HANDBOOK OF SEMICONDUCTOR CIRCUITS

TAB BOOKS
BLUE RIDGE SUMMIT, PA. 17214

This handbook was originally published by the U.S. Government Printing Office for the Armed Forces Supply Support Center under the title "Military Standardization Handbook, Selected Semiconductor Circuits," MIL-HDBK-215.

Printed in the United States of America

CONTENTS

	Page
Introduction	11

PART 1

Abbreviations and Symbols 17

PART 2 - DIRECT-COUPLED AMPLIFIERS

Section 1. Design Philosophy

A.	Introduction	23
B.	Direct-Coupled Transistor Amplifiers (Single Unbalanced Stages)	23
C.	Balanced Configurations	28
D.	Balanced Difference Amplifiers	30
E.	Modulators	31

Bibliography 33

Section 2. List of Selected Circuits

2-1	Difference Amplifier	36
2-2	Compensated D.C. Amplifier (Temperature Adjusted)	39
2-3	Low Drift High Impedance Straight D.C. Current Amplifier	41
2-4	Balanced Circuit with Independent Zero Set and Temperature Balance Adjustments	44
2-5	Silicon Diode Ring Modulator	47
2-6	Semiconductor-Capacitor Chopper	49
2-7	High Voltage Direct-Coupled Amplifier......	52

PART 3 - LOW-FREQUENCY AMPLIFIERS

Section 1. Design Philosophy

A. Transistor Characteristics
1.	Input Characteristics.........................	54
2.	Transfer Characteristics.....................	55
3.	Output Characteristics.......................	55
4.	Equivalent Circuit............................	56
5.	Noise..	57

CONTENTS (Continued)

		Page
B.	Circuit Considerations	
	1. Bias Stabilization	58
	a. Series Emitter Stabilization	58
	b. Collector-to-Base Shunt Stabilization	59
	c. Bias Compensation	59
C.	Thermal Considerations	60
D.	Distortion	
	1. Nonlinear Input Characteristics	62
	2. Nonlinear Current Amplification Factor	63
E.	Negative Feedback	
	1. Series Degeneration	63
	2. Shunt Degeneration	64
F.	Basic Amplifiers	
	1. RC Amplifiers	65
	2. Transformer-Coupled Amplifiers	66
	3. Power Amplifiers	67

Bibliography 69

Section 2. List of Selected Circuits

3-1	Low Level Audio Amplifier	70
3-2	Transformer-Coupled Output Stage	73
3-3	Input Circuit for Capacitive Transducer	75
3-4	Input Circuit for Inductive Transducer	77
3-5	Low Noise Transistor Preamplifier	79
3-6	High Input Impedance Preamplifier	82
3-7	Wide Temperature Range Preamplifier	84
3-8	High Input Impedance Wide Temperature Range Amplifier	86
3-9	Diode Coupled A.C. Preamplifier	89
3-10	Twin Tee Rejection Amplifier	91
3-11	Transformerless Quasi-Complementary Audio Amplifier	93
3-12	Complementary Symmetry Amplifier	95
3-13	Self-Balancing Push-Pull Class A Output Stage	98
3-14	75 Watt Audio Amplifier	100
3-15	High Gain Servo Amplifier	103
3-16	High Efficiency Servo Output Stage	105

PART 4 - HIGH-FREQUENCY AMPLIFIERS

Section 1. Design Philosophy

A.	Introduction	108

CONTENTS (Continued)

		Page
B.	Equivalent Circuits	108
C.	Power Gain and Stability	110
D.	Unilateralization and Neutralization	115
E.	High Frequency Noise (23 to 31)	117
F.	Automatic Gain Controls.....................	119
G.	Tuned Amplifier Interstages	120
H.	Video Amplifiers............................	121

References and Bibliography 123

Section 2. List of Selected Circuits

4-1	One Stage Neutralized 455 KC I.F. Amplifier	127
4-2	Two Stage 455 KC I.F. Strip	129
4-3	Stable Wide Band Amplifier	131
4-4	R.F. Stage for 3-Band Receiver	134
4-5	60 MC I.F. Amplifier Using Silicon Tetrodes	136
4-6	Narrow-Band Selective Amplifiers..........	138
4-7	Emitter-Tuned I.F. Amplifier..............	140
4-8	30 MC I.F. Amplifier	142
4-9	Wide Band Transistor Feedback Amplifier...	143
4-10	Video Amplifier	146
4-11	Wide Band Video Amplifier.................	148
4-12	Wide-Band High-Frequency I.F. Amplifier...	150

PART 5 - OSCILLATORS

Section 1. Design Philosophy

A.	Introduction.....................................	152
B.	Frequency Stability	152
C.	Types of Oscillators	156
D.	Summary	159

References 161

Bibliography 162

Section 2. List of Selected Circuits

5-1	High-Stability Self-Excited Oscillator	
	5-1.1 - Self-Limiting, Self-Excited 1 MC Oscillator	163
	5-1.2 - Self-Excited 1 MC Oscillator with AVC.........................	164
5-2	Wide-Range Constant-Amplitude Oscillator...	167

5

CONTENTS (Continued)

		Page
5-3	VHF Oscillator................................	169
5-4	100 KC Reference Oscillator..............	171
5-5	450 KC Crystal Oscillator................	173
5-6	1 MC Crystal Standard Oscillator........	175
5-7	9 MC Crystal Standard Oscillator........	177
5-8	108 MC Crystal Oscillator................	178
5-9	20 KC Audio Crystal Oscillator..........	180
5-10	Audio-Frequency Oscillator..............	182
5-11	Phase-Shift Oscillator...................	184
5-12	Twin Tee Oscillator......................	186
5-13	Voltage-Controlled Wien Bridge Oscillator................................	189
5-14	Series-Tuned Emitter, Tuned-Base Oscillator................................	191

PART 6 - SWITCHING CIRCUITS

Section 1. Design Philosophy

A.	Transistor Characteristics....................	192
B.	Circuit Design Methods.......................	198
C.	Basic Flip-Flop Design Procedures............	198
	1. Saturated Flip-Flop......................	198
	2. Nonsaturating Flip-Flop..................	200
	3. Clamping Methods.........................	204
D.	Symmetrical Circuits..........................	206
E.	Triggering Methods............................	207
F.	Blocking Oscillator Circuit Design Procedures...................................	209
G.	Direct-Coupled Transistor Methods............	212
H.	Negative Resistance Circuit Methods..........	213
I.	Other Circuit Methods.........................	214

References.. 214

Bibliography.. 215

Section 2. List of Selected Circuits

6-1	Saturated RC Coupled Flip-Flop..........	217
6-2	Nonsaturated RC Coupled Flip-Flop.......	221
6-3	Direct-Coupled Transistor Flip-Flop.....	223
6-4	Current Switching Flip-Flop.............	225
6-5	Base Gated Direct-Coupled Flip-Flop.....	227
6-6	Saturated Clamped Flip-Flop.............	229
6-7	Saturated Clamped High Speed Flip-Flop..	232
6-8	High Speed Graded Base Transistor Flip-Flop................................	234
6-9	Medium Speed Silicon Flip-Flop..........	237

CONTENTS (Continued)

		Page
6-10	Emitter Follower Cross Coupled Saturated Flip-Flop...................	238
6-11	Negative Resistance Diode Flip-Flop......	239
6-12	Astable Circuit for Direct-Coupled Transistors........................	242
6-13	Basic Saturated Astable Multivibrator.....	243
6-14	Monostable Multivibrator..................	246
6-15	Astable Unijunction Transistor Multivibrator........................	248
6-16	Direct-Coupled Silicon Monostable Multivibrator........................	250
6-17	Fast Recovery Monostable Multivibrator....	252
6-18	Schmitt Trigger Circuit...................	254
6-19	Blocking Oscillator.......................	257
6-20	Blocking Oscillator with Square Loop Transformer.........................	259
6-21	Blocking Oscillator with Delay Line Width Control.............................	261
6-22	Dual Inverter Circuit.....................	263
6-23	Symmetrical Buffer Inverter...............	265
6-24	Differentiator and Pulse Shaper...........	267

PART 7 - LOGIC CIRCUITS

Section 1. Design Philosophy

A.	Logic Circuits............................	268
B.	Basic Gates, "And" - "Or"..................	268
C.	Counters and Shift Registers..............	277
D.	High Speed Gating.........................	279

References... 281

Bibliography... 281

Section 2. List of Selected Circuits

7-1	Current Steering Gate.....................	283
7-2	Diode-Gated Amplifier.....................	285
7-3	Diode-Transformer Gated Amplifier, Exclusive "Or".......................	287
7-4	Transistor Resistor Gate..................	289
7-5	Current Type Dual Gate....................	291
7-6	Exclusive "Or"............................	293
7-7	Direct-Coupled Counter Stage with Gating.	295
7-8	RC Coupled Flip-Flop with Direct-Coupled Gating...............................	297
7-9	Direct-Coupled Type Ring Counter..........	299

CONTENTS (Continued)

		Page
7-10	Unijunction Ring Counter................	300
7-11	Direct-Coupled Type Ring Counter with Conditional Steering...............	302
7-12	Seven-Stage Ring Counter................	303
7-13	Shift Register with Conditional Steering.	306
7-14	Direct-Coupled Half Adder...............	307
7-15	Current Switch Half Adder Circuit........	309
7-16	Diode-Transistor Matrix Switch...........	311
7-17	Shift Register with Resistor-Capacitor Diode Gates..........................	313
7-18	Low Power Ring Counter..................	315

PART 8 - A.C. TO D.C. POWER SUPPLIES

Section 1. Design Philosophy

A. Unregulated Power Supplies
1. Introduction........................... 317
2. Capacitor-Input-Filter Rectifiers......... 318
3. Choke-Input-Filter Rectifiers............. 326
4. Filtering............................... 330

B. Regulated Power Supplies
1. Shunt Voltage Regulators.................. 330
2. Series Voltage Regulators................. 332
3. Constant-Current Regulators............... 335
4. Switching Regulators...................... 337
5. Regulating Amplifiers..................... 341
6. Overload Protection....................... 345

References.. 347

Bibliography.. 348

Section 2. List of Selected Circuits

8-1	Breakdown-Diode Regulated Power Supply....	349
8-2	Emitter-Follower Regulator...............	350
8-3	Six-Volt Silicon Shunt Regulator..........	352
8-4	Current-Limited Series Voltage Regulator..	354
8-5	150-Volt Silicon Series Regulator.........	357
8-6	6.3-Volt, 5-Ampere Series Regulator.......	359
8-7	Wide-Range Regulator with 0.005% Regulation............................	361
8-8	Chopper-Stabilized Strain-Gauge Supply....	365
8-9	Small Lab Supply with Current Limiting....	368
8-10	Convection-Cooled, Wide-Range Power Supply	371
8-11	High-Temperature, 120-Volt Silicon Regulator.............................	375

CONTENTS (Continued)

                                                                                                                Page

8-12	Silicon Controlled Rectifier Regulator...	377
8-13	Dual-Regulator, Wide-Range Power Supply..	380
8-14	Transistor Constant-Current Regulator....	383
8-15	High Voltage Regulator with Series Stack.	385
8-16	500-Volt Regulated Supply...............	389
8-17	Minature Silicon Regulator Module........	392

PART 9 - POWER CONVERTERS

Section 1. Design Philosophy

A.	Basic Transistor Oscillator..................	394
B.	Other Connections for Basic Circuits.........	396
C.	Starting Circuits............................	397
D.	Basic D.C. to D.C. and D.C. to A.C. Power Converters..................................	399
E.	De-Spiking Networks..........................	401
F.	Transistor Selection.........................	402
G.	Transformer Selection and Design.............	403
H.	Modified and Improved Circuits...............	405
I.	Regulators and Protective Circuitry..........	407
J.	Power Converter Thermal Considerations.......	409

References.. 410

Bibliography.. 410

Section 2. List of Selected Circuits

9-1	D.C. to D.C. Power Converter.............	411
9-2	Magnetic Amplifier Regulated D.C. to D.C. Converter...........................	413
9-3	D.C. to D.C. Converter Using the Base Control Method of Output Regulation..	415
9-4	D.C. to A.C. Converter with 400 c.p.s. Square Wave Output....................	417
9-5	D.C. to D.C. Converter, 130 Watts Output, with Zener Diode Transient Surge Protection............................	419
9-6	D.C. to D.C. Converter - 50Watts, 400 c.p.s...................................	421
9-7	Series Connected D.C. to D.C. Converter..	423

PART 10 - SMALL-SIGNAL NONLINEAR CIRCUITS

Section 1. Design Philosophy

A.	Introduction................................	425
B.	Modulators..................................	425
C.	Mixers and Converters.......................	426

CONTENTS (Continued)

		Page
D. Detectors		427
E. Frequency Multipliers		428
F. Frequency Dividers		429
Bibliography		430

Section 2. List of Selected Circuits

10-1	Locked-Oscillator Frequency Divider	431
10-2	Low-Level Modulator	433
10-3	Frequency Modulator and Oscillator	435
10-4	Frequency Multiplier	436
10-5	Autodyne Converter	438
10-6	VHF Mixer with RF Amplifier-Oscillator	440
10-7	Regenerative Frequency Divider	443
10-8	Second Detector with AGC	445
10-9	Second Detector with Delayed AGC	447

INTRODUCTION

The transistor has become one of the major significant breakthroughs in the electronics field. New techniques of circuit design, miniaturization, and reliability are now emerging, due to the utilization of the transistor and other simiconductor devices. Latest design trends for military equipment have been predominantly towards the use of these devices for the active elements.

It is probable that every engineer will be involved in some phase of transistor circuit design. From a cursory investigation, it would seem not too difficult a task to utilize previous electronic background and experience in the design of transistor circuits. However, this is not entirely the case. Established circuit design based on vacuum tubes was a guide to the early transistor circuit designer; however, due to the imagination and interest created by this new device, new applications and methods of design have resulted. Practical experience has shown that it requires a modified design approach, which is made mandatory by the unique characteristics and parameters of the semiconductor devices.

During the developmental years of the transistor, many techniques of design were invented to overcome some transistor shortcomings. In many cases, these techniques are still useful, although the deficiencies associated with the early type transistors have nearly vanished. At present, the emphasis is on circuit refinements and new systems approaches, more so than on new circuit techniques. It is for this reason that the Handbook of Semiconductor Circuits has been written.

The purpose of this handbook is to provide the transistor circuit engineer with a reference of reliable, well-designed examples of contemporary circuits suitable for use in military equipment, as well as general usage. To make this a valuable handbook in years to come, when the selected circuits may have been outdated, the design philosophies included with each part have been given great emphasis.

The design philosophies have been presented in a practical and informative manner, as a basis for the design of circuits other than those shown in the text. It was felt that a mere presentation of selected circuits would not have been sufficient, as the design philosophies will give the circuit designer valuable information in further developing the selected circuits, as well as the design of his own circuits.

The circuits were selected on the basis of inventiveness, reliability, or advanced state of the art design. We wish to stress here that this is not a handbook of preferred circuits, but rather a group exemplifying good design and reliability. This is a book employing building block circuits, rather than systems. There is a circuit description with any pertinent or unique design or operational data on all selected circuits, followed by a schematic. The contributor is listed with each circuit. In some cases, the contributor is not the inventor of the idea or circuit, but has modified or further developed the original circuit into a more reliable-type circuit.

The areas of circuit design covered are complete and are listed here for reference.

Part 2 - Direct-Coupled Amplifiers: The problems associated with transistors in d.c. amplifier applications presented perhaps the greatest challenge to the circuit designer. The leakage current was high in early transistors and the need for stabilization under these conditions lead to excellent design techniques that presently provide a high degree of reliability. The selected circuits included in this part are considered to be the best available and will find wide application. The design philosophy of this section illustrates not only the problems related to d.c. amplifier design, but provides an excellent discussion of transistor fundamentals.

Part 3 - Low-Frequency Amplifiers: This section is written in basic terms and gives concise design principles. The feedback techniques are illustrated in both the design philosophy and selected circuits and show clearly the various characteristics of the different circuit configurations. Numerous graphical representations of the transistor characteristics provide a better understanding of the design considerations. Both high and low power amplifiers are included.

Part 4 - High-Frequency Amplifiers: This section covers video amplifiers, narrow band and wide band amplifiers, as well as many other important high frequency circuits. The design procedure illustrates recent techniques that will prove invaluable to the engineer. In addition, the discussion is written in a classical manner to provide a wealth of advanced circuit design techniques.

Part 5 - Oscillators: This is the first work on this subject in recent years to give practical circuits in addition to a theoretical analysis. The various types of oscillators shown provide a wide choice of designs for current circuit applications.

Part 6 - Switching Circuits: Described in detail are the many circuit approaches that exist in this area. The various circuit methods such as direct-coupled (DCTL), resistor transistor (RTL), and others presently in use are compared in terms of efficiency, complexity, and ease of design. The design philosophy illustrates the design

methods commonly used in switching circuits and many examples
are included. The text is written in simple terms with many complete
design examples. Basic switching circuits capable of operation in
the 5 to 10 megacycle range are included.

   Part 7 - Logic Circuits: This section is written in a manner that
complements the switching circuits section. All the presently used
logic methods are discussed, and associated circuits are included.
The logic section contains gates, ring counters, shift registers, half-
adders and matrix switches.

   Part 8 - A.C. to D.C. Power Supplies: This section is unique among
material of this sort in that it describes the important design considerations with emphasis on practical applications. The circuits
covered include unregulated, regulated, shunt voltage, series voltage,
series current and transistor magnetic regulators. Overload protection and regulated amplifiers are also discussed fully. Economical
design considerations are a major feature of this section.

   Part 9 - Power Converters: The converter section covers both
d.c. to d.c. and d.c. to a.c. type. The selected circuits have been
chosen to represent the most classical work in this area. The design
philosophy is written to provide a firm basis for continued design
work.

   Part 10 - Small Signal Nonlinear Circuits: This section describes
mixers, converters, detectors, modulators, multipliers, and dividers.
The important transistor characteristics for these circuits are
discussed in the light of recent design procedures.

   This handbook is not an attempt to set up any type of preferred
circuitry. Rather we hope to provide a handy reference, with a large
number of circuits, exemplifying good design and reliable-type
circuitry, as well as to encourage the standardization of semiconductor circuit nomenclature and symbols.

HANDBOOK OF
SEMICONDUCTOR CIRCUITS

PART 1

ABBREVIATIONS AND SYMBOLS

10. SCOPE

10.1 This section covers the abbreviations and symbols for use with semiconductor devices.

20. GENERAL

20.1 The abbreviations and symbols used in this specification are those generally accepted by the electronic and electrical industries. The following is an explanation of the elements of these abbreviations and symbols:

- (a) Lower case letters are used as primary symbols for small signal internal device parameters and for instantaneous time variable current, voltage and power.
- (b) Lower case letters are used as subscripts to indicate small signal parameters and varying component values.
- (c) Upper case letters are used as primary symbols for external circuit parameters and components, large signal device parameters, and for maximum, average, and root-mean-square values of current, voltage, and power.
- (d) Upper case letters are used as subscripts to indicate static values, large signal parameters and d.c., average and instantaneous total values.
- (e) For electrical quantities, the first subscript designates the electrode at which the measurement is made.
- (f) For device parameters, the first subscript designates the element of the four pole matrix, for example, I or i for input, O or o for output, F or f for forward transfer, and R or r for reverse-transfer.
- (g) The second subscript normally designates the reference electrode.
- (h) Supply voltages are indicated by repeating the associated device electrode subscript.

30. LISTING

30.1 Example and special abbreviations and symbols are as follows: (J or j is used as a general reference electrode indicator and K or k is used as a general measurement electrode indicator).

A	Ampere (a.c. r.m.s. or d.c.)
a	Ampere (peak)
a.c.	Alternating current
Aac	Ampere (a.c. r.m.s.)
Adc	Ampere (d.c.)
AQL	Acceptable quality level
AVG or avg . .	Average
B, b	Base electrode
BV	Breakdown voltage
BV_{KJO}	Breakdown voltage, open circuit
BV_{KJR}	Breakdown voltage, resistance return
BV_{KJS}	Breakdown voltage, short circuit
bz	Small signal breakdown impedance
BZ	Large signal breakdown impedance
C, c	Capacitance and collector electrode
C(dep)	Depletion layer capacitance
C(dif)	Diffusion capacitance
°C	Degree Centigrade
C_{ij}	Input capacitance
C_L	Load capacitance
C_{oj}	Output capacitance
cm	Centimeter
c.p.s.	Cycle per second
CRO	Cathode-ray oscilloscope
db	Decibel
d.c.	Direct current
Δ (delta)	A change in the value of the indicated variable
ΔBV	Breakdown voltage regulation
E, e	Emitter electrode
f	Frequency
F(ref)	Reference frequency
f_{aj}	Small-Signal short-circuit forward-current transfer ratio cutoff frequency
f_{pg}	Power gain cutoff frequency
f_{osc}	Maximum frequency of oscillation
G	Acceleration of gravity
g_{MJ}	Static transconductance
g_{mj}	Small signal transconductance
G_{MJ}	Large signal transconductance

H_{FJ}	Large signal short circuit forward current transfer ratio
h_{FJ}	Static forward current transfer ratio
h_{fj}	Small-signal short-circuit forward-current transfer ratio
h_{IJ}	Static input impedance
h_{ij}	Small-signal short-circuit input resistance
h_{oj}	Small-signal open-circuit output admittance
h_{rj}	Small-signal open-circuit reverse-voltage transfer ratio
I	Current
i (surge)	Surge current
I_{KJO}	Cutoff current, open circuit
I_{KJR}	Cutoff current, resistance return
I_{KJS}	Cutoff current, short circuit
I_{KJV}	Cutoff current, reverse voltage
I_{KJX}	Cutoff current, specified condition
i.f.	Intermediate frequency
I_F	Forward current, d.c.
i_f	Peak forward current
i_f (surge)	Peak forward surge current
in	Inch
INV	Inverse
I_o	d.c. output current
I_R	Reverse current, d.c.
i_r (surge)	Peak reverse surge current
I_s	Saturation current
J, j	Junction and reference electrode, general
K, k	Measurement electrode, general and kilo (thousand)
°K.	Degrees kelvin
kc.	Kilocycle per second
L_c	Conversion loss (ratio of available signal power to the available intermediate frequency power
ma.	Milliampere (peak)
mAac	Milliampere (a.c. r.m.s.)
mAdc	Milliampere (d.c.)
Max or max.	Maximum
mc.	Megacycles per second
Meg.	Million
μ	Micro (millionth)
μa	Microampere (peak)
μAac	Microampere (a.c. r.m.s.)
μAdc	Microampere (d.c.)

μf	Microfarad
μh	Microhenry
μmho	Micromho
μsec	Microsecond
μμf	Micromicrofarad
μVac	Microvolt (a.c. r.m.s.)
μVdc	Microvolt (d.c.)
μW	Microwatt
m	Mila (thousandth)
mμsec	Millimicrosecond
mm	Millimeter
msec	Millisecond
mVac	Millivolts (a.c. r.m.s.)
mVdc	Millivolts (d.c.)
mW	Milliwatt (max., ave. or r.m.s.)
mw	Milliwatt (peak)
MIN or min. .	Minimum
NF	Noise figure
NF_o	Overall noise figure (power ratio)
NR_o	Output noise ratio, or noise-temperature ratio (power ratio)
N region	Region of semiconductor device where electrons are the majority carriers
Ω	Ohms
P	Power (max., ave. or r.m.s.)
P_g	Power gain, small signal
P_G	Power gain, large signal
P_j	Average value of the power dissipation at the junction associated with the jth electrode
P region	Region of semiconductor where holes are the majority carriers
PRE	Power rectification efficiency
R_B	External base resistance
r_b	Base spreading resistance
RE	Rectification efficiency (voltage)
RE_{hij}	Real part of small signal short circuit input impedance
R_C	External collector resistance
R_e	Reference resistor for noise-ratio measurements
R_E	External emitter resistance
R_F	Radio frequency

R_L . . . . Load resistance
r.m.s. . . . . . . Root mean square
r_{KJ} (sat) . . . . Saturation resistance
SCD . . . . Semiconductor device
⊥ . . . . Symbol for semiconductor diode, arrow points in forward direction
γ_r . . . . Thermal response time
γ_t . . . . Thermal time constant
T . . . . Temperature
t . . . . Time
T_A . . . . Ambient temperature
T_C . . . . Case temperature
TCBV . . . . . Temperature coefficient of breakdown voltage
t_d . . . . Pulse delay time
t_f . . . . Pulse fall time
t_{fr} . . . . Forward recovery time
T_j . . . . Junction temperature
T_{max} . . . . . . Absolute maximum temperature
t_p . . . . Pulse time
θ (theta) . . . . Thermal resistance
θ_{J-A} . . . . . . Thermal resistance, junction to ambient
θ_{J-C} . . . . . . Thermal resistance, junction to case
t_r . . . . Pulse rise time
t_{rr} . . . . Reverse recovery time
t_s . . . . Pulse storage time
T_{stg} . . . . . . Storage temperature
t_w . . . . Pulse average time

V . . . . Voltage (max., ave. or r.m.s.)
v . . . . Volt (peak)
Vac . . . . Volt (a.c. r.m.s.)
V_{BB} . . . . Base voltage (d.c.) supply
V_{BJ} . . . . Base voltage
V_{CC} . . . . Collector voltage (d.c.) supply
V_{CJ} . . . . Collector voltage
V_{dc} . . . . Volt (d.c.)
V_{EE} . . . . Emitter voltage (d.c.) supply
V_{EJ} . . . . Emitter voltage
V_F . . . . Forward voltage, d.c.
V_I . . . . Input voltage, d.c.
V_{JJ} . . . . Supply voltage, d.c.
V_{KJF} . . . . . . Floating potential

V_{KJ} (sat). . . . Saturation voltage, d.c.
V_O . . . . Output voltage, d.c.
V_R . . . . Reverse voltage, d.c.
v_r . . . . Peak inverse voltage
V_{RT} . . . . Voltage reach through

W . . . . Watt (max., ave. or r.m.s.)
w . . . . Watt (peak)

X_{XX}(INV) . . . Inverse electrical characteristics

y_{fj} . . . . Small-signal short-circuit forward transfer admittance
y_{ij} . . . . Small-signal short-circuit input admittance
y_{oj} . . . . Small-signal short-circuit output admittance
y_{rj} . . . . Small-signal short circuit reverse-transfer admittance

Z_F . . . . Large signal forward impedance
z_f . . . . Small signal forward impedance
z_{fj} . . . . Small-signal open-circuit forward-transfer impedance
Z(IF) . . . . . . Intermediate frequency impedance
z_{ij} . . . . Small-signal open-circuit input impedance
z_{oj} . . . . Small-signal open-circuit output impedance
z_{rj} . . . . Small-signal open-circuit reverse-transfer impedance
Z(V). . . . . . . Video impedance

PART 2

DIRECT-COUPLED AMPLIFIERS

Section 1. Design Philosophy

A. Introduction

Direct-coupled amplifiers are among the basic needs of analog computers, measurement instruments, and industrial control equipment. It is the purpose of this section to present in some detail the inherent limitations imposed by the temperature dependencies of transistor and diode parameters.

One of the major considerations of amplifiers is the minimum detectable signal. The ultimate limitation is determined by the random noise behavior of the amplifier. However, in most practical amplifiers operating over a temperature range as small as ten degrees centigrade, the minimum detectable signal is determined by drift of the quiescent operating point.

There are a number of techniques that are employed to achieve d-c gain. Although a major emphasis of this section will be devoted to resistive coupling of transistors, the use of choppers or modulators will be covered also.

B. Direct-Coupled Transistor Amplifiers (Single Unbalanced Stages)

In this section, the design of direct-coupled amplifiers that use germanium or silicon transistors will be discussed. The two-diode equivalent circuit of a p-n-p transistor is shown in Figure 2-0.1. The diodes in the circuit follow the exponential equation

$$I = I_{CO}\left(e^{-\frac{qV}{KT}} - 1\right) \quad 2\text{-}0.1$$

where I is the current flowing through the diode, I_{CO} is a saturation current (this value is highly dependent on temperature), q is electronic charge, V is the voltage applied across the diode, K is Boltzman's constant, and T is absolute temperature. A typical characteristic for a germanium

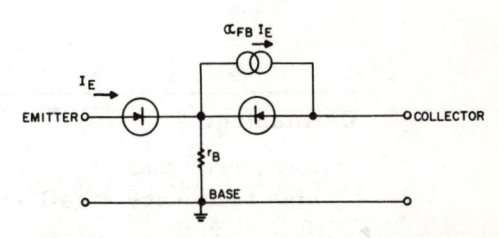

Fig. 2-0.1 - Two diode equivalent circuit of a PNP transistor

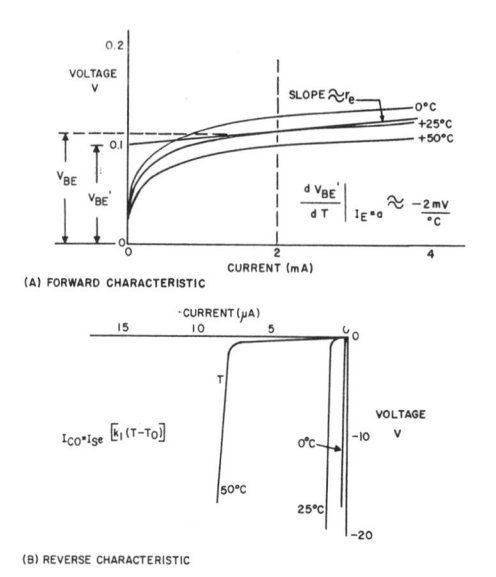

Fig. 2-0.2 - Typical germanium diode characteristics

(A) FORWARD CHARACTERISTIC

(B) REVERSE CHARACTERISTIC

diode is shown in Figure 2-0.2. The forward bias characteristic is temperature dependent. This temperature variation arises from two factors, the change in I_{CO} with temperature and the dependence of the exponential term on absolute temperature. The variation is most conveniently expressed as a change in voltage with temperature at a constant forward bias current. The variation of emitter diode voltage with temperature falls in the range of -1.5 mVdc/°C to -3.0 mVdc/°C for both germanium and silicon transistors. The coefficient is slightly dependent on diode current but for a given type of unit the use of a constant value is a valid approximation.

For d-c analysis, the emitter to base diode can be approximated by a battery with value V'_{BE} and a resistance r_e equal to the incremental value at the operating point.

The saturation current of the reverse characteristic is highly temperature dependent. The dependence follows an exponential relation given by

$$I_{CO} = A_0 \exp [k(T - T_0)] \qquad 2\text{-}0.2$$

where I_{CO} is the reverse saturation current, A_0 is a constant measured at a reference temperature T_0, k is a constant and T is the diode junction temperature in degrees centigrade. The values of A_0 and k will vary from unit to unit of a given type transistor as well as with different types. Typical values of A_0 for germanium and silicon transistors are given in Table

	A_0 μamps
Germanium:	
low power audio	3 - 20
high frequency small area units	0.1 - 1.0
Silicon - low power	< 0.01 - 0.0001

The values of k are considerably more constant than the values of A_0. Typical values of k for germanium units are in the range of 0.08 ± 0.01, whereas the coefficient tends to be somewhat larger for silicon units. In silicon transistors that have a very small value of A_0, it is difficult to separate the true saturation current of the body from a surface leakage current from the collector to the base of the transistor. Although the I_{CO} of silicon transistors generally varies more rapidly than the I_{CO} of germanium, the difference in initial values is so great that germanium units are quite inferior from this aspect.

The forward current transfer ratio is also temperature dependent. For germanium units the gain can either increase or decrease with temperature, whereas the gain of silicon units generally increases with temperature. The percentage variation in gain with temperature varies greatly with operating point. Many units show a change in sign as well as in magnitude. At room temperature, gain variations of silicon units may typically be from two to ten times larger than of germanium.

The parameters of a small signal equivalent circuit are also temperature dependent. It is usually possible to swamp out these variations by appropriate choice of external resistors. On the other hand, the first order effect of a variation in parameters will be a variation in gain. This variation can be lumped into the variation of h_{ab} with temperature and the combined variation treated experimentally.

An equivalent circuit that represents the d.c. behavior of a p-n-p transistor is shown in Figure 2-0.3. The use of the small signal current gain α_{fb} rather than a total d.c. gain is an approximation that may cause some confusion. Since the discussion that follows is concerned with incremental gain at a quiescent current rather than total d.c. gain, the incremental notation is justified although it may be misleading.

In most amplifiers, the load resistance R_L is very small compared with the collector resistance r_c. Thus, it will be assumed that r_c approaches an infinite value.

From Figure 2-0.3, the emitter current may be expressed as

$$I_E = \frac{V_{EE} - V_{BE}' - v_G + I_{CO}(R_g + r_b)}{R_e + r_e + (R_g + r_b)(1 - \alpha_{fb})} \quad 2\text{-}0.3$$

but the collector current is equal to

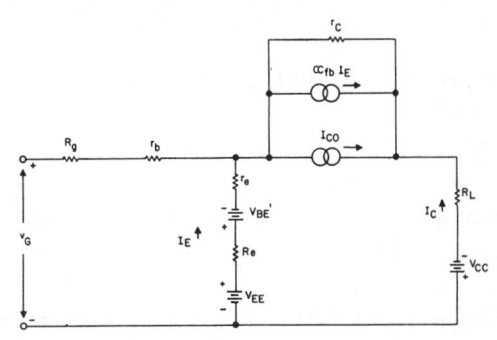

Fig. 2-0.3 - Equivalent circuit for d.c. analysis of grounded emitter stage

$$I_C = -(\alpha_{fb} I_E + I_{CO}) \qquad 2\text{-}0.4$$

Thus

$$I_C = -\alpha_{fb} \left[\frac{V_{EE} - V'_{BE} - v_G + I_{CO}(R_g + r_b + R_e + r_e)}{R_e + r_e + (R_g + r_b)(1 - \alpha_{fb})} \right] \qquad 2\text{-}0.5$$

Thus, we see that there are several terms in the solution for the collector current. To evaluate an amplifier design it is convenient to break the expression into two parts, a signal term, and a drift term. The signal term is given by

$$i_c = +\alpha_{fb} \frac{v_G}{R_e + r_e + (R_g + r_b)(1 - \alpha_{fb})}. \qquad 2\text{-}0.6$$

The drift term is best expressed in terms of a differential variation of collector current as a function of temperature. Since there are three major terms that vary with temperature, V'_{BE}, I_{CO}, and α_{fb}, a partial differentiation of equation 2-0.5 yields

$$\frac{dI_C}{dT} = \frac{\alpha_{fb}}{(\text{Den})} \frac{dV'_{BE}}{dT} - \alpha_{fb} \frac{R_T}{(\text{Den})} \frac{dI_{CO}}{dT} - \left[\frac{V_{EE} - V'_{BE} + I_{CO} R_T}{(\text{Den})} \right] \left[\frac{R_T}{(\text{Den})} \right] \frac{d\alpha_{fb}}{dT} \qquad 2\text{-}0.7$$

where $(\text{Den}) = R_e + r_e + (R_g + r_b)(1 - \alpha_{fb})$, and $R_T = R_e + r_e + R_g + r_b$. In equation 2-0.7, the quiescent emitter current can be recognized as

$$I_E = \frac{V_{EE} - V'_{BE} + I_{CO} R_T}{\text{Den}}. \qquad 2\text{-}0.8$$

Thus, the drift current may be written as

$$i_D = dI_C = \left[\frac{\alpha_{fb} \frac{dV'_{BE}}{dT} - \alpha_{fb} R_T \frac{dI_{CO}}{dT} - I_E R_T \frac{d\alpha_{fb}}{dT}}{R_e + r_e + (R_g + r_b)(1 - \alpha_{fb})} \right] \Delta T. \qquad 2\text{-}0.9$$

The ratio of signal to drift current is of considerable interest, and is given by

$$\frac{i_C}{i_D} = \frac{v_G}{\left[\frac{dV'_{BE}}{dT} - R_T \frac{dI_{CO}}{dT} - \frac{I_E R_T}{\alpha_{fb}} \frac{d\alpha_{fb}}{dT} \right] \Delta T}. \qquad 2\text{-}0.10$$

If we wish to determine the minimum detectable signal, the ratio of i_C/i_D is set equal to unity. Conversely, if the collector current is to be kept constant, a correcting voltage (v_D) may be applied to the input. From equation 2-0.10, it can be shown that the correcting voltage v_D required to keep the collector current constant is

$$v_D = -\left[\frac{dV'_{BE}}{dT} - R_T \frac{dI_{CO}}{dT} - \frac{I_E R_T}{\alpha_{fb}} \frac{d\alpha_{fb}}{dT}\right] \Delta T . \qquad 2\text{-}0.11$$

At this stage, it is fruitful to take a typical set of transistor parameters and evaluate the magnitudes of the three terms that appear in equation 2-0.11. The first term, dV'_{BE}/dT has a value of approximately -2.5 mVdc/°C for either germanium or silicon. Thus at room temperature with a control of ± 8°C, the emitter-base diode imposes a limitation of approximately ∓20.0 mVdc, (for emphasis ∓0.02 volts).

The second term is $R_T\, dI_{CO}/dT$. If we use values of R_g = 1000 ohms, R_E = 100 ohms, the value of R_T may be approximated as 2000 ohms. If a small-area high frequency germanium transistor is employed, the value of I_{CO} at room temperature can be approximately 0.25 μamp. For a ± 8°C change, the variation in I_{CO} can be approximated as + 0.25 μamp and - 0.125 μamp. The drift voltage from this term will be + 500 μvolts, - 250 μvolts. Taking this same transistor to 55°C will give results of + 5 mVdc and - 2.5 mVdc. The use of a transistor with a larger junction area can easily increase the above results by one to two orders of magnitude.

The I_{CO} of a silicon transistor can on the other hand be in the range of 0.001 μamp to 0.0001 μamp. The equivalent drifts at room temperature will be approximately + 2.0 μvolts to + 0.2 μvolts. At 55°C the drift voltage may be in the range of 2 to 25 μvolts and at even 85°C the drift voltage may be as low as 40 to 500 μvolts.

The temperature coefficient of current gain $d\alpha_{fb}/dT$ of germanium transistors lies in the range of - 0.003/°C to + 0.005/°C. The drift voltage caused by this parameter is

$$v_D = \frac{I_E R_T}{\alpha_{fb}} \frac{d\alpha_{fb}}{dT} \Delta T . \qquad 2\text{-}0.12$$

If R_T = 2000 ohms, $d\alpha_{fb}/dT$ = + 0.005/°C ΔT = ± 8°C, the drift voltage becomes

$$v_D = (80)(I_E) . \qquad 2\text{-}0.13$$

Since the value of R_T = 2000 ohms may be valid for a quiescent current greater than 200 μamps, the smallest value of v_D becomes ± 16 mVdc. The variation in gain for silicon transistors can be anywhere in the range of + 0.0002/°C to 0.005/°C. For a 2N338, a high gain silicon transistor, a typical value might be +0.0004/°C. Thus, drift voltages due to gain variations for the silicon transistor over a ±8°C range may be ±1.3 mVdc.

The drift voltage caused by gain variations with temperature are a direct function of the quiescent emitter current. However, as the emitter current is reduced, the incremental resistance r_e increases. An order of magnitude limit may be obtained for the drift voltage at very small values of I_E by letting the resistors R_g, r_b, and R_e be small compared with r_e. The incremental resistance r_e is approximated by

$$r_e = \frac{0.027 \text{ (volts)}}{I_E \text{ (amps)}} \qquad 2\text{-}0.14$$

thus the lower limit of drift term due to $d\alpha_{fb}/dT$ can be approximated by

$$v_D \geq 0.027 \frac{d\alpha_{fb}}{dT} dT. \qquad 2\text{-}0.15$$

For a temperature coefficient of gain equal to 0.0004/°C the smallest drift voltage from this cause would be in excess of 10 µvolts/°C.

The above discussion can be summarized in the following manner. First, the use of germanium transistors must be restricted to those units that have an I_{CO} less than 5.0 µamps for room temperature operation, and less than 1.0 µamps for operation at elevated temperature. Second, the variation in voltage of the emitter to base diode is of major importance. In a single unbalanced unit the drift voltage is limited to -2.5 mVdc/°C. This limit is unaffected by the external parameters and varies only with the transistor used and to a lesser extent with the quiescent current. Third, the drift voltage caused by gain variation is a direct function of emitter current. For best results the current should be made as small as possible. Transistors with constant gain should be selected. Four, at elevated temperatures silicon transistors are superior but at room temperature, germanium units may yield better results since the gain variation with temperature is smaller.

C. Balanced Configurations

The major portion of the drift voltage was shown to be the variation of the emitter to base voltage. The emitter diode characteristic is quite uniform from one transistor to another. With virtually no selection, it is possible to obtain two units that have temperature dependent variations that track each other within better than 10%. Furthermore, it is possible to select units that track each other within 1% and with extreme care over a small temperature range to less than 0.25% or 1 part in 400 parts. For general discussion, we will employ a balancing of units to 4% of the normal variation. Thus, with two balanced units the emitter voltage variation can be reduced to ± 100 μvolts/°C. Two circuits that accomplish this reduction are shown in Figures 2-0.4 and 2-0.5. In Figure 2-0.4, the balancing action is obtained by the use of a diode in series with the emitter of the transistor. The diode is polarized to balance out the variations of the emitter to base voltage. The resistance R_3 should be large to simulate a current source.

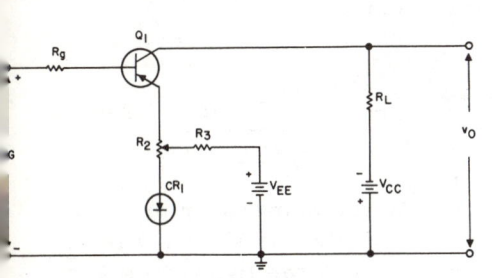

Fig. 2-0.4 - Balanced circuit with an auxiliary diode

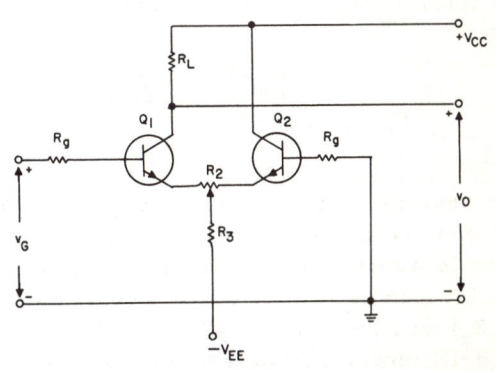

Fig. 2-0.5 - Circuit for balancing the variation in emitter-base voltage

The potentiometer R_2 can be employed to achieve a fine control on the balancing action. The circuit of Figure 2-0.5 is similar in action, except that a second transistor is employed. The circuit is a single-ended amplifier and balances out only the emitter to base voltage of Q_1. The two transistors should be of similar type. With either of these circuits, the order of magnitude of the drift voltage is approximately ± 100 μVdc/°C caused by the variation of V'_{BE}. With low values of emitter current (approximately 50 μamp) the drift voltage due to the variation of a_{fb} is approximately 30 μVdc/°C to 300 μVdc/°C. Neither of the circuits shown in Figures 2-0.4 and 2-0.5 are balanced for power supply variations or for gain variations. If the drift level is reduced to the range of 100 microvolts at the input, the stability of the collector supply may require some care. If the stage gain is 10, then a power supply stability better than 1 mVdc is required. Before proceeding to more thoroughly balanced circuits, a discussion of minimum detectable signal power is in order.

Up to this point, the criterion established for design evaluation has been a signal to drift ratio or drift voltage at the input. The equations that were developed included the generator resistance. Thus, it is possible to consider a minimization technique of drift power taken from the source. If the I_{CO} of the transistor can be neglected, then the drift voltage referred to the input is from equation 2-0.10

$$v_D \simeq \left[\frac{\pm \Delta V'_{BE}}{dT} - I_E R_T \frac{d\alpha_{fb}}{dT}\right] \Delta T \qquad 2\text{-}0.16$$

where $\Delta V'_{BE}$ indicates the unbalance voltage of a balanced circuit. The incremental input resistance is

$$R_{in} = r_b + \frac{r_e + R_e}{1 - \alpha_{fb}}. \qquad 2\text{-}0.17$$

The total drift power taken from the source is then

$$P_{drift} = \frac{\left[\frac{\mp \Delta V'_{BE}}{dT} + I_E (R_g + r_b + R_e + r_e) \frac{d\alpha_{fb}}{dT}\right]^2 (1 - \alpha_{fb})}{(R_g + r_b)(1 - \alpha_{fb}) + r_e + R_e}. \qquad 2\text{-}0.18$$

It is obvious that there is a possibility of equating the numerator to zero if the signs of the two terms are dissimilar. It can also be shown that there is a minimum value of drift power if the signs of the two terms are similar. In either case, the resulting optimum value of the resistance R_e (or R_g) will be dependent on the differential drift voltage $\Delta V'_{BE}$. In any case, the drift power can be reduced by the use of a high gain transistor. This aspect will not be pursued further except to emphasize that the optimum value of R_e can be finite and of large value.

D. Balanced Difference Amplifiers

The two previous circuits afforded a first order balance for emitter diode variations. By the use of a difference amplifier technique, reduction of the drift that is caused by vatiations in gain, I_{CO}, and power supply can be achieved. A circuit diagram is shown in Figure 2-0.6. The expression for drift voltage referred to the input can be shown to be similar to equation 2-0.11, except that in each term that is variable with temperature the difference between the parameter variations of the two transistors appears. This circuit is an excellent circuit for a low level input stage. To reduce common mode effects, the resistor R_1 should be very large. In some designs the resistor R_1

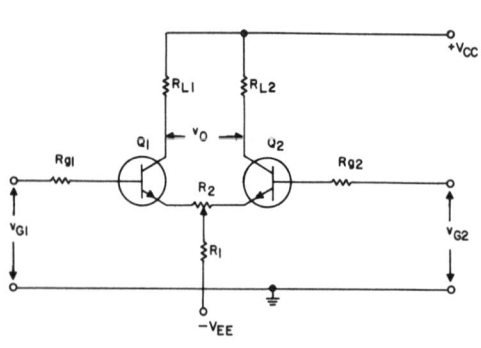

Fig. 2-0.6 - Balanced difference amplifier

and battery V_E are replaced by a transistor biased to provide a constant current. Amplifiers of this type have been designed to yield a drift voltage of less than 10 μvolts/°C over a small temperature range and less than several millivolts over a temperature range of 50°C. The minimum detectable signal power of a well designed differential amplifier can be as low as 10^{-14} watts/°C.

The drift of second or third stages that follow an output stage is usually not difficult to reduce. The importance of second stage drift is reduced by the gain of the first stage. Since the quiescent bias current of the second stage must be larger than the first, the gain dependent drift can be the determining factor. However, a careful choice of bias currents and gain can be employed to effect a cancellation of drift in the first and second stages.

E. Modulators

The design of low level d.c. amplifiers requires extreme care and careful selection of transistors. With very good design, the minimum detectable signal power approaches 10^{-14} to 10^{-15} watts/°C. The thermal noise power in a 1000 c.p.s. bandwidth at room temperature is approximately 2×10^{-17} watts. The noise figure of the transistor will limit the minimum detectable power to a value greater than 2×10^{-17} watts. The use of a modulator to convert the d.c. signal to an a.c. signal is an alternative technique that may in the future produce results superior to direct coupling.

There are a large number of converters that can be employed. First, a mechanical chopper with conventional techniques can yield a noise or drift voltage of the order of 10 to 100 microvolts over a very wide temperature range into a high impedance. In many applications, the finite life of a chopper with mechanically moving parts proves a great, if not impossible, disadvantage. Magnetic converters that function on a second harmonic principle can provide minimum detectable signals of 10^{-9} amp with an input impedance of 1000 ohms. The disadvantage of this modulator is its size, the power required to operate it, and its large volume in comparison with transistors. In both of the previous types, the "chopping" or modulating frequency is restricted to 1000 c.p.s. or less. Thus, only narrow bandwidth amplifiers can be obtained by use of a modulate-demodulate technique.

A single transistor can be employed as a switching device to convert low-level d.c. signals to an a.c. waveform. A simplified circuit is shown in Figure 2-0.7 and the operating characteristics in Figure 2-0.8. If the voltage applied to the base is negative, both the emitter and collector diodes are shorted and the output voltage becomes approximately V_1 of

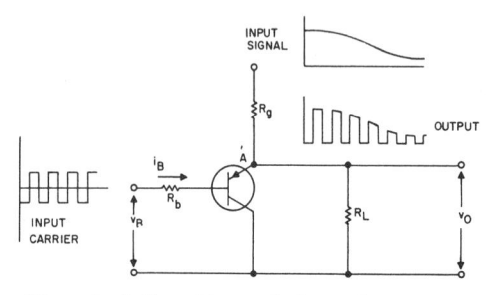

Fig. 2-0.7 - Transistor chopper

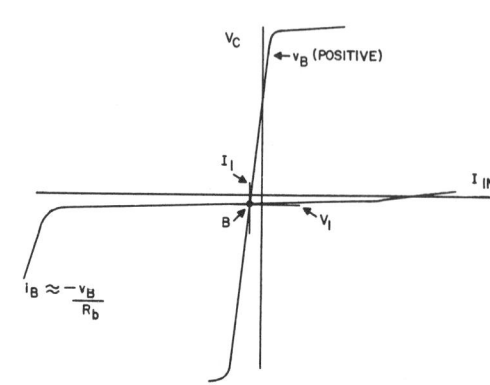

Fig. 2-0.8 - Operating characteristic of a transistor chopper

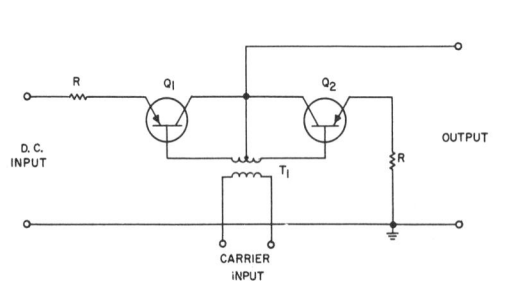

Fig. 2-0.9 - Balanced transistor chopper

Figure 2-0.8. If the base voltage is positive, the diodes are open circuited. The resistance of the switch from point A to ground is in the megohm region and the signal applied at the input is not attenuated. If no voltage is applied at the input the voltage at point A during the high impedance portion of the cycle is approximately $-I_1(R_g R_L/R_g R_L)$. Typical values of the coordinates of the point B are $V_1 = 1$ mVdc $I_1 < 1$ μamp. The variation of the coordinates of this point with temperature are small. For silicon transistors I_1 may be less than 1 microamp at temperatures as high as 125°C. Some of the temperature dependent drift of the voltage V_1 may be greatly reduced by the use of two transistors as shown in Figure 2-0.9. The coordinate values of matched silicon units remain between ± 0.2 mVdc and ± 0.2 microamp over a very wide temperature range. Undesired drifts due to the current I_1 can be minimized by keeping either R_g or R_L a small value. For very small signals the switching transients may limit the minimum detectable signal.

There are now commercially available silicon transistor choppers that have an equivalent drift voltage at the input of less than ± 200 μVdc over a temperature range of -55°C to +135°C. The input impedance can be made very high with some degradation of drift but the generator resistance must be kept low. A minimum detectable signal power less than 10^{-1} watts/°C can be achieved. The conversion power gain of a resistance chopper must be less than unity. Thus the chopped signal must be efficiently matched to the a-c amplifier that follows the chopper.

Another type of semiconductor modulator employs the voltage sensitive capacitance of silicon junction diodes. This modulator is essentially a parametric amplifier. The modulator can provide power gain. Reasonably low drift voltage can be achieved over small temperature ranges. The basic operation is illustrated in Figure 2-0.10. If the capacitors are voltage sensitive and have the form

$$C_1 = C_o + K f(v) \qquad 2-0.19$$

the magnitude of the output voltage can be shown to be proportional to the product of the magnitude of the carrier signal times the d.c. input signal. To achieve a balance with zero d.c. signal applied, the capacitors must be equal or the voltage applied to the capacitors varied. To maintain a balance with temperature, the diode characteristics must be identical. In practical circuits, silicon junction capacitors have been employed in a bridge circuit. The drift voltage of the modulator is limited by a balance of the contact potentials of the two diodes. Since the temperature coefficient of contact potential is approximately -1 mVdc/°C for heavily doped materials, the diodes must be carefully selected to reduce drift.

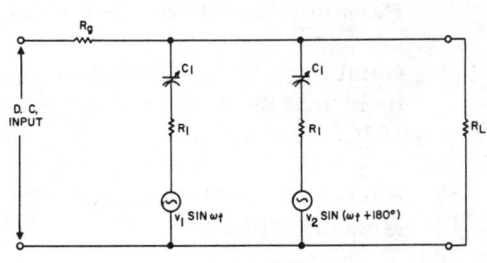

Fig. 2-0.10 - Capacitance modulator for low-level d.c. conversion

Conversion gain can be maximized by conjugate matching the load resistance R_L to the output impedance of the bridge.

BIBLIOGRAPHY

1. Adler, R. B., "A Large Signal Equivalent Circuit for Transistor Static Characteristics," M.I.T., RLE Transistor Group Report T-2, August 1951.

2. Biard, J. R. and Matzen, W. T., "Differential Amplifier Features D-C Stability," Electronics, Vol. 32, No. 3, pp 60-62, January 16, 1959.

3. Bode, H. W., "Network Analysis and Feedback Amplifier Design," D. Van Nostrand Company, Inc., New York, 1947.

4. Bright, R. L., and Kruper, A. R., "Transistor Choppers for Stable DC Amplifiers," Electronics, April 1955.

5. DeBolt, H. E., "A High Sensitivity Semiconductor Diode Modulator for DC Measurement." Internal report of Fairchild Camera and Instrument Corporation.

6. Ebers, J. J. and Moll, J. L., "Large-signal Behavior of Junction Transistors," Proc. IRE, Vol. 42, No. 12, p. 1761, December 1954.

7. Gartner, W. W., "Temperature Dependence of Junction Transistor Parameters," Proc. IRE, Vol. 45, No. 5, pp 662-680, May 1957.

8. Goldberg, E. A., "Stabilization of Wide Band Direct-current Amplifiers for Zero and Gain," RCA Review, Vol. 11, No. 2, p. 296, June 1950.

9. Hunter, L. P., "Handbook of Semiconductor Electronics," Section 13, McGraw-Hill, 1956.

10. Hurtig, C. R., "Bias Stabilization of Junction Transistors," M.I.T., RLE Transistor Group Report, August 1953.

11. Lin, H. C. and Barco, A. A., "Temperature Effects in Circuits Using Junction Transistors," Transistors I, RCA Labs., Princeton, N.J., pp 369-402, 1956.

12. Manley, J. M., "Some General Properties of Magnetic Amplifiers," Proc. IRE, March 1951.

13. Moody, N. F., "A Silicon Junction Diode Modulator of 10^{-8} Ampere Sensitivity for Use in Junction Transistor Direct-Current Amplifiers Proc. NEC, Vol. 11, p. 441, 1955.

14. Rote, W. A., "Magnetic Converter DC Amplifier," Electronics, December 1953.

15. Roy, R., "Transistorized High Frequency Chopper Design," Electron Design, August 6, 1958.

16. Schaffner, J. S. and Shea, R. F., Letter to the Editor, Proc. IRE, p. 101, January 1956.

17. Shea, R., "Transistor Circuits Engineering," John Wiley and Sons, Inc., New York, 1957.

18. Slaughter, D. W., "Feedback Stabilized Transistor Amplifier," Electronics, May 1955.

9. Slaughter, D. W., "The Emitter-Coupled Differential Amplifier," Trans. IRE, PGCT, Vol. CT-3, No. 1, p. 51, March 1956.

10. Stanton, J. W., "A Transistorized D.C. Amplifier," Trans. IRE Vol. CT-3, pp 65-66, March 1956.

11. Warren J. P., "A New Approach to the Design of Low Drift D.C. Amplifiers." Technical Digest, Solid State Circuits Conference, pp 36-37, February 1959.

CIRCUIT 2-1

DIFFERENCE AMPLIFIER

Transistor Applications, Inc.

One of the most commonly used techniques for balancing the temperature dependent drifts of resistively-coupled transistor amplifiers employs the emitter-coupled difference amplifier. This class of amplifier originates with D. W. Slauthter of California Institute of Technology, and has been additionally developed by Transitron Corporation and Texas Instruments Corporation. A circuit similar to Slaughter's original work is shown in Circuit 2-1. Special attention has been focused on the selection of operating points of the transistors. As described in the design philosophy, the first stage of the transistor amplifier should be operated at a very low collector current, and the emitter to base voltage variation must be balanced out. The objective of this circuit design is to obtain low drift rates without extreme selection of transistors and also without the use of a temperature-dependent balance. Accordingly the only selection that is made is the matching of transistors very carefully for gain at the proper operating current. In Circuit 2-1, the operating currents are as follows: collector current of the

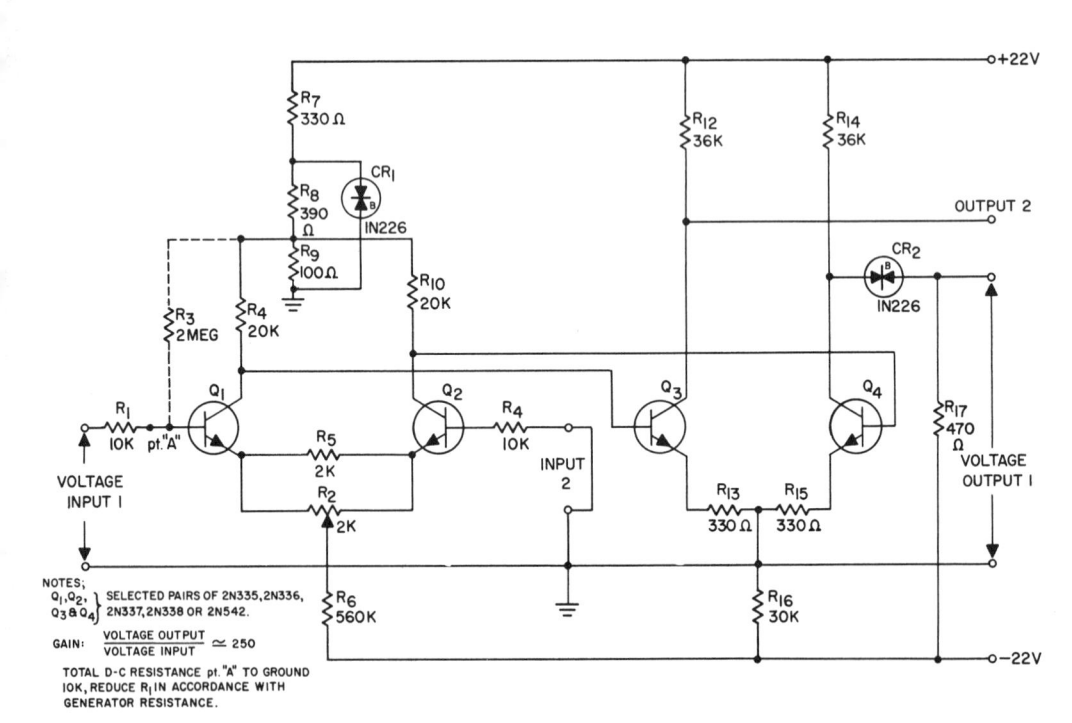

Circuit 2-1 - Difference amplifier

Circuit 2-1, Difference Amplifier (cont'd.)

first pair approximately 20 μamps per unit, collector current of second pair approximately 350 μamps. The gain selection can be easily and quickly made with a transistor plotter. The transistors are selected for current gain of the grounded emitter connection (α_{FE}) into groups with a tolerance of ± 5%. The highest gain units are employed in the first stage. Approximately 25% of the 2N335 transistors had a current gain in excess of 20 at an emitter current of 20 μamps.

The difference amplifier pair affords a reduction in variations caused by emitter to base voltage, saturation current, temperature dependence of the current gain, and power supply variations. Since considerable attention has been given to common mode rejection in the existing literature, this discussion will be confined to the operation of the unit and the results achieved.

The potentiometer R_2 is employed for a zero adjustment of the output voltage at point #1. The diode CR_2 may require selection or a change to a unit with a different breakdown voltage if the gain selection yields results significantly different from those described. If the amplifier is operated with a single-ended input the drift voltage for an average pair of units is ± 120 μvolts/°C. This change is quite linear and will hold with reasonable accuracy over a temperature range from 10°C to 65°C. At temperatures above 65°C, the drift departs from a linear variation. This is due in part to the variation of the saturation current with temperature. Since the units were not selected for equal saturation currents at an elevated temperature, the balancing action is not necessarily obtained. Depending on which transistor has the larger saturation current the variation in drift slope can either increase or decrease.

At a bias current of 20 μamps, the incremental input impedance of the pair is approximately 70,000 ohms. Thus the minimum detectable signal power of a typical pair is approximately 2×10^{-13} watts/°C. The spread of values for different pairs places drift signals in the range of 40 μVdc/°C to 600 μVdc/°C. The corresponding minimum detectable power levels are 2×10^{-14} watts/°C to 5×10^{-12} watts/°C.

At this point it is worthwhile to point out one very fundamental practical construction technique. The variation of emitter to base voltage is approximately 2.5 mVdc/°C. The resulting drift of the current pair shown in Figure 2-1, which did not require an excessively difficult selection of transistors is 0.120 mVdc/°C. This is an effective reduction of 20:1. Thus a differential temperature drift of the two units of 0.05°C can result in an error as large as the unbalance. It is therefore of utmost concern to operate the transistors at the same temperature within very close tolerances.

Circuit 2-1, Difference Amplifier (cont'd.)

The transistors should therefore be mounted in close proximity to each other in a material of high heat conductivity. To avoid transients caused by sudden changes in temperature, the amplifier should be packaged in a manner that produces a unit with high heat capacity.

CIRCUIT 2-2

COMPENSATED D.C. AMPLIFIER (TEMPERATURE ADJUSTED)

General Electric Company   Contributed by J. W. Stanton

Circuit 2-2 employs germanium transistors where the major temperature dependence arises from the saturation currents of the units. Transistors Q_1 and Q_3 form a forward gain amplifier whereas the third unit Q_2 is employed for temperature compensation. Transistors Q_1 and Q_3 are operated in a straight grounded emitter connection. The current gain of the two-stage amplifier is limited by the resistors R_2 and R_7 to the approximate value or R_5/R_7, which in this case is approximately 100. Transistor Q_2 is operated in a grounded emitter connection (with feedback) at essentially a low current level that is determined by the saturation current of the transistor. To achieve temperature compensation, the saturation current I_{CO_2} must be large compared with I_{CO_1}. Introduction of the

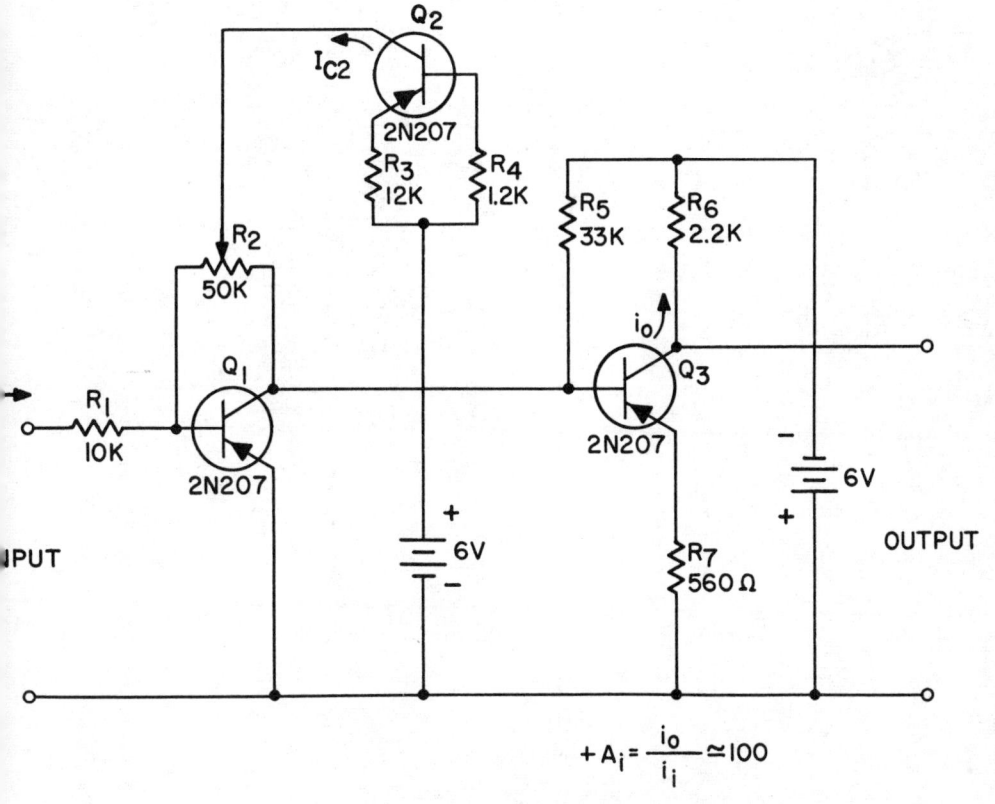

Circuit 2-2 - Compensated d.c. amplifier

Circuit 2-2, Compensated D.C. Amplifier (Temperature Adjusted) (cont'd.)

compensating current into the base of transistor Q_1 results in the maximum degree of compensation. Potentiometer R_1 is adjusted as a function of temperature to maintain a minimum drift.

A high degree of stability is difficult to achieve with this circuit. Since the saturation currents vary exponentially with temperature, and the saturation currents are not small in value, it is difficult to maintain a zero balance over a wide temperature range. Over a temperature range from 20°C to 60°C it is possible to achieve temperature drifts as low as 0.5×10^{-6} amps at the input. The drift, however, can be as large as 5×10^{-6} amps. With the better degree of balance the minimum detectable signal power is approximately 3×10^{-9} watts. Since the drift current is not linear with temperature, it is not possible to specify a temperature gradient of the drift signal. However, the major portion of the drift occurs at temperatures above 45°C.

CIRCUIT 2-3

LOW DRIFT HIGH IMPEDANCE
STRAIGHT D.C. CURRENT AMPLIFIER

Transitron Electronic Corporation	Contributed by Nicholas DeWolf
and John Warren

    Circuit 2-3 is designed to amplify very low currents, not exceeding 50 millimicroamperes (0.05 μAdc), from high impedance sources, preferably at least 10 megohms. The amplifier is of the "straight" d.c. type, and has a current gain somewhat over 60 db. This is a temperature adjusted type amplifier.

    The circuit consists of two direct-coupled grounded emitter stages, Q_1, Q_2. The input transistor is a special type designed to operate at extremely low currents. At these low operating levels, the absolute changes will be small for typical changes in transistor parameters with time or with varying temperature.

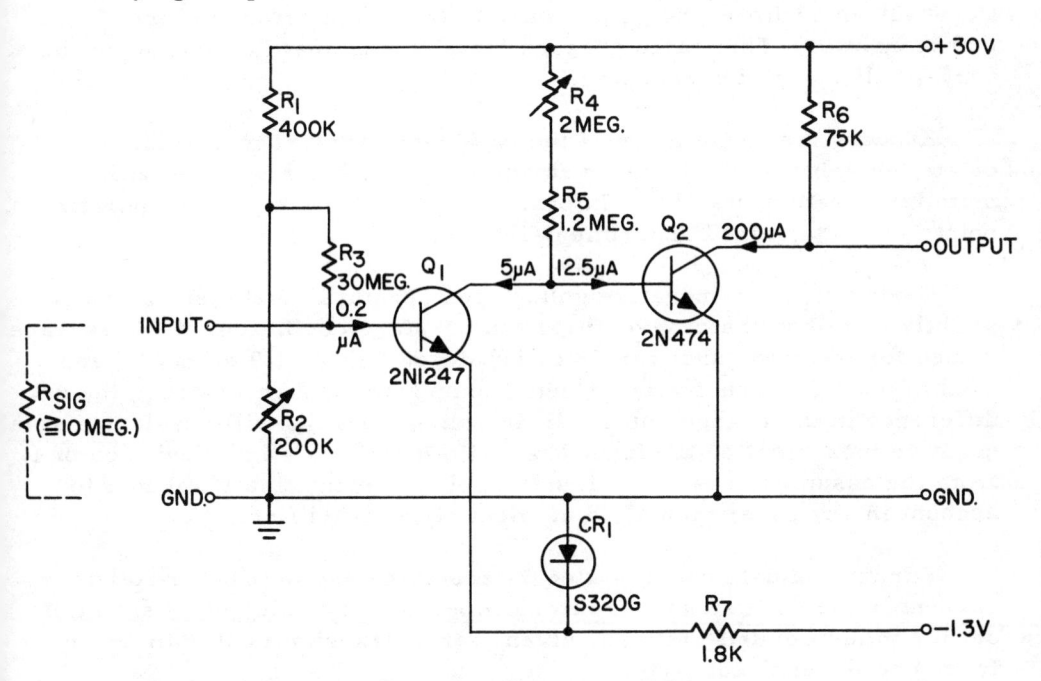

Circuit 2-3 - Low drift, high impedance,
d.c. current amplifier transitron

Circuit 2-3, Low Drift Straight D.C. Current Amplifier (cont'd.)

Temperature coefficients of transistor d.c. current gains, defined in such a way as to include I_{CO} variations, must be taken into account. In a current amplifier, these coefficients affect the overall performance more than the base-to-emitter voltage temperature coefficients conventionally considered. Relatively simple circuitry has been evolved, in which careful selection of circuit elements permits a close approach to the cancellation of all temperature coefficients present.

The base bias current of Q_1 is very little dependent upon temperature being derived through a high resistance from a relatively high voltage. Any increase in ambient temperature causes an increase in the collector current of Q_1, due to the temperature coefficient of the d.c. current transfer ratio. This collector current I_{C_1}, and the base current of Q_2, I_{B_2}, are drawn from another current source, composed of the high resistance in the collector circuit and the high supply voltage. Accordingly, if the collector current increases, the following base current must decrease by approximately the same amount. The bias conditions are arranged so that this decrease in I_{B_2}, and the increase in the d.c. current transfer ratio of Q_2, result in as little change as possible in the collector current, I_{C_2}, of that transistor. This same process tends to eliminate variations in the very small I_{CO} of the first stage.

Knowing the parameters of the two transistors, it is possible to calculate the value of I_{B_2} for a given value of I_{C_1}, which will result in optimum temperature stability. In practice, however, the proper operating points are determined experimentally.

Diode CR_1 has a negative voltage temperature coefficient which is slightly less than the base-emitter temperature coefficient of Q_1, average values for the diode and transistor employed being - 1.9 mVdc/°C and - 2.3 mVdc/°C respectively. When a temperature change occurs the net difference in the changes of the diode voltage and V_{BE} will result in a small negative increment on the input base. However this slight deviation of I_{B_1} from the assumed constant current supply is automatically taken into account in any experimental adjustment of the amplifier.

For this adjustment, resistor R_2 should be set so that the collector current in the first stage is 5 microamperes. The value of R_4 should be set for minimum drift with any given pair of transistors if optimum performance is required.

The circuit is designed to operate from 25° to 65°C ambient, and over this range input drifts as low as 0.1 millimicroamperes/°C (0.0001 µAdc) and 1.0 millimicroampere per day may be achieved. If even smaller drifts are required, the designer has recourse to oven operation.

Circuit 2-3, Low Drift Straight D.C. Current Amplifier (cont'd.)

One limitation occurring in amplifiers of this type is the restriction of frequency range because of the high impedance levels; the combination of these with typical collector capacitances may be enough to reduce cutoff frequency to the order of 1 kc in extreme cases.

The minimum detectable signal power of this circuit is approximately 10^{-13} watts/°C for the recommended generator resistance.

CIRCUIT 2-4

BALANCED CIRCUIT WITH INDEPENDENT ZERO SET AND TEMPERATURE BALANCE ADJUSTMENTS

Hermes Electronic Company   Contributed by C. R. Hurtig

    In many circuits, such as the balanced difference amplifiers, the resulting temperature dependent drift signal is a linear function of temperature. This fact arises from the difference in the temperature coefficients of the emitter to base voltages of the transistor pairs. To achieve improved performance, the transistors may be selected for "matched" temperature coefficients or a temperature dependent balance. The first technique is limited by either cost or the inability to perfectly match transistors. Thus, temperature dependent balances are quite often employed to reduce costs or to improve performance. Two techniques of accomplishing a temperature dependent balance have been shown in Circuits 2-2 and 2-3. Another means of accomplishing a temperature dependent balance is shown in Figure 2-4.2. Basically the circuit of Figure 2-4.2 is a

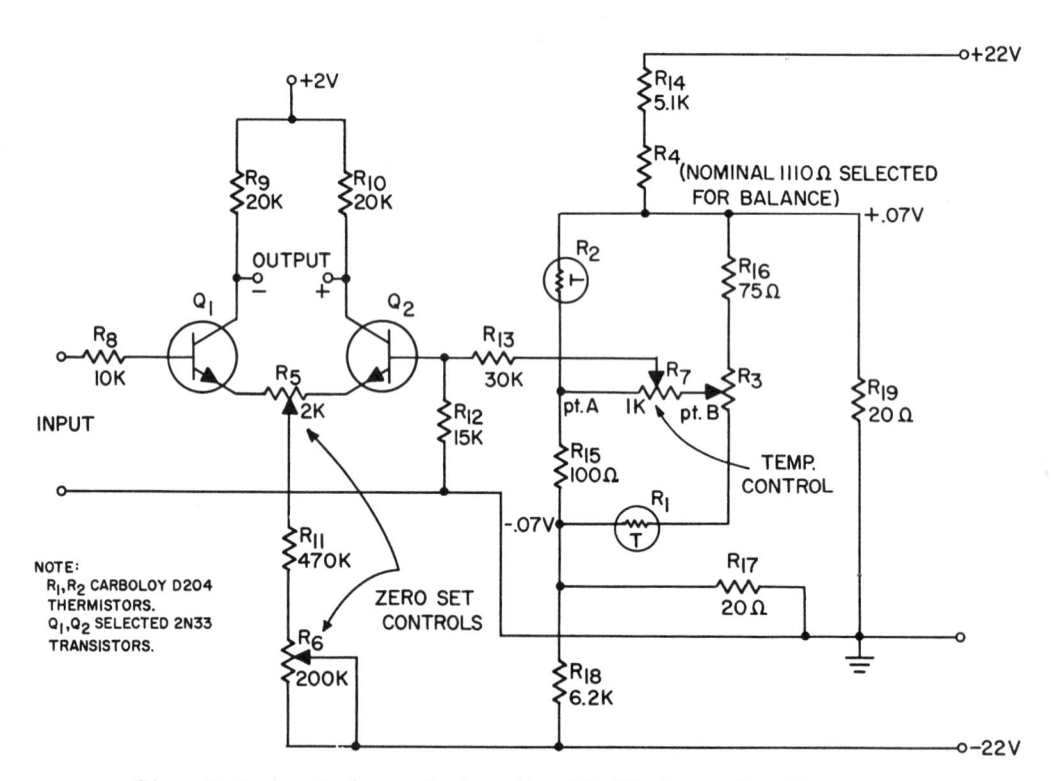

Circuit 2-4 - Balanced circuit with "independent" zero set and temperature balance adjustments

Circuit 2-4, Balanced Circuit with Independent Zero Set and
        Temperature Balance Adjustments (cont'd.)

difference amplifier with the addition of a transistor Q_3 that acts as a current source. The resistance R_1, moreover, is a temperature sensitive component such as a Texas Instrument sensitor. This type of resistor provides a temperature coefficient of resistance that has a value of +0.7%/°C over a wide temperature range. Over a 15°C temperature range the change in resistance approaches a linear change with temperature. The output current of transistor Q_3 can be designed to vary approximately as a linear function of temperature. As temperature increases, the value of R_1 increases and causes the current Q_3 to decrease. If the arm of potentiometer R_2 is centered and the transistor Q_1 and Q_2 are matched units, the differential output voltage will not be influenced by the variation in I_3. Depending on the position of the arm of potentiometer R_2 the output voltage can be made to increase or decrease with temperature. Proper adjustment can result in a high degree of stability of the output voltage. With the use of an auxiliary balance and matched units, an equivalent input drift voltage as low as ± 6 μvolts/°C have been achieved. The drift power can be as low as 5×10^{-15} watts/°C.

In Circuits 2-2, 2-3 and 2-4, temperature dependent adjustments are made. In all of these circuits, the adjustment employed to compensate for drift unfortunately also varies the quiescent operating point (the zero adjustment) of the amplifier. Thus to achieve a low value of drift signals several temperature cycles must be made. Two circuits will be presented which are designed to afford an independence of the temperature balance and zero set adjustments. Although these circuits are not perfect, a first order of magnitude of independent adjustments can be achieved.

The first circuit, as shown in Figure 2-4.2, employs a thermistor bridge in conjunction with a single-ended amplifier. The first stage of the amplifier employs two transistors in a differential balancing circuit. The base lead of the balancing transistor is connected to the thermistor bridge.

The thermistor characteristic is shown in Figure 2-4.1. The resistance is a highly nonlinear function of temperature. However, the combination of a thermistor and resistor results in a ratio that is a close approximation to a linear function of temperature. Over a ± 10°C range the departure of the ratio $R_1/(R_1+100)$ from a linear function is less than 10%.

The thermistor bridge of Circuit 2-4 is arranged so that points A and B are at zero potential. The voltage at point A increases with temperature at a rate of + 0.7 mVdc/°C, whereas the potential at point B decreases at a rate of -0.7 mVdc/°C. If the potentiometer R_7 is set at its mid point, the voltage applied to the base of Q_2 is constant with temperature. At either extreme of the potentiometer, the temperature correcting voltage has a

Circuit 2-4; Balanced Circuit with Independent Zero Set and
Temperature Balance Adjustments (cont'd.)

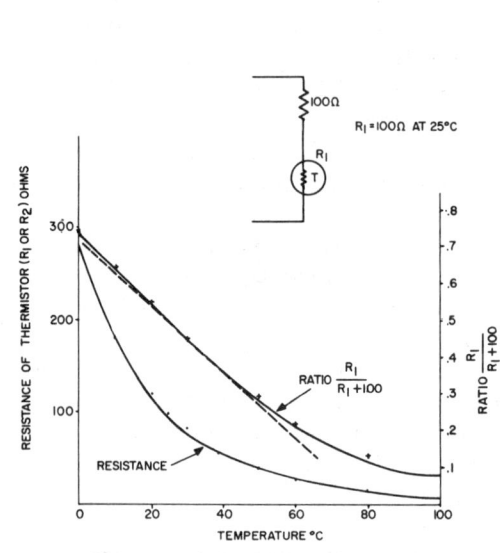

Fig. 2-4.1 - Thermistor characteristics

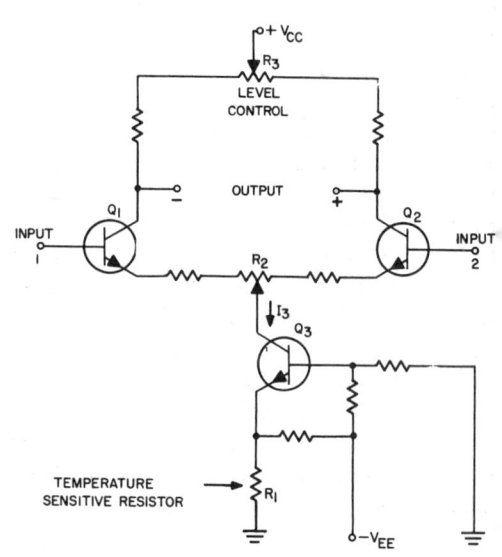

Fig. 2-4.2 - Example of a difference amplifier incorporating a temperature balance

magnitude of ± 230 μvolts/°C. The adjustments required are as follows: R_5 corrects for unequal gain of the transistors; R_6 provides a zero set adjustment; R_4 provides equalization of the voltages applied to the bridge; R_3 balances the bridge.

Over a limited temperature range ± 10°C it is possible to obtain an equivalent input drift voltage less than ± 20 μVdc/°C. With carefully matched components and extreme care in the thermal mounting of the transistors and thermistor bridge, drift signals of ± 10 μVdc/°C can be realized. The corresponding drift power is less than 10^{-15} watts/°C. The drift resulting from power supply variation is not negligible. A 1.0 mVdc differential unbalance of the + 22 volt and - 22 volt supplies will create an error signal at the base of Q_2 of approximately 1 μvolt.

A similar technique can be designed using forward biased junction diodes to obtain a correcting signal that is linear over a wide temperature range. However, the sensitivity to power supply variations is increased by a factor of 20.

Slightly improved results can be obtained by placing a bridge incorporating breakdown diodes in the emitter circuit of the first pair.

CIRCUIT 2-5

SILICON DIODE RING MODULATOR

Defense Research Board, Telecommunications Establishment, Ottawa, Canada

Contributed by N. F. Moody

This circuit represents a large variety of modulator circuits. Rather than employ a resistively coupled amplifier it is possible to modulate the low-level d.c. signal, amplify the modulated signal, and then demodulate to obtain the d.c. signal at high level. The diode bridge of Circuit 2-5 employs silicon diodes and has a stability of 10^{-8} amp at room temperature.

The modulator is a conventional diode bridge. The carrier signal is applied by a balanced transformer and the output is fed from a balanced transformer. The center tap of the transformer is critical to successful operation. The silicon junction diodes require matched forward characteristics and a reverse current smaller than 10^{-8} amp. The output waveform is basically a square wave filtered by the output transformer. The amplitude of the output signal is proportional to the magnitude of the d.c. signal. The phase of the output signal with respect to the carrier signal is proportional to the sign of the d.c. signal.

The input impedance of the modulator to signal current is approximately 70,000. The minimum detectable current is 10^{-8} amp at room

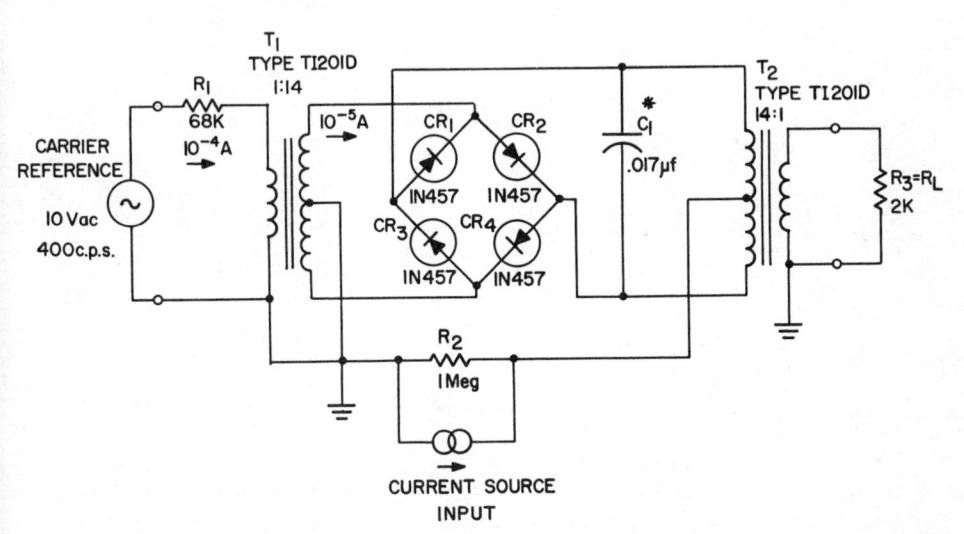

Circuit 2-5 - Diode ring modulator

Circuit 2-5, Silicon Diode Ring Modulator (cont'd.)

temperature corresponding to approximately 10^{-10} watts. The drift current is approximately 10^{-7} amp for a temperature range from 25°C to 80°C. The drift signal is an exponential function since the saturation current of the diodes is the limiting factor.

There are many other types of modulator circuits. In particular the second harmonic magnetic converter provides very low values of drift signals.

(From the article "A Silicon Junction Diode Modulator," Electronic Eng., 28, 94-100, March 1956, by N. F. Moody)

CIRCUIT 2-6

SEMICONDUCTOR-CAPACITOR CHOPPER

Hermes Electronic Company　　　　　　Contributed by C. R. Hurtig

Circuit 2-6 is an experimental balanced modulator that is useful for converting very low-level d.c. signals to a.c. signals. The circuit consists of an RC bridge as shown in simplified form in Figure 2-6.1. The capacitors are zero-biased silicon junction diodes. The transition capacitance of the diodes employed is an inverse square-root function of the sum of the applied modulating voltage and the contact potential of the diodes. The bridge is balanced under the condition

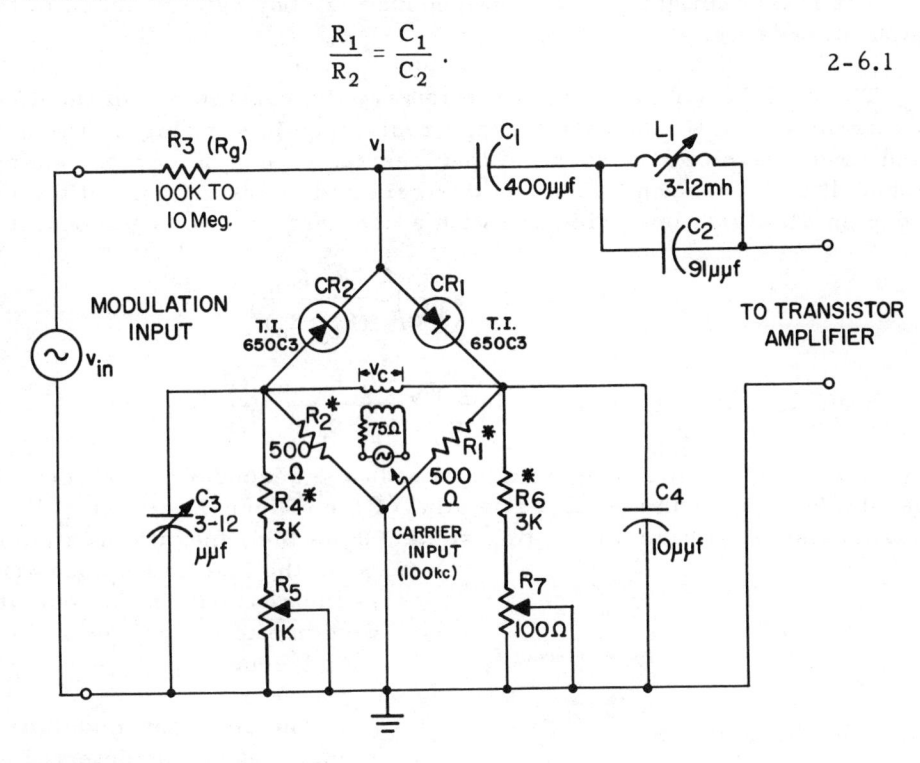

$$\frac{R_1}{R_2} = \frac{C_1}{C_2}.$$ 2-6.1

* 1% WIREWOUND RESISTORS

Circuit 2-6 - Semiconductor-capacitor chopper

Circuit 2-6, Semiconductor-Capacitor Chopper (cont'd.)

The application of a small d.c. voltage at the modulation terminals causes the capacitance of the forward-biased diode to increase and of the reverse-biased diode to decrease. The bridge is, therefore, unbalanced and a signal at the carrier frequency is developed across the output terminals. The phase of the output signal varies by 180° as a function of the polarity of the modulation signal. The output impedance of the bridge at the carrier frequency is approximately

$$Z_{out} = \frac{R_1 R_2}{R_1 + R_2} + \frac{C_1 + C_2}{j\omega C_1 C_2}. \qquad 2\text{-}6.2$$

For maximum gain the load impedance is conjugate-matched to this output impedance.

The modulation power required to vary the capacitance of the diodes is extremely small since the leakage resistance is very high. The sideband power developed across the load resistance can be correspondingly large. It can be shown that the power gain and operating gain of the bridge under small signal operation and with a conjugate match at the output are

$$\text{Power gain} = \left(\frac{v_c}{4V_C}\right)^2 \frac{R_L R_{in}}{(R_L + R_p)^2} \qquad 2\text{-}6.3$$

$$\text{Operating gain} = \left(\frac{v_c}{4V_C}\right)^2 \frac{R_L R_g}{(R_L + R_p)^2} \frac{(R_{in})^2}{R_{in} + R_g} \qquad 2\text{-}6.4$$

where v_c is the sum of rms voltage across the secondary of the coupling transformer, V_C is the contact potential of the silicon diodes. R_{in} is the input resistance to the modulating signal, R_g is the generator resistance, R_L is the load resistance and R_p is the real part of the output resistance of the bridge at the carrier frequency.

The smallest modulating signal that can be detected at the output is limited by one of three basic factors: first, the temperature dependence of the balance of the bridge; second, the effect of stray coupling of the carrier, which tends to produce a residual output; and third, by the noise performance of the overall circuit.

Circuit 2-6.1 - Simplified circuit of the balanced modulator

Circuit 2-6, Semiconductor-Capacitor Chopper (cont'd.)

The temperature dependence of the bridge balance is to a major degree dependent on the temperature dependence of the diode capacitance. In general, the temperature dependence is determined almost entirely by the variation of the internal contact potential of the diode with temperature. The temperature coefficient of the transition capacitances of the diodes used is approximately $650 \times 10^{-6}/°C$ at zero bias. The reverse breakdown voltage of the diodes is approximately 4.0 volts. With some selection of diodes the effective drift voltage referred to the input is approximately 5×10^{-6} volts/°C. With this order of magnitude of temperature stability, the short time limitation on the minimum input signal is caused by a modulation of the diode capacitance by thermal noise and residual carrier present at the input. Noise figure can be defined as the ratio of total noise at the carrier frequency at the output to the noise at the output caused by modulation of the capacitors. It has been possible to achieve noise figures as low as 10 db.

The diodes were first chosen for equal breakdown voltage and then in addition for optimum temperature performance. Since the input impedance is greater than 10^6 ohms the minimum detectable signal power is less than 5×10^{-18} watts/°C. Thus the device is limited in the practical case by thermal and shot noise.

CIRCUIT 2-7

HIGH VOLTAGE DIRECT-COUPLED AMPLIFIER

Philco Corporation   Contributed by J. B. Angell

The high voltage, temperature uncompensated, direct-coupled amplifier shown in Circuit 2-7 may be used effectively where large voltage swings are desired.

The arrangement of Q_1, Q_2, Q_3, and Q_4 permits the supply of voltage, V_{CC}, of -100 volts to be divided among the transistors, such that the collector voltage of each is within the rated value of -30V for the type 2N384 transistor. Since a single transistor does not permit a voltage swing in excess of its rated maximum collector voltage, the high instantaneous output voltage across R_L must be, and is, shared at all times by the transistor connected in series. The transistors used do not have to be matched for identical characteristics and any number, in addition to the input transistor, may be used.

The voltage swing for each transistor appears across the 18K resistor connected in the respective base circuit, and the overall voltage swing appearing across the load resistor R_L is the sum of the voltage swings of the individual stages.

The current gain of the overall amplifier is that of the input stage Q_1, where the connection can be either grounded-base or grounded-emitter. When large current gains are desired, the grounded-emitter connection, as shown, should be used.

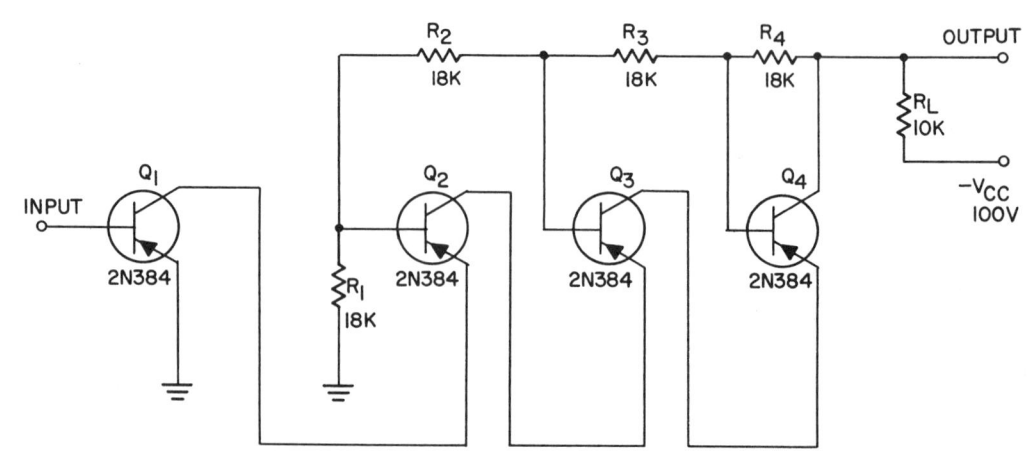

Circuit 2-7 - High voltage direct-coupled amplifier

PART 3

LOW-FREQUENCY AMPLIFIERS

Section 1. Design Philosophy

Transistors find wide applications in low frequency operations because of their high efficiency, freedom from microphonics, small size, and circuit economy achieved with the use of complementary transistors. However, there are many problems associated with the operation of transistors as low frequency amplifiers.

For efficient operation, it is desirable to alternately swing the transistor down to a very low voltage and high current level or high voltage, low current level. On the other hand, at a low current level there may be distortion due to the increased input resistance. At high currents, there may be distortion due to the fall off in current amplification. The proper choice of operating point will affect the distortion considerably. Linear operation of course depends on low distortion, and this will be an important consideration in the amplifier design.

Negative feedback can be used to reduce distortion, and the choice of shunt or series negative feedback depends on the source impedance and the nature of the distortion.

In all amplifiers, it is important to stabilize the operating current against temperature variations, since distortion, dissipation, gain and many other characteristics are related to the temperature variation. The technique of stabilization depends on whether the stage is direct coupled, RC or transformer coupled, and whether it is a low level or high power stage.

In high power stages, high temperature may be created by the internal dissipation of the transistor. This may not only deteriorate the performance, but also cause the transistor to run away due to thermal instability. Thermal stability requires good circuit design, as well as good thermal design, and a knowledge of proper heat sink design and derating techniques.

A good circuit design should have provision for the interchangeability of transistors. Generally transistors of same type may vary widely in current gain, reverse current and input resistance. Since transistor circuits with good temperature stability are well guarded against any variations in these respects, such circuits also provide good interchangeability of transistors.

In low-level input stages, the noise may pose a problem. The operating current and voltage should be chosen to be low enough to yield low noise. However, the choice of operating point should be compromised with consideration of distortion, gain and input impedance.

Transistor RC-coupled or direct-coupled amplifiers are essentially current amplifiers. The high output impedance is connected to the low input impedance of the following stage. For current amplifiers, the input should be fed to the base. The output may be taken from either the collector or the emitter. The selection depends on the input impedance desired, and whether the following transistor is complementary or not.

For maximum transfer of energy, transformer coupling may be used to match the high output impedance to the low input impedance of the following stage. Transformers may also be used to transform the high impedance of a generator to a lower value for low noise operation. In general, the use of a transformer often results in the reduction of the number of stages. On the other hand, transformers impose problems which do not exist in RC-coupled amplifiers. For example, in the Class B amplifier the leakage inductance in the transformer causes transients during the switching off period of a transistor which create an undesirable cross-over distortion. In addition, transformers introduce phase shift which limits the amount of negative feedback which may be applied.

High efficiency is an important consideration in the power stages. Class B amplifiers have higher efficiency than Class A amplifiers; i.e., lower dissipation for the same power output than Class A amplifiers but they have lower power gain. Transistor amplifiers can approach the theoretical efficiency of Class B operation of 78%, and that of Class A operation of 50%.

Before the description of the selected circuits, the considerations just enumerated shall be treated individually to provide a better background and understanding of the selected circuits. The basic configurations for different functions will also be shown.

A. Transistor Characteristics

   1. Input Characteristics

   The input characteristics of transistors are sensitive to the operating current and temperature.

   The input characteristics of a transistor in the common-emitter configuration are shown in Figure 3-0.1. The inverse slope of the characteristic curve is the input resistance of the transistor. Note that at low currents, the input resistance is higher than at higher currents and the characteristics are less linear. The a.c. input resistance r_{in} is approximately

$$r_{in} = r_b' + 25/I_B \qquad 3-0.1$$

where r_b' is the base spreading resistance and I_B is the d.c. base current in mAdc. In the region where $25/I_B \gg r_b'$, the characteristic is nonlinear and inversely proportional to the d.c. base current. As current is increased, r_b' becomes greater than $25/I_B$ and the characteristic becomes more linear.

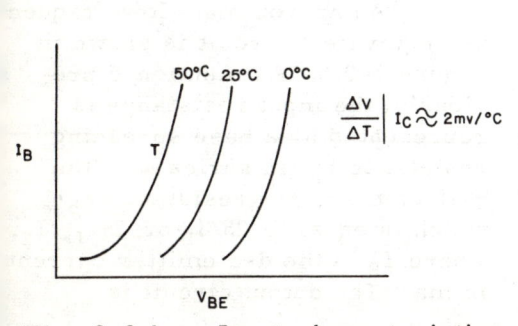

The input characteristic curves move toward the left as the temperature is increased with a rate that is approximately 2 mVdc/°C. Whereas the d.c. characteristics are quite sensitive to temperature, the incremental or a.c. characteristics are not sensitive to temperature as long as the d.c. operating current remains the same.

2. Transfer Characteristics

The output current vs. input voltage characteristics are similar to the input characteristics as shown in Figure 3-0.2. The output current is equal to the current amplification factor, α_{fe}, times the input current. The temperature effect is also similar to that of the input characteristics.

The current amplification characteristics are shown in Figure 3-0.3. Note that as the d.c. emitter current is increased, the current amplification factor first increases, reaches a maximum, and then decreases.

3. Output Characteristics

The output characteristics of a transistor are shown in Figure 3-0.4. There are three regions

Fig. 3-0.1 - Input characteristics

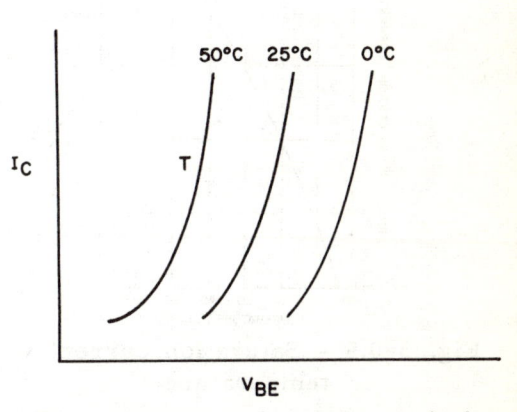

Fig. 3-0.2 - Transfer characteristics

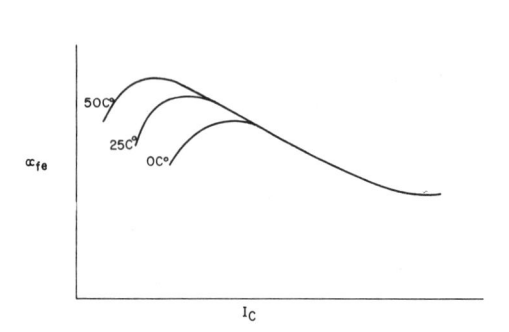

Fig. 3-0.3 - Common emitter current gain vs collector current

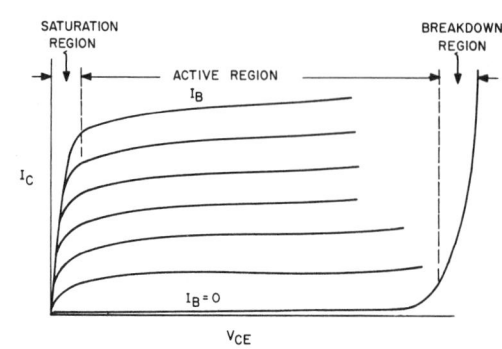

Fig. 3-0.4 - Collector characteristics

of interest. In the active region, the collector current is essentially proportional to the base current, but does not vary appreciably with collector voltage. The output voltage is therefore very high. In the saturation region, the voltage is very low, usually only a fraction of a volt, and the collector current does not increase with increasing base current after the base current reaches a certain value. In the breakdown region, the collector current rises rapidly with voltage and it is no longer controllable by the base current.

As the temperature is increased the curves move upward by an amount equal to the increase of cutoff current I_{CEO}. I_{CEO} is equal to $\alpha_{fe} I_{CBO}$, where I_{CBO} is the collector base reverse current with the emitter open-circuited. This current is very sensitive to temperature and varies almost exponentially, as shown in Figure 3-0.5 increasing at a rate of 8 to 10 per cent per degree centigrade.

4. Equivalent Circuit

An approximate low-frequen a-c equivalent circuit is shown in Figure 3-0.6. As mentioned previously, the input resistance is represented by a base spreading resistance r_b' in series with the reflected emitter resistance $r_b'_e$ which is equal to $25/I_B$ or $25\alpha_{fe}/I_E$, where I_E is the d-c emitter current in ma. The output circuit is

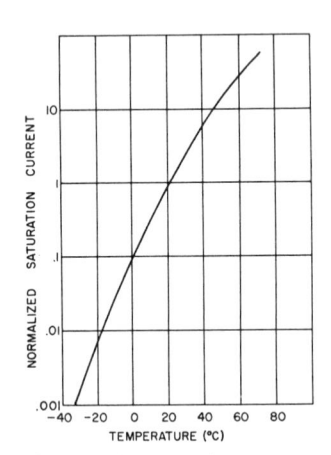

Fig. 3-0.5 - Saturation current vs temperature

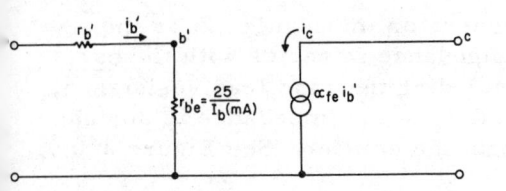

Fig. 3-0.6 - Common emitter low frequency equivalent circuit

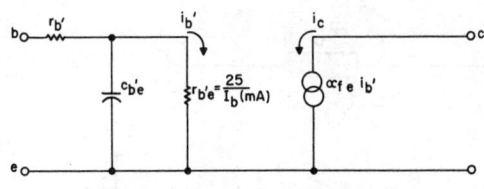

Fig. 3-0.7 - High frequency equivalent circuit

represented by a current generator, generating α_{fe} times the current flowing through $r_{b'e}$.

At the high frequency end of the audio spectrum, the low-frequency equivalent circuit may not hold true for some transistors. The equivalent circuit should then be modified as shown in Figure 3-0.7. The only difference here is that a diffusion capacitance $C_{b'e}$ is shunted across $r_{b'e}$. The value of $C_{b'e}$ is $I_E/80\pi f_{ab}$, where f_{ab} is the alpha cutoff frequency.

5. Noise

There are three kinds of noise in the transistor, namely: white noise, 1/f surface noise, and 1/f leakage noise. In the equivalent circuit as shown in Figure 3-0.8, the dominant white noise and 1/f surface noise can be represented by the current generator located between b' and e, whereas the 1/f leakage noise can be represented by a current generator located between b' and c. The white noise power is proportional to the d.c. base current. The 1/f surface noise power is proportional to the square of the d.c. emitter current. The 1/f leakage noise power is proportional to the collector leakage current, which in turn varies as the d.c. collector voltage. If the leakage noise is negligible, the total noise power generally increases by 3 to 6 db for each 2:1 increase in d.c. current.

For low noise operation, the external base and emitter impedances should be low, of the order of the base lead resistance. When $i_{b'e}$ is the only dominant noise source, in Figure 3-0.8, the signal to noise output current ratio for a base input or emitter input circuit is equal to

$$\left(\frac{S}{N}\right)_i = \frac{V_g}{(Z_g + Z_b + Z_e) i_{b'e}} \quad 3-0.2$$

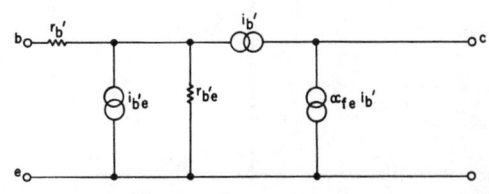

Fig. 3-0.8 - Noise generators in equivalent circuit

where V_g is the generator voltage of the desired signal, Z_g is the

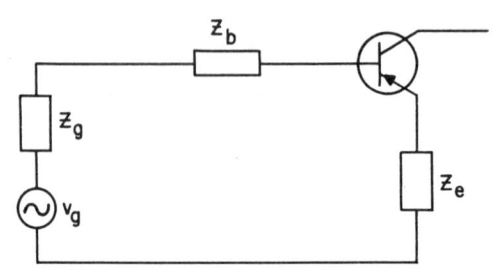

Fig. 3-0.9 - Transistor input circuit

generator impedance, Z_b is the impedance in series with the base including the base lead resistance, and Z_e is the impedance in series with the emitter. See Figure 3-0.9.

B. Circuit Considerations

1. Bias Stabilization

The operating point of a transistor is sensitive to temperature variations unless some means of stabilization is provided. The temperature sensitivity is due to two main reasons: (1) the shift of d.c. input or transfer characteristics (see Figures 3-0.1 or 3-0.2), and (2) the exponential increase of cutoff current with temperature, see Figure 3-0.5. The following circuits are often used for biasing a transistor stage to achieve stabilization.

a. Series Emitter Stabilization

The basic emitter degeneration circuit for temperature stabilization is shown in Figure 3-0.10. The d-c voltage drop across the external emitter resistance, R_E, is much higher than the temperature sensitive base-to-emitter voltage and the voltage drop across the external base resistance. As the temperature varies, the emitter current is held substantially constant. Since the collector current is nearly equal to the emitter current, the collector current also remains insensitive to temperature variations.

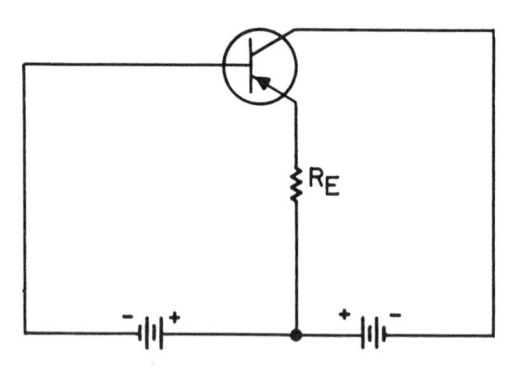

Fig. 3-0.10 - Temperature stabilized by series emitter degenerization

V_{BE} varies nearly 2 mVdc/°C as described earlier so that one requirement for R_E is that the voltage drop $I_E R_E$ should be much greater than $2\Delta T$ mVdc, where ΔT is the temperature range over which the transistor is expected to operate. In general, $I_E R_E$ is greater than 1 volt in a good design.

The value of R_B should be small enough so that the forward bias created due to the flow of temperature sensitive saturation current in R_B is small compared with the

$I_E R_E$ drop. This is usually satisfied if R_B is smaller than $\alpha_{fe} R_E$.

b. Collector-to-Base Shunt Stabilization

The d.c. shunt feedback circuit shown in Figure 3-0.11 is another common method for stabilizing a transistor circuit against temperature variations. In this circuit, any increase in collector current due to a temperature increase will result in lower collector voltage. This lower collector voltage reduces the forward bias at the base of the transistor through R_F, thereby decreasing the increase in collector current. The negative feedback is effective when R_F is low, and R_C and R_B are high. The external parallel resistance as seen by the base

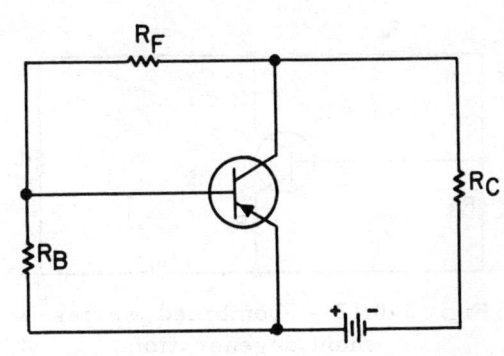

Fig. 3-0.11 - Temperature stabilization by collector-base shunt degeneration

$$\left[\frac{R_B (R_F + R_C)}{R_B + (R_F + R_C)} \right], \qquad 3\text{-}0.3$$

however, should not be too high, because at high temperature the flow of saturation current in this resistance will create a forward bias to increase the collector current. Since R_F is usually quite high, this means that R_B should only be moderately high. Good stabilization is achieved when

$$R_F < \alpha_{fe} R_C \qquad 3\text{-}0.4$$

$$1/40\ I_C < \frac{R_B}{R_B/(R_F + R_C)} < \frac{1}{40\ I_{CBO}} \qquad 3\text{-}0.4a$$

where I_{CBO} is the reverse collector base current at the highest temperature of interest.

A combination of emitter degeneration and shunt degeneration may be combined to achieve a higher degree of stabilization as shown in Figure 3-0.12.

c. Bias Compensation

In circuits where one cannot afford any d.c. voltage drop such as the $I_E R_E$ voltage of Figure 3-0.10, or the $I_C R_C$ voltage of Figure 3-0.12,

59

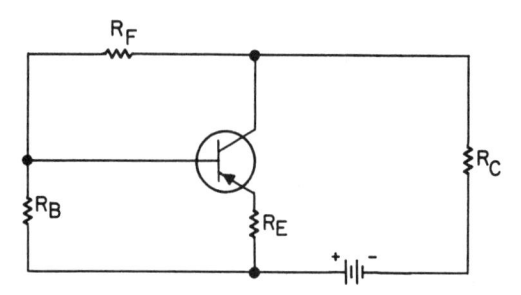

Fig. 3-0.12 - Combined series & shunt degeneration

temperature compensating bias may be used. In this method, constant collector or emitter current is maintained by decreasing the base-to-emitter forward bias at a rate of approximately 2 mVdc/°C as the temperature is increased.

This compensating bias may be derived from a thermistor as shown in Figure 3-0.13 or a constant current fed junction diode as shown in Figure 3-0.14. Most thermistors need padding resistors such as R_p to "track" the characteristics of the transistor. Diodes on the other hand, need no padding resistor, so long as the diode is made of the same kind of material as the transistor to be compensated.

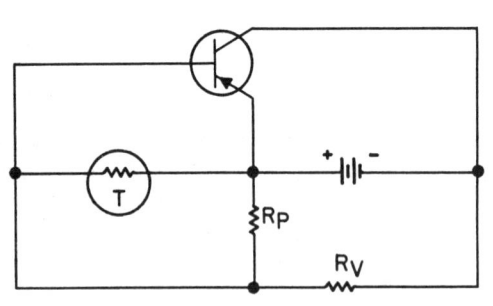

Fig. 3-0.13 - Thermistor temperature compensation

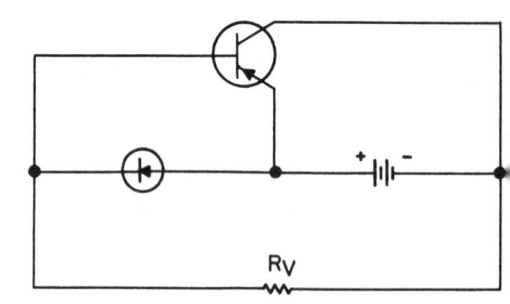

Fig. 3-0.14 - Diode temperature compensation

C. Thermal Considerations

In the operation of power transistors, a large amount of power may be dissipated in the transistor. For a Class A amplifier, maximum dissipation occurs when the signal output is zero. Under this condition, the dissipation is equal to the input power or twice the maximum undistorted sine wave signal output. In Class B operation, the maximum dissipation occurs when there is a square wave output signal equal to half the maximum peak amplitude in current and voltage. In this case, the dissipation in each transistor is equal to one quarter of the maximum sine-wave output.

The maximum power which can be dissipated in a transistor is limited by two factors: the maximum junction temperature, and thermal stability.

The total temperature rise from the ambient to the junction equals the sum of the individual temperature rise of the chassis (heat sink), the mounting hardware and the internal transistor. Thus the junction temperature, T_j, can be expressed as

$$T_j = P(\theta_c + \theta_m + \theta_t) + T_a \qquad 3-0.5$$

where P is the power dissipation, θ_c is the chassis thermal resistance, θ_m is the mounting resistance and θ_t is the thermal resistance from the junction to the case of the transistor. For high power operation, the thermal resistance should be minimized. Figure 3-0.15 is a plot of the thermal resistance as a function of the radiating surface. The mounting thermal resistance results from the insulating hardware (such as mica or anodized aluminum washers) used to electrically insulate the transistor from the chassis. Varnish or grease should be used and the transistor should be tightened against the mounting to reduce the mounting thermal resistance.

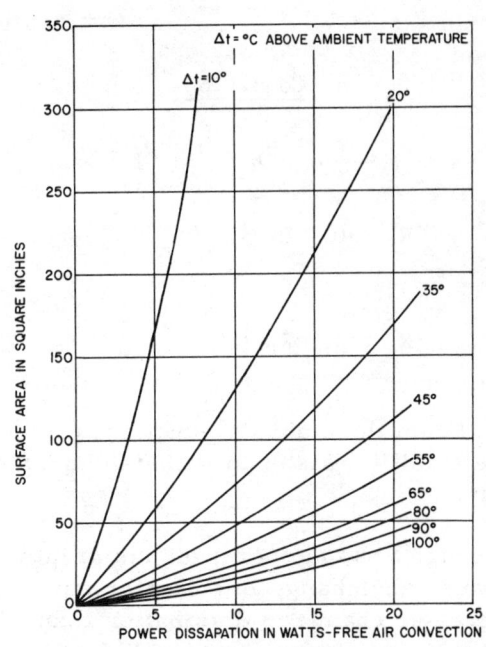

Fig. 3-0.15 - Temperature rise vs power dissipation

When power dissipation and temperature increase regeneratively without reaching equilibrium, the transistor is said to be in a runaway condition. The thermal stability of a transistor depends on many factors. The maximum power dissipation without causing thermal instability is

$$P = \frac{23}{\theta} \log \frac{10}{\theta \, S \, I_{CBO} \, V_{CE}} \qquad 3-0.6$$

where θ is the total thermal resistance, I_{CBO} is the reverse current at ambient temperature, V_{CE} is the collector-emitter voltage, and S is the stability factor which is given in Table I for the generalized circuit Figure 3-0.16.

*** Table I ***

Condition	S
$R_C = R_B = R_E = 0$, $R_F = \infty$	α_{fe}
$R_E \neq 0$, $R_F = \infty$	$\dfrac{(R_B + R_E)\,\alpha_{fe}}{R_B + \alpha_{fe}\,R_E}$
$R_E \neq 0$, $R_F \neq 0$	$\dfrac{\alpha_{fe}}{R_B R_C \alpha_{fe}/(R_B + R_F)\,(R_F + R_C)}$

From the stability equation, circuits with low S, i.e., good temperature stability, also can withstand greater dissipation without causing runaway.

High voltage operation has greater tendency to cause runaway, because any change in collector current due to temperature rise results in greater change in dissipation than with lower voltage. The runaway criteria are shown graphically in Figure 3-0.17.

D. Distortion

1. Nonlinear Input Characteristics

In the input characteristics shown in Figure 3-0.1, the d.c. current varies exponentially with the d.c. voltage in the low current region, and the input resistance varies inversely as the d.c. base current. When the source impedance is low compared

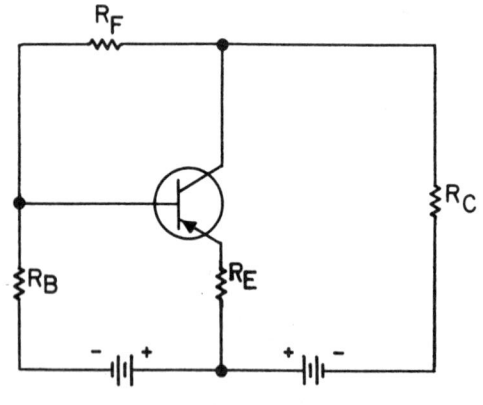

Fig. 3-0.16 - Generalized D. C. circuit

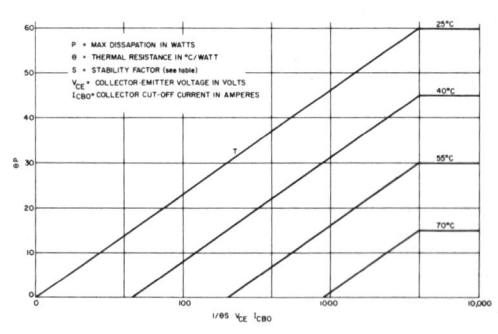

Fig. 3-0.17 - Power dissipation derating curves

with the input resistance, nonlinear input resistance gives rise to distortion.

The harmonics generated on an exponential curve increase as the input voltage is increased. The amplitude of the second, third and fourth harmonics are plotted against the amplitude of the signal in Figure 3-0.18.

As the d-c base current is increased to a value such that $r_{b'e}$ is much less than the series impedance of the generator and the base spreading resistance, the input characteristic becomes linearized. Thus, the transistor should be operated with a low signal input or a high d.c. base current, if distortion due to the nonlinear input characteristic is to be minimized.

2. Nonlinear Current Amplification Factor

At high current densities, the current amplification factor decreases as the d.c. emitter current increases. When the source impedance is high, this fall-off in current amplification may introduce high distortion.

E. Negative Feedback

1. Series Degeneration

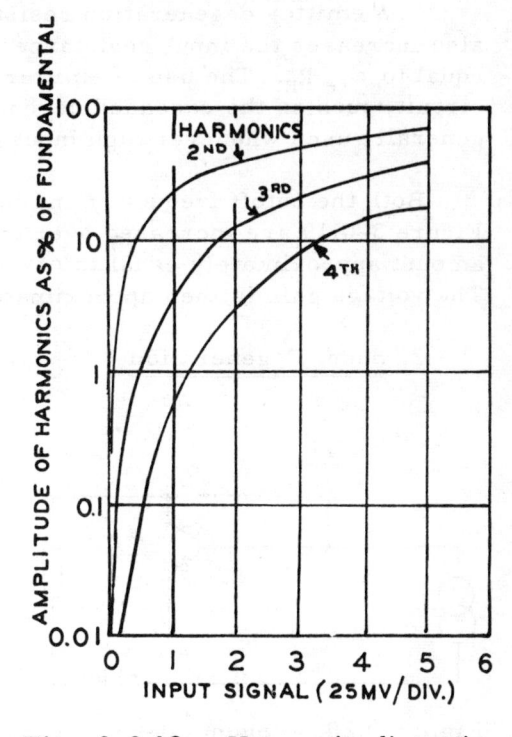

Fig. 3-0.18 - Harmonic distortion vs input voltage

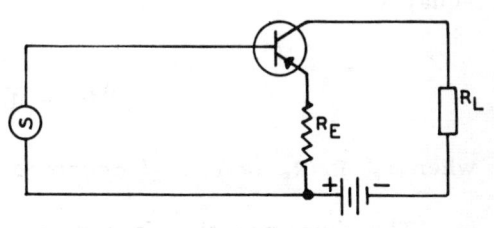

Fig. 3-0.19 - Series degeneration

The basic series negative feedback circuit is shown in Figure 3-0.19. This type of negative feedback is most effective in reducing the distortion when the source impedance is lower than the base-to-emitter resistance. The value of the emitter resistor, R_E, should be so chosen that $\alpha_{fe} R_E \gg r_{b'e}$. Higher current gain transistors give a greater amount of degeneration and are therefore preferred.

The emitter degeneration resistance not only reduces distortion but also increases the input resistance. The increase in input resistance is equal to $\alpha_{fe} R_E$. The use of emitter degeneration or modifications of this circuit, such as the cascaded degeneration circuit (Darlington circuit), ar generally used whenever high input resistance is required.

Both the cutoff frequency and the output impedance of the circuit in Figure 3-0.19 are increased over the common emitter circuit by an amount approximately equal to $(1 + \alpha_{fe} R_E/R_g)$ times when $\alpha_{fe} \gg r_{b'e}$. The voltage gain is then approximately equal to R_L/R_E.

2. Shunt Degeneration

The shunt degeneration circuit is shown in Figure 3-0.20. This type of degeneration is most effective in reducing the distortion when the source impedance is high compared to the input impedance. For a high source impedance, distortion is mostly due to the fall-off of the current amplification factor with increasing emitter current, see Figure 3-0.3.

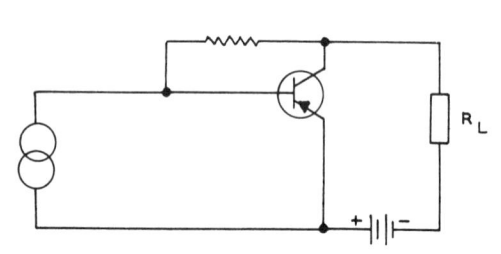

Fig. 3-0.20 - Shunt degeneration

The amount of negative feedback is approximately equal to $\alpha_{fe} R_L/R_F$ and the effective current amplification factor, α_{fe}', is approximately equal to

$$\alpha_{fe}' = \frac{\alpha_{fe}}{1 + \alpha_{fe} R_L/R_F} \qquad 3-0.$$

when $\alpha_{fe} R_L/R_F \gg 1$, α_{fe}' becomes R_F/R_L.

The shunt negative feedback reduces the input and output resistance. The reduction factor for the input resistance is approximately equal to $(1 + \alpha_{fe} R_L/R_F)$. The output resistance is approximately equal to $R_F \alpha_{fe}$.

The cutoff frequency is also extended by the use of shunt negative feedback by a factor approximately equal to $(1 + \alpha_{fe} R_L/R_F)$.

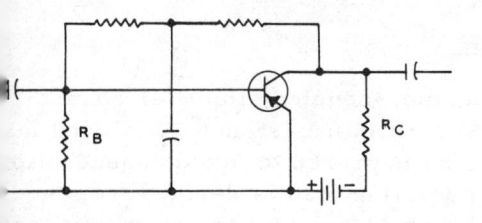

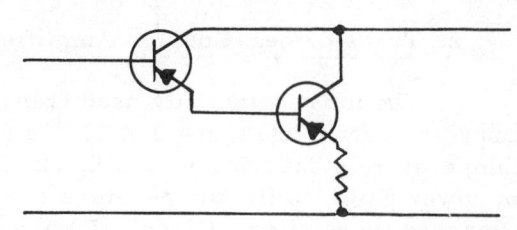

Fig. 3-0.21 - R-C coupled circuit

Fig. 3-0.22 - Darlington circuit

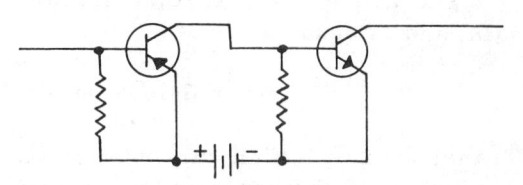

Fig. 3-0.23 - Cascade complementary direct coupled circuit

Basic Amplifiers

1. RC Amplifiers

RC amplifiers can be connected as shown in Figure 3-0.21. Due to the high output resistance and the low input resistance of a common emitter stage, the RC amplifiers are essentially current amplifiers. The figure of merit for such an amplifier is simply the current amplification factor.

The amplifier can be biased more conveniently by the shunt degeneration as shown in Figure 3-0.11, particularly since R_L is to be used. In addition, if a bypass capacitor is to be used the required capacitance is usually smaller for the shunt stabilization than the series degeneration as shown in Figure 3-0.10.

The value of R_B should be much greater than the input resistance of the transistor to avoid shunting the signal. The value of R_C should also be much greater than the input resistance of the next stage for the same reason.

Direct-coupling can be used for low-frequency amplifiers. Direct coupling sometimes has the advantage of saving components and reducing signal degeneration due to the elimination of load resistance. One commonly used circuit is the cascaded emitter follower amplifier sometimes referred to as the Darlington circuit (see Figure 3-0.22). Another convenient direct-coupled arrangement is the cascaded complementary amplifier as shown in Figure 3-0.23. In either of these cases, a shunt resistor is used between the base and the emitter for temperature stabilization. Further temperature stabilization may be obtained by negative feedback over more than one stage.

2. Transformer-Coupled Amplifiers

The most commonly used transformer-coupled amplifier connections are shown in Figure 3-0.24. In (A) emitter resistance R_E is used for temperature stabilization, and R_E should be bypassed to avoid degeneration in power gain. In (B) temperature compensating bias is derived from a bypassed thermistor. Circuit (B) has the advantage that the supply voltage is not sacrificed and is therefore useful in applications where the supply voltage is limited.

The figure of merit for transformer-coupled amplifiers is the power gain, and is equal to

$$\text{Power Gain} = \text{Output Power}/\text{Input Power}. \qquad 3\text{-}0.8$$

Transistors with high current gain and low input resistance are therefore desirable. High supply voltage permits the use of high load resistance R_L and hence high power gain.

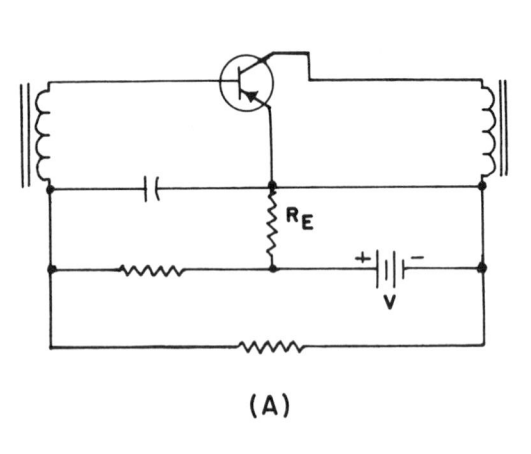

Push-pull amplifiers can be operated either Class A, Class AB, or Class B for audio purposes. A typical circuit is shown in Figure 3-0.25. The class of operation depends on the amount of forward bias.

If zero bias is used in Class B operation, the output signal will have cross-over distortion and look something like that shown in Figure 3-0.26(A). This distortion is due to the lack of initial bias required to operate the transfer characteristic

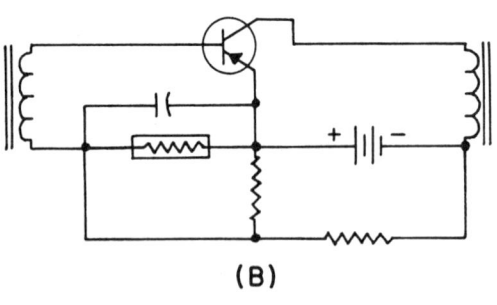

Fig. 3-0.24 - Transformer coupled amplifier

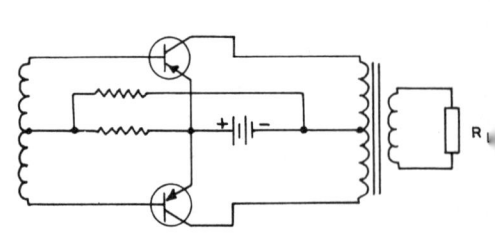

Fig. 3-0.25 - Push pull amplifier

in the linear region as shown in
Figure 3-0.26(B). This kind of
cross-over distortion can be eliminated by applying a small forward
bias. This required forward bias
changes at different temperatures
due to the variation in the d.c.
transfer characteristics. In Class
AB and Class B operation, no bypass
capacitor can be used across the
biasing resistor R_{BE} to avoid crossover distortion.

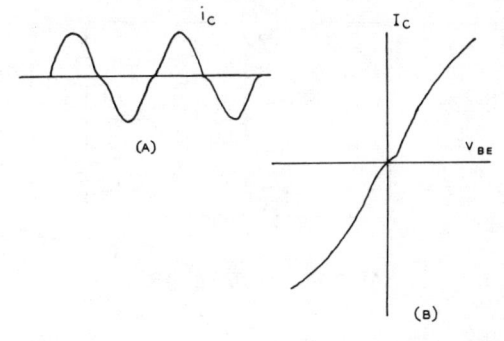

Fig. 3-0.26 - Cross over distortion

3. Power Amplifiers

Power Amplifiers may be operated in Class A, Class AB, or Class B. For linear operation, a single transistor must be operated in Class A. If two transistors are used, the more efficient class of operation may be used.

The common emitter configuration is used most widely in power amplifiers because of high power gain. If distortion must be reduced and the frequency response must be extended, it is generally better to apply negative feedback to the common emitter circuit than resorting to other kinds of configurations with low power gain.

There are two basic push-pull circuit arrangements. One has a push-pull output and the other has a single-ended output. The push-pull output circuit is shown in Figure 3-0.25. The single-ended output circuits are shown in Figures 3-0.27(A) and (B).

The single-ended output circuit can be obtained with either complementary type of transistors as shown in Figure 3-0.27(A) or the same type of transistors as shown in Figure 3-0.27(B). The advantage of the single-ended output stage is the elimination of output transformers.

The complementary circuit shown in Figure 3-0.27(A) has a further advantage over the series connection of the same type of transistors in that the former needs only a single-ended input, whereas the latter needs a split phase input. The split-phase can be obtained by various schemes as shown in Figure 3-0.28(A), (B), and (C). Figure 3-0.28(A) uses an input transformer with two secondary windings. Figure 3-0.28(B) uses a complementary pair as phase splitter. Figure 3-0.28(C) cannot be used for Class B operation unless the two diodes shown in dotted lines are used to furnish a low impedance path when the transistor in shunt is not conducting.

(A) COMPLEMENTARY CIRCUIT

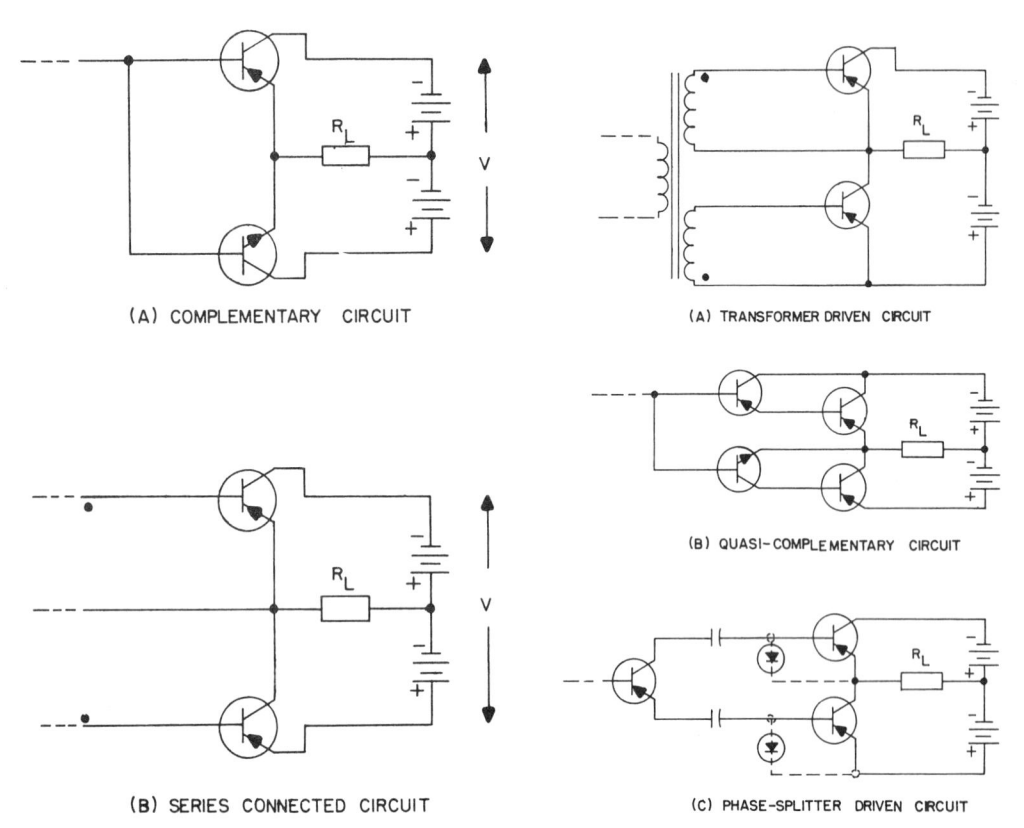

(B) SERIES CONNECTED CIRCUIT

Fig. 3-0.27 - Single-ended output circuits

(A) TRANSFORMER DRIVEN CIRCUIT

(B) QUASI-COMPLEMENTARY CIRCUIT

(C) PHASE-SPLITTER DRIVEN CIRCUIT

Fig. 3-0.28 - Methods for driving single-ended push-pull amplifier

The discussion on circuits is devoted only to base-input configurations. In practice the common emitter configuration is the most commonly used because of high gain. The common collector is used where high input impedance is desired, while the common base configuration is seldom used in low frequency amplifiers. Although the common base configuration gives lower distortion in certain instances, the low distortion can generally be obtained by other base-input configurations with negative feedback.

BIBLIOGRAPHY

1. R. L. Wallace and W. J. Pientenpol, "Some Circuit Properties and Applications of n-p-n Transistors," Proc. IRE, Vol. 39, July, 1951.

2. R. J. Kircher, "Properties of Junction Transistors," IRE Transaction on Audio, Vol. AU-3, July-August, 1955.

3. W. M. Webster, "Saturation Current in Alloy Junctions," Proc. IRE, Vol. 43, March, 1955.

4. W. H. Fonger, "A Determination of 1/f Noise in Semiconductor Diodes and Triodes," Transistor I, RCA Laboratories, 1956.

5. H. C. Lin and A. A. Barco, "Temperature Effects in Circuits using Junction Transistors," Transistor I, RCA Laboratories, 1956.

6. H. T. Moores, "Design Procedures for Power Transistors," Electronic Design, July, September & October, 1955.

7. N. DeWolf, "Rating Transistors to prevent Runaway," Electronic Design, Vol. 3, Feb., 1955.

8. W. M. Webster, "On the Variation of Junction Transistor Current Amplification Factor with Emitter Current," Proc. IRE, Vol. 42, June, 1954.

9. Transistor Electronics by Lo, Endres, Zawels, Waldhauer, Cheng.

10. Transistor Audio Amplifiers by R. F. Shea.

11. G. C. Sziklai, "Symmetrical Properties of Transistors and their Applications," Proc. IRE, Vol. 41, June, 1953.

12. M. B. Herscher, "Designing Transistor A-F Power Amplifier," Electronics, April, 1958.

CIRCUIT 3-1

LOW LEVEL AUDIO AMPLIFIER

Radio Corporation of America

Circuit 3-1 represents a straightforward well-designed low level amplifier. The operating current of this amplifier is 1 mAdc, a point where the current amplification factor is at its maximum. The d.c. current is held constant against temperature variation by both the shunt degenerative action of R_3 and the series emitter degeneration action of R_2. Note that there is a voltage drop of 1.5 V across R_2. The operating range of temperature is from -60°C to +75°C.

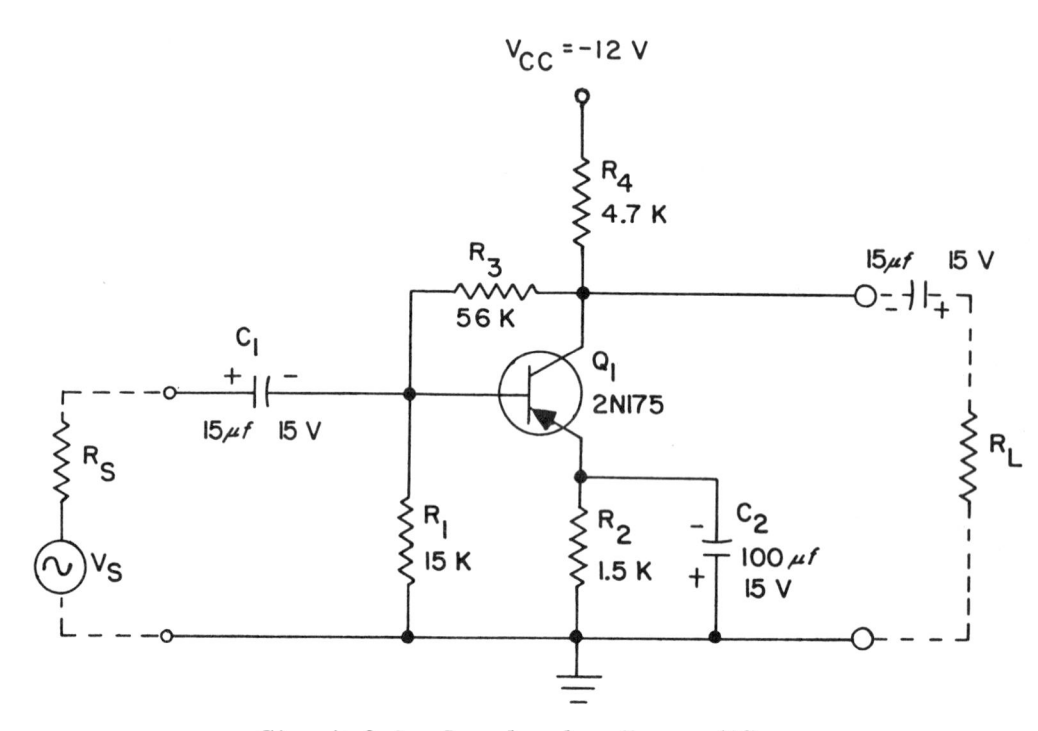

Circuit 3-1 - Low level audio amplifier

Circuit 3-1, Low Level Audio Amplifier (cont'd.)

The d.c. voltage drop across R_4 is such that the quiescent collector voltage is approximately half the supply voltage. This bias will give maximum dynamic range, because the instantaneous voltage may now swing from zero up to the supply voltage.

R_2 is bypassed to avoid a-c degeneration. However, R_3 is not bypassed. This shunt degeneration linearizes but reduces the current gain, reduces the input and output impedances, and extends the frequency response. Greater amount of negative feedback is effected when the load resistance R_L or the source impedance is higher. As a voltage amplifier, the source impedance is low and hence the shunt feedback becomes less effective.

This amplifier is suitable as a general purpose amplifier building block. The performance of this amplifier is shown in the following table:

Frequency Response: A. Current Gain vs. Frequency
(10 to 40,000 c.p.s. @ 3 db down)

B. Voltage Gain vs. Frequency
(55 to 10,000 c.p.s. @ 3 db down)

Minimum Input Level: 0.1 mVac

Maximum Input Level: 20 mVac

Distortion: Maximum % Total Harmonic Distortion

Input Voltage:	R_s 5000 ohms	R_s 1000 ohms
A. 1 mVac	0.7%	1.0%
B. 10 mVac	3.9%	6.2%
C. 20 mVac	8.3%	13.0%

Gain at 30°C	Current Gain	Voltage Gain
A. R_L = 2000 ohms	13	36
B. R_L = 500 ohms	41	16

Circuit 3-1, Low Level Audio Amplifier (cont'd.)

    Characteristic Impedance at 30°C

        A. Output Impedance = 1620 ohms when R_s = 1000 ohms,
                                    980 ohms when R_s = 5000 ohms.

        B. Input Impedance = 1620 ohms when R_L = 500 ohms,
                                    900 ohms when R_L = 2200 ohms.

    Power Requirements: -12V ± 10% @ 1 mAdc 12 mw

    Temperature Range: -60°C to +75°C

CIRCUIT 3-2

TRANSFORMER-COUPLED OUTPUT STAGE

C.B.S.

A transformer-coupled Class A amplifier is used where high power gain is required. Circuit 3-2 shows a typical transformer-coupled amplifier such as those used in the automotive radio output stage. The input transformer is used to match the high output impedance of the driver (transistor or tube). The output transformer may be an auto-transformer for matching the voice-coil impedance.

Temperature stabilization is achieved by the thermistor, R_T, which provides a compensating bias at the base. The series emitter degeneration for temperature stabilization is not practical, because an emitter resistance would reduce the d.c. collector-emitter voltage, which in turn would reduce the power gain.

Due to the variation in the d.c. input characteristics from transistor to transistor, a variable resistor, R_2, should be used in series with R_3 to adjust the forward bias of the transistor to the desired operating point. A padding resistor, R_1, is connected across the thermistor to obtain the desired variation in forward bias with temperature.

In a Class A amplifier, the dissipation is more than twice the maximum power output. A good heat sink is required if the amplifier is to be operated satisfactorily at high temperatures.

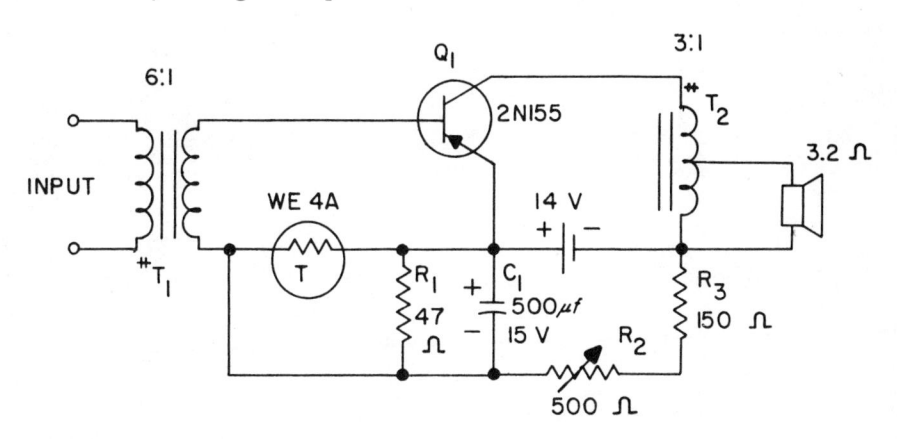

T_1 - THORDARSON - TR-20

T_2 - THORDARSON - TR-58

Circuit 3-2 - Transformer-coupled output stage

Circuit 3-2, Transformer-Coupled Output Stage (cont'd.)

The typical performance is as follows:

Supply voltage:	14V
Collector current:	420 mAdc
Power output:	2W
Input impedance:	30 ohms
Load impedance:	30 ohms
Power gain:	36 db
Distortion:	5%
Frequency response:	7 kc
Operating temperature:	-40°C to +70°C

CIRCUIT 3-3

INPUT CIRCUIT FOR CAPACITIVE TRANSDUCER

C.B.S.   Contributed by H. C. Lin*

    The impedance of a capacitive transducer decreases as the frequency is increased. If the input impedance is lower than that of the transducer, the input current will be higher at higher frequencies. An input circuit should equalize the frequency response, give good signal-to-noise ratio and have low distortion. An input circuit for capacitive pickup (crystal or ceramic) satisfying all of these requirements is shown in Circuit 3-3.

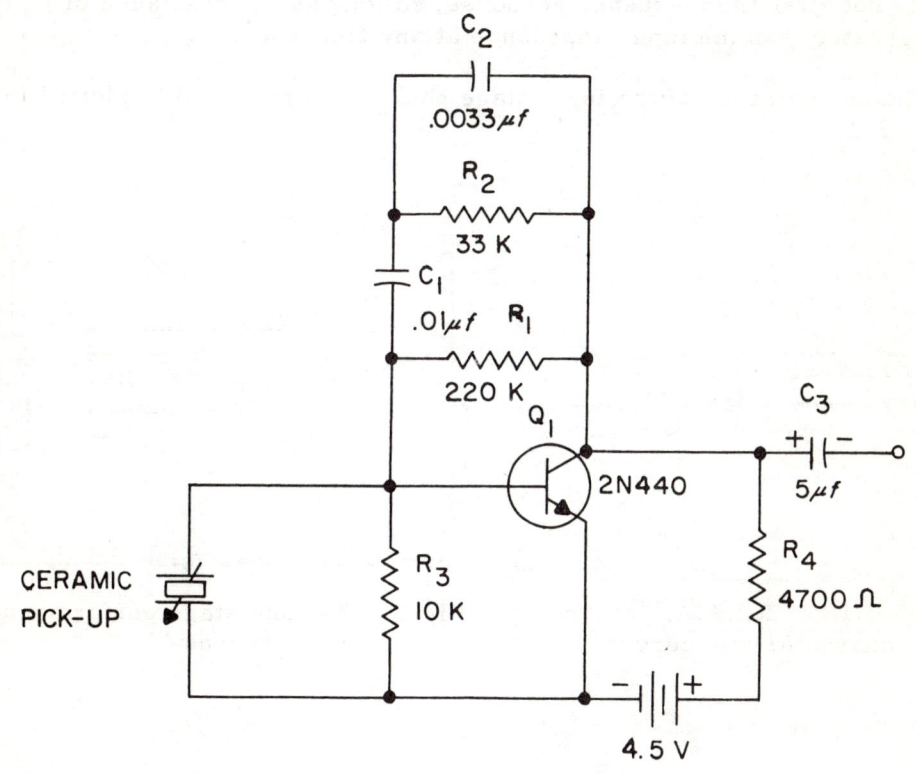

Circuit 3-3 - Input circuit for capacitive transducer

---

*Now with Westinghouse Electric Corp.

Circuit 3-3, Input Circuit for Capacitive Transducer (cont'd.)

The input impedance is made much smaller than the source impedance and the current gain is made to vary inversely as the velocity response of the RIAA recording characteristic shown in Figure 3-3.1. These functions are obtained by means of negative feedback. C_1 is effective between 50 and 500 c.p.s.; R_2 between 500 and 2000 c.p.s.; C_2 above 2000 c.p.s.

The large amount of negative feedback reduces the distortion and permits the use of low operating current, which is essential for low noise operation. The fact that no equalizing network is connected in series with the base also helps to reduce the noise.

This circuit has the advantage that interchanging pickups of different C_o does not alter the frequency response, so long as the reactance of C_o is much greater than the input impedance at any frequency.

The performance of the input stage shown in Circuit 3-3 is plotted in Figure 3-3.2.

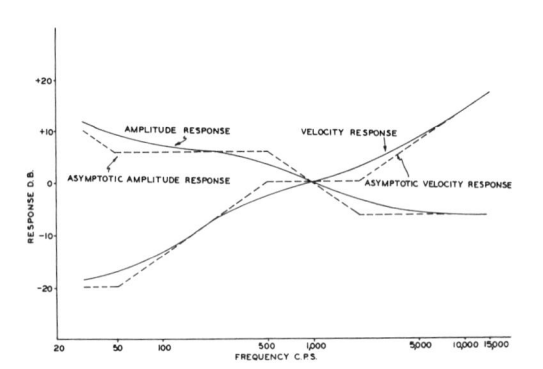

Fig. 3-3.1 - R.I.A.A. recording characteristic curves

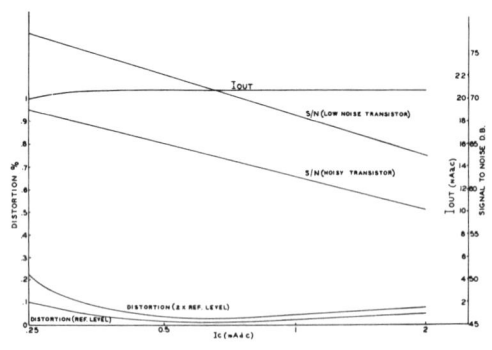

Fig. 3-3.2 - Input stage performance curves

CIRCUIT 3-4

INPUT CIRCUIT FOR INDUCTIVE TRANSDUCER

C.B.S.   Contributed by H. C. Lin*

The impedance and the signal voltage of an inductive transducer generally increase as the frequency is increased. Like the input circuit for capacitive transducer, the input circuit for inductive transducer should equalize the frequency response, and have low noise and distortion.

These requirements are satisfied in the input circuit for inductive pickup in Circuit 3-4. In this circuit the input impedance increases with the frequency as the velocity response of the recording characteristic (see Figure 3-3.1). This is accomplished by negative feedback applied to

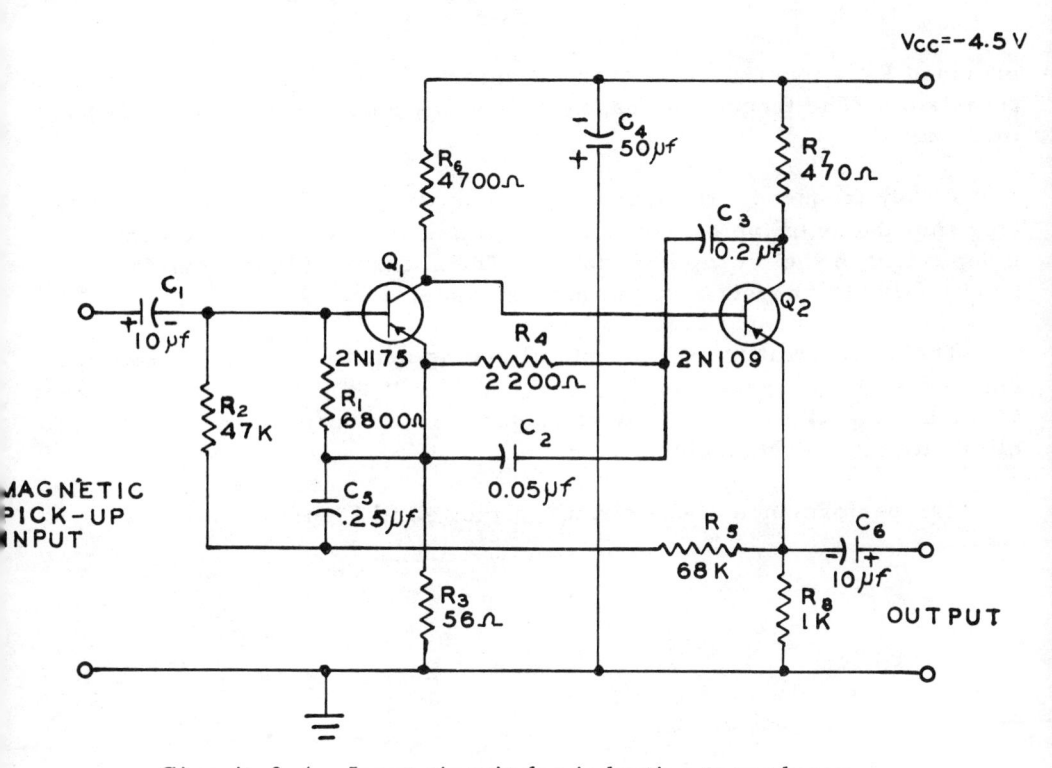

Circuit 3-4 - Input circuit for inductive transducer

---

*Now with Westinghouse Electric Corp.

Circuit 3-4, Input Circuit for Inductive Transducer (cont'd.)

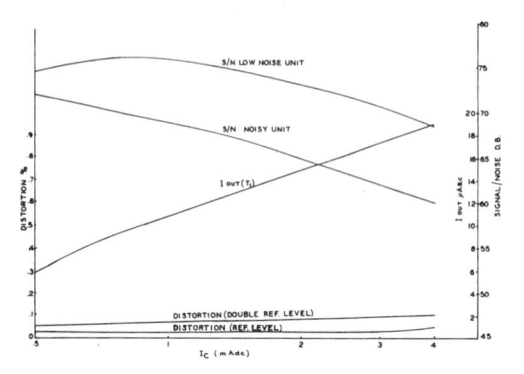

Fig. 3-4.1 - Circuit performance curves

the emitter of the first transistor. For $R_7 \gg R_3$, the amount of feedback is equal approximately to

$$\alpha_{fe_2} R_7/Z_F \qquad 3\text{-}4.1$$

where α_{fe_2} is the α_{fe} of transistor Q_2, and Z_F is the impedance of the feedback network C_2, C_3 and R_4. The input impedance is equal to approximately

$$(\alpha_{fe_1} R_3)(1 + \alpha_{fe_2} R_7/Z_F) \qquad 3\text{-}4.2$$

provided R_3 is much greater than the internal emitter resistance r_e of the transistor. The frequency response deviates 3 db below 50 c.p.s. if R_3 is made equal to r_e.

At any frequency, the input impedance is much higher than the generator impedance of typical inductive pickups. Hence, this circuit is independent of the pickup impedance. The absence of base resistance is beneficial with respect to noise performance.

The d.c. operating point is stabilized by negative feedback from the emitter of Q_2 to the base of Q_1 through R_2 and R_5. In addition, R_1 is used to further stabilize the individual stages. Bypass capacitor C_3 is used to eliminate a.c. degeneration.

The performance of the circuit is plotted in Figure 3-4.1.

CIRCUIT 3-5

LOW NOISE TRANSISTOR PREAMPLIFIER

Minneapolis-Honeywell   Contributed by Richard S. Burwen

    Circuit 3-5 is a general purpose preamplifier featuring an extremely low noise figure. The amplifier as shown has a gain of 100 or 1000 and a frequency response of 3 cycles to 30 kilocycles. The noise figure is less than 5 db under a wide range of circuit modifications, and is as low as 2.4 db under optimum conditions. Large amounts of d.c. and a.c. feedback provide excellent circuit stability, even with large changes in temperature and large variations among transistors.

    The circuit uses four transistors. The first two stages employ the common emitter connection with direct-coupling between them. The first stage is operated at very low collector voltage and current to facilitate low noise performance. The collector of second transistor Q_2 is direct-coupled to the base of Q_3, an n-p-n unit, which in turn drives the output transistor, Q_4, a p-n-p emitter follower. The use of the n-p-n transistor after a p-n-p transistor simplifies the circuit to a considerable extent. The feedback network is connected between the emitter of the output stage

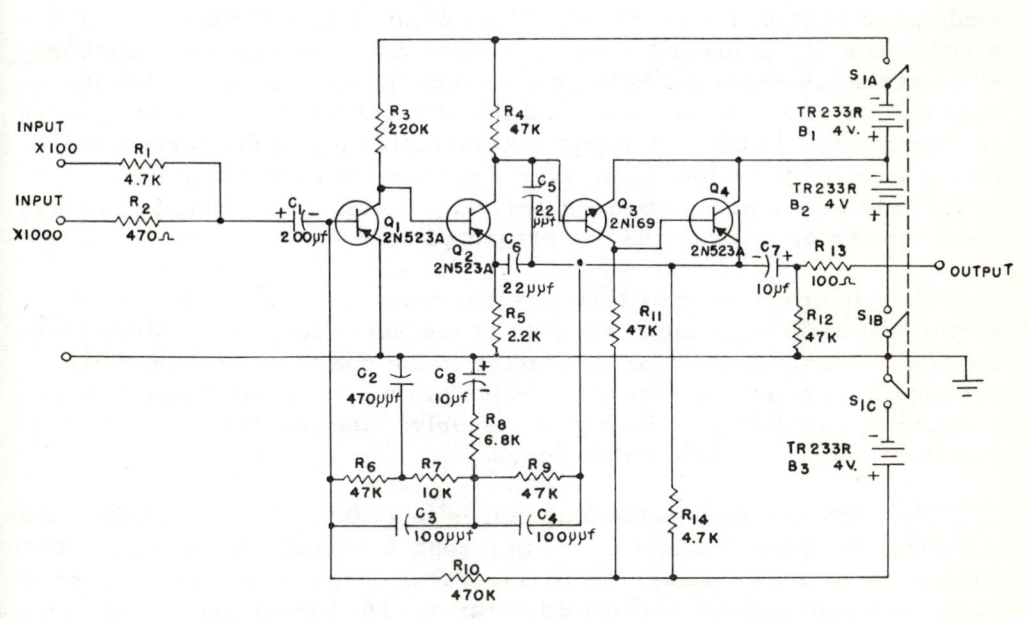

Circuit 3-5 - Low noise transistor preamplifier

Circuit 3-5, Low Noise Transistor Preamplifier (cont'd.)

and the base of the input stage. Several additional small capacitors are used to insure high frequency stability.

The input impedance of the preamplifier is quite low. The impedance is determined primarily by the input network, R_1 or R_2 and C_1. The heavy feedback reduces the base to ground impedance of transistor Q_1 to a very low value. The input impedance at mid-frequencies is then 4.7K for a gain of 100 and 470 ohms for a gain of 1000. The circuit is intended for use with low impedance sources which will not be adversely affected by the loading of the input.

The noise figure of the amplifier using the X100 input is approximatel 2.4 db; the measured input noise is 2.0 μV. Using the X1000 input results in a noise figure of 3.2 db and the measured noise is 0.7 μV. These noise figures are typical of the capabilities of this type of input circuit. Over a range of input resistors from 100 ohms to 10K, the noise figure should not exceed 5 db.

It should be noted at this point that the wideband noise figure obtained in this circuit will not correspond directly to manufacturer's noise specifications, as the latter are measured at 1000 cycles at very narrow bandwidth, and at higher than optimum collector currents. This procedure tends to emphasize the so-called 1/f, or semiconductor, noise, which is predominant at the lower frequencies. The noise voltage per cycle bandwidth at the low frequencies is greater than the average noise density over the entire audio spectrum, and the high collector current introduces additional noise in the test setup; consequently the manufacturer's noise figure specifications tend to be overly pessimistic, and results much better than the reference figures generally quoted can be obtained in a wideband circuit such as the one presented here.

The use of mercury batteries rather than an a-c power supply is partly responsible for the attainment of low noise operation. Noise from the power supply is an important factor in the operation of a sensitive preamplifier, and the mercury batteries are much quieter than the corresponding rectifier and filter power supply. The life of the batteries specified is approximately 1500 hours.

The open loop gain of the four stages is 40,000. This high gain makes stabilization of the preamplifier with feedback is somewhat critical. Without the use of d-c coupling, stabilizing a four-stage amplifier with heavy feedback would present serious difficulties. The time constants associated with the feedback network and the output network cannot cause the amplifier to be unstable at small inputs. However, under overload conditions

Circuit 3-5, Low Noise Transistor Preamplifier (cont'd.)

the normally low output impedance of the common emitter output stage Q_4 becomes appreciably higher, and the time constant of the output coupling capacitor C_7 and the load resistance is consequently placed in the feedback loop. The additional low-pass RC network appearing under overload conditions introduces more phase shift in the feedback loop, rendering the circuit unstable at some low frequency, and oscillation results. Fortunately, the instability is conditional; as the circuit oscillates the emitter follower impedance swings from quite low to quite high, and if the overload signal is removed from the input, the oscillation ceases upon the first swing to the low impedance condition. Thus, if the overload input signal is removed, stability is restored within one cycle of the oscillation.

In addition to possessing desirable low noise characteristics, this preamplifier also provides low distortion amplification. The midband distortion at 1 volt rms output is 0.005% with a gain of 100 and 0.05% with a gain of 1000. The open loop distortion is approximately 2%, most of which arises in the third stage because the static collector current of Q_3 is not too much larger than the variation with large signals. However, the heavy feedback minimizes any distortion arising in the circuit.

A desirable feature of this circuit is its low component count, due to direct-coupling and the use of cascade complementary arrangement. Aside from the overall d.c. stabilization and ease of stabilization, the d-c coupling also results in the saving of two or three large capacitors per stage. This means a tremendous decrease in size and a tremendous increase in circuit reliability.

This preamplifier is useful in a large number of audio applications. It has excellent noise figure, frequency response, and component economy. It is temperature stabilized and should provide reliable performance under a wide range of temperatures, although the use of germanium transistors does limit the upper temperature. A similar circuit using silicon transistors should be feasible, although the noise figure will not be as good. The circuit herein presented should be satisfactory under the most stringent of low-noise requirements.

CIRCUIT 3-6

HIGH INPUT IMPEDANCE PREAMPLIFIER

Burr-Brown Contributed by R. P. Burr

This preamplifier employs the Darlington, or "super-alpha" connection of the input transistors to provide a high input impedance. By connecting two common collector stages in cascade, the a.c. input impedance is nearly equal to (β^2) times the load, R_5 in parallel with R_6. The actual input impedance at room temperature is greater than 2 megohms.

In spite of the high a.c. input impedance, the d.c. resistance in series with the base of Q_1 (R_1 plus the parallel resistance of R_2 and R_3) is considerably lower. This low d.c. resistance in series with the base is

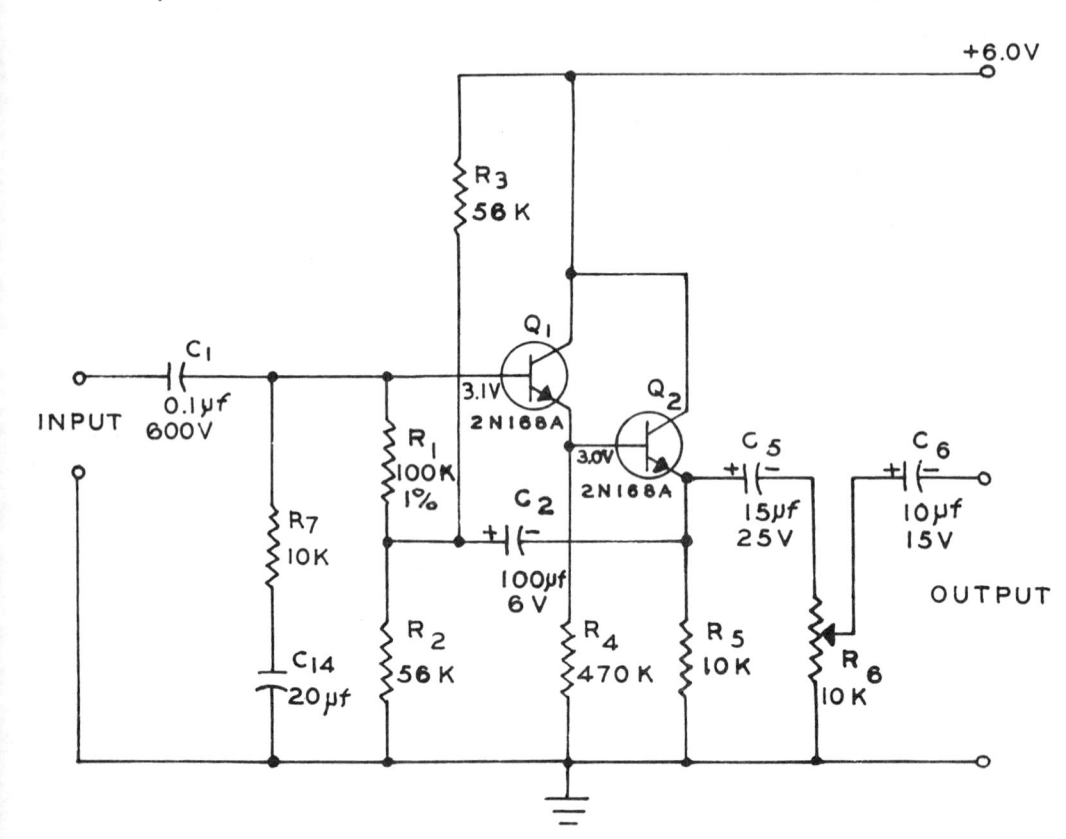

Circuit 3-6 - High input impedance preamplifier

Circuit 3-6, High Input Impedance Preamplifier (cont'd.)

essential for temperature stabilization. The high ratio of a.c. input impedance to d.c. input resistance is made possible by the use of capacitor C_2 connected between the emitter of Q_2 and the junction of R_1 and R_2. This couples the output voltage back to the junction. Since the output voltage is nearly the same as the input voltage, the actual difference voltage across R_1 is very small and the effective value of R_1 is increased by several orders of magnitude. The upper limit of R_1 is set by temperature considerations. R_4 adds sufficient current to Q_1 to maintain high β and frequency response.

The circuit as shown functions well over the temperature range of -25°C to +55°C. Frequency response is flat within 0.5 db over the frequency range of 10 c.p.s. to 250 kc. The circuit is used primarily as an instrument or oscilloscope preamplifier, but is easily adapted to any use requiring high input impedance. The use of germanium transistors limits the temperature range, but the circuit is well proven within its rating.

CIRCUIT 3-7

WIDE TEMPERATURE RANGE PREAMPLIFIER

General Electric Company   Contributed by S. K. Ghandhi

Circuit 3-7 is designed for wide temperature operation. Silicon transistors are employed to extend the operating capabilities in the high temperature range.

The three n-p-n silicon transistor stages are direct coupled. The emitter potentials of Q_1, Q_2 and Q_3 are progressively more positive. This is accomplished by connecting breakdown diodes CR_1 and CR_2 in series with the emitters of Q_2 and Q_3 respectively. The purpose of the progressively increased emitter potential is to insure that the collector of the preceding stage is not saturated.

Overall negative feedback is applied from the collector of the output transistor to the base of the input transistor through R_{11}. This feedback stabilizes the d-c operating point and maintains a constant gain against temperature variations. The initial operating point is adjusted by R_2 so that 12.5V appears at the collector of Q_3. By connecting R_1 in series with the base, the negative feedback will not lose its effectiveness even if the source impedance is very low. A local negative feedback circuit,

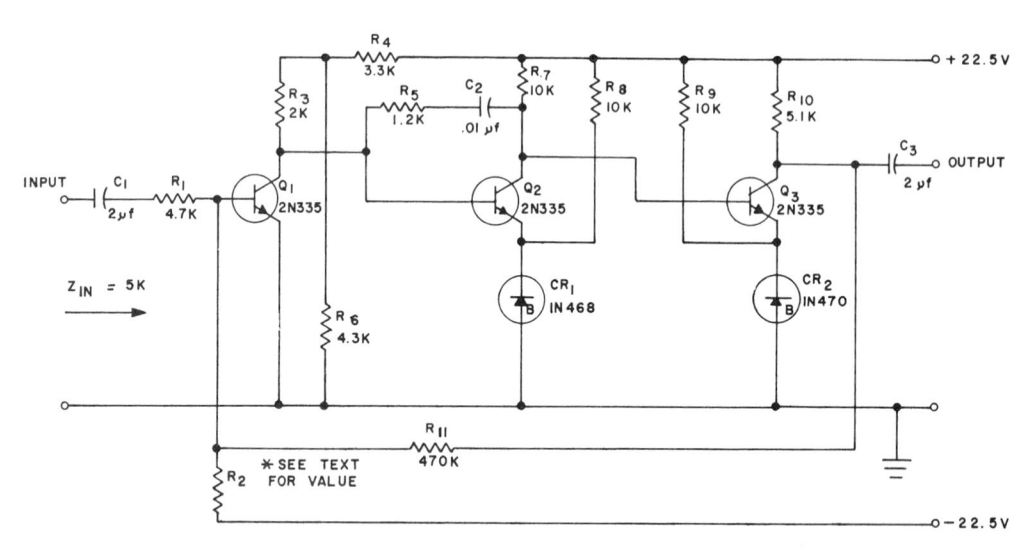

Circuit 3-7 - Wide temperature range preamplifier

Circuit 3-7, Wide Temperature Range Preamplifier (cont'd.)

consisting of C_2 and R_5, is connected between the collector and the base of Q_2 to prevent oscillations at high frequencies due to heavy overall feedback.

The circuit has a voltage gain of approximately 40 ± 0.3 db over a temperature range of -55°C to 125°C. The frequency response extends from 20 c.p.s. (limited only by C_1 and C_3) up to greater than 25 kc. No electrolytic capacitors are used in the circuit proper. This feature improves the reliability and component economy. Besides, the component tolerances are not critical, and almost any n-p-n silicon transistor may be used provided that it is operated within its ratings.

CIRCUIT 3-8

HIGH INPUT IMPEDANCE WIDE TEMPERATURE RANGE AMPLIFIER

Texas Instruments   Contributed by Arthur D. Evans

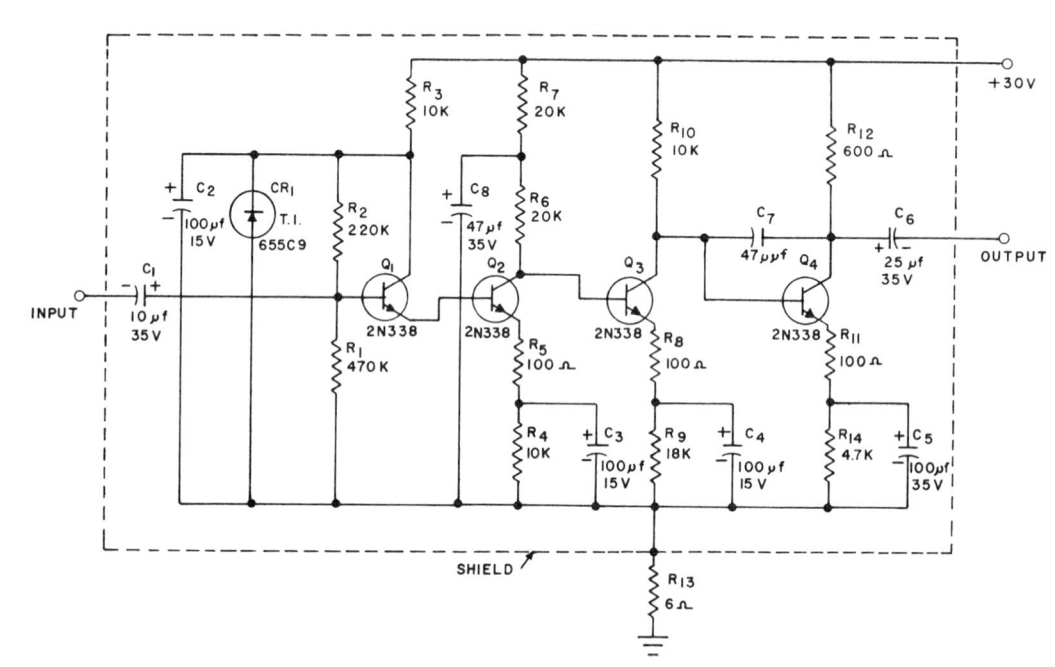

Circuit 3-8 - High input-impedance wide temperature range amplifier

This circuit is extraordinary in that it provides a very high input impedance over an extended temperature range. The input impedance at 1 kc is in excess of 7 to 8 megohms over the temperature range of -55°C to 125°C, while the gain is constant to within 0.2 db over the same range.

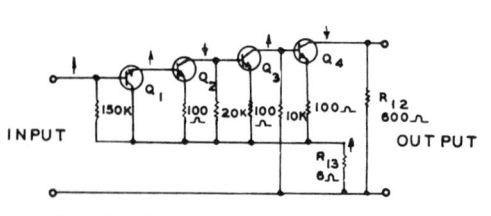

Fig. 3-8.1 - Simplified circuit

Figure 3-8.1 shows the simplified equivalent circuit of the amplifier. It consists of a common-collector stage followed by three common-emitter stages. By employing negative feedback, voltage gain is exchanged for input impedance, and overall gain is stabilized. This feedback is

Circuit 3-8, High Input Impedance Wide Temperature
          Range Amplifier (cont'd.)

accomplished by placing feedback resistor R_{13} in the common return lead of the entire amplifier. The a-c current through R_{13} is approximately equal to the product of the input current and the current gain of the amplifier. Thus the effective value of R_{13} referred to the input terminals is AR_{13}, where A is the current gain of the amplifier. If the amplifier current gain is 10^6 and R_{13} is 6 ohms, then the equivalent input impedance is approximately 6 megohms.

The complete amplifier is shown in Circuit 3-8. The four stages are direct-coupled. Bias stability is assured by the use of large emitter resistors R_4, R_9, and R_{14}; these are bypassed for signal frequencies. Bias voltage for the first stage is developed across the 9 volt breakdown diode CR_1.

The output impedance is set by the value of R_{12}, which is 600 ohms. The voltage gain is set by the ratio of the output resistor R_{12} to the feedback resistor R_{13} and is almost independent of transistor parameters. As shown above, the input impedance is equal to the product of the current gain of the amplifier and the feedback resistor R_{13}. Since the current gain is a function of temperature and frequency, the input impedance is affected by these parameters; however, the circuit shown is stable to ±0.1 db over the temperature range and to within ±1 db from 6 c.p.s. to 300 kc. The input impedance is greater than 1 megohm between the frequencies of 25 c.p.s. and 350 kc and greater than 8 megohms between 400 c.p.s. and 30 kc. Input impedance as a function of temperature and frequency is shown graphically in Figures 3-8.2 and 3-8.3.

The equivalent input noise ranges from about 24 microvolts with input shorted to about 540 microvolts with input open. Careful consideration should be given to the layout of the amplifier, particularly with respect to stray capacitance to ground. In the circuit shown, excellent results are obtained by mounting all components on a metal sheet which is insulated from the main chassis and electrically connected to the ungrounded side of feedback resistor R_{13} as shown in the circuit diagram. This prevents deterioration of the high frequency characteristics of the amplifier by reducing the effect of stray capacitance between the components and ground.

This circuit will provide excellent results in many applications requiring a very high input impedance over a large temperature range. It can be used reliably as a preamplifier following high impedance devices such as infra-red detectors and photocells. This amplifier offers

Circuit 3-8, High Input Impedance Wide Temperature Range Amplifier (cont'd.)

extremely desirable and stable characteristics under adverse environmental conditions, and rates serious consideration for applications where high impedance is a necessity.

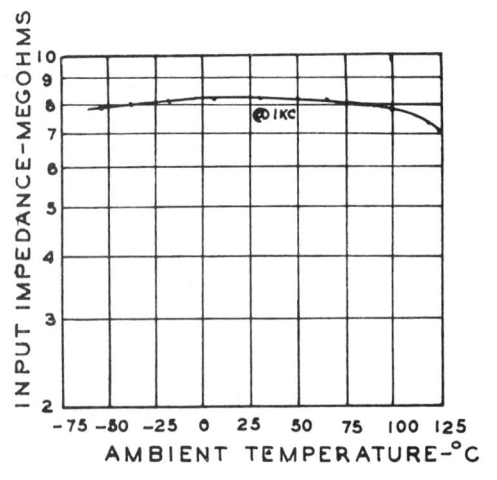

Fig. 3-8.2 - Input impedance vs ambient temperature

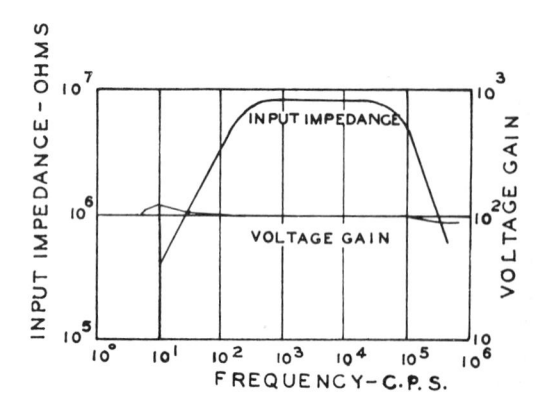

Fig. 3-8.3 - Input impedance vs frequency

CIRCUIT 3-9

DIODE COUPLED A.C. PREAMPLIFIER

Transitron Electronic Corporation   Contributed by Nicholas DeWolf and John Warren

The preamplifier shown in Circuit 3-9 provides stable operation over a wide temperature range, while allowing for a very considerable spread of the semi-conductor component parameters. The interstage coupling is unusual in that forward biased diodes are employed as coupling elements. All silicon transistors and diodes are used in a circuit which performs over the temperature range -55°C to +150°C.

The amplifier is a three stage design with all stages used in the common emitter connection. The input to the base of transistor Q_1 is through a coupling capacitor C_1. Each stage uses a bare minimum of components.

The d.c. operating conditions are stabilized by an overall feedback loop including R_4 and R_8 and the d.c. voltage at the output collector varies only 0.125 volts per microampere change in the first stage base current, due for example to any increase of leakage current. The d.c. feedback

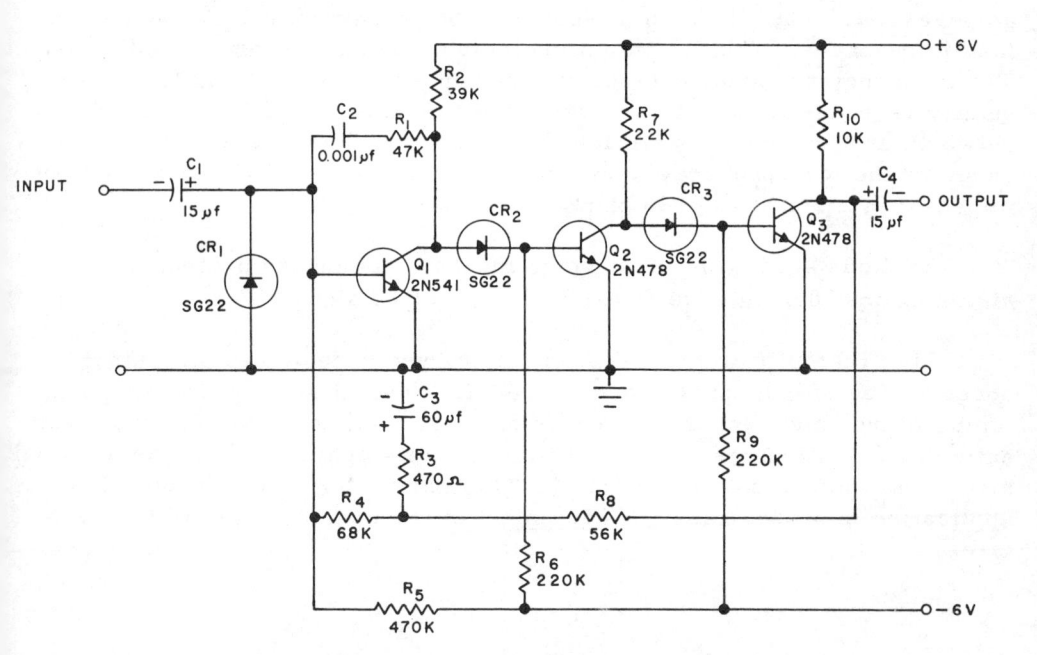

Circuit 3-9 - Diode coupled A.C. preamplifier

Circuit 3-9, Diode Coupled A.C. Preamplifier (cont'd.)

loop is partially bypassed by C_3 so that the amount of a.c. feedback is considerably less. The amplifier is designed for operation with source impedances above 10K, and load impedances above 1K. The negative feedback would be less effective with lower values of source and load impedances. A voltage gain is 58 db with 10K source and load.

The interstage diodes being forward-biased accomplish a.c. coupling without loss of gain. Any need for large coupling or bypass capacitors is eliminated. Due to the d.c. voltage drop across these diodes, the collector-emitter voltages of the first two transistors are set at approximately 1 volt which results in reliable low noise operation. The objectionable effect of saturation current at high temperatures is minimized because of the low external base resistance furnished through these diodes.

Due to the heavy a.c. feedback, the circuit would be unstable at a frequency of the order of 80 kc unless the high frequency gain were reduced. The RC feedback network between the base and collector of the input stage lowers the gain at these frequencies and eliminates this possibility of oscillation.

The circuit as presented provides 1 volt a.c. output per 0.125 microamperes a-c input. This corresponds to 58 db gain with 10K source and load impedances. The maximum undistorted output signal is 1 volt rms. The frequency response extends from 4 c.p.s. to 100 kc. The low frequency response is determined solely by the capacitor C_3 in the feedback network; any tendency toward low frequency instability is eliminated by reducing the low frequency loop gain, i.e., large C_3. The input and output impedances are 200 and 250 ohms respectively.

The diode across the first stage input is intended to protect the transistor against damage due to excessive negative signal.

This preamplifier represents a unique yet reliable design featuring conservation of components and space. It is useful over a wide range of temperatures and over a wide range of source and load impedances. The entire circuit requires only 5 milliwatts for its operation, and the physical size of the unit is extremely small. This particular circuit should find application in a wide range of equipment subject to extremes of temperature.

CIRCUIT 3-10

TWIN TEE REJECTION AMPLIFIER

Baird-Atomic, Inc.   Contributed by William R. Lamb

Circuit 3-10 was designed to pass a band of frequencies centered about a 400 cycle carrier. The required bandwidth is forty cycles. The phase shift in this region must be as small as practical. Furthermore, the amplifier must have low transmission at the rejection frequencies, 200 and 600 cycles.

These design requirements were met by an amplifier consisting of two cascaded twin tee rejection networks with suitable isolation and impedance matching between them; the two networks being tuned to 200 and 600 cycles, respectively.

The basic twin-tee circuit is characterized by infinite rejection at the null frequency. Because a perfect balance is difficult to obtain, especially when component variations due to aging and temperature are taken into account, a rejection of not more than 45 db should be expected. The phase shift through the network becomes rather violent as the null frequency is approached, sweeping from plus 90° to minus 90° as the null is passed. There is a rigid relationship between the driving and load impedances which must be maintained if the network is to have symmetrical skirt characteristics. However, these impedances have reasonable values and

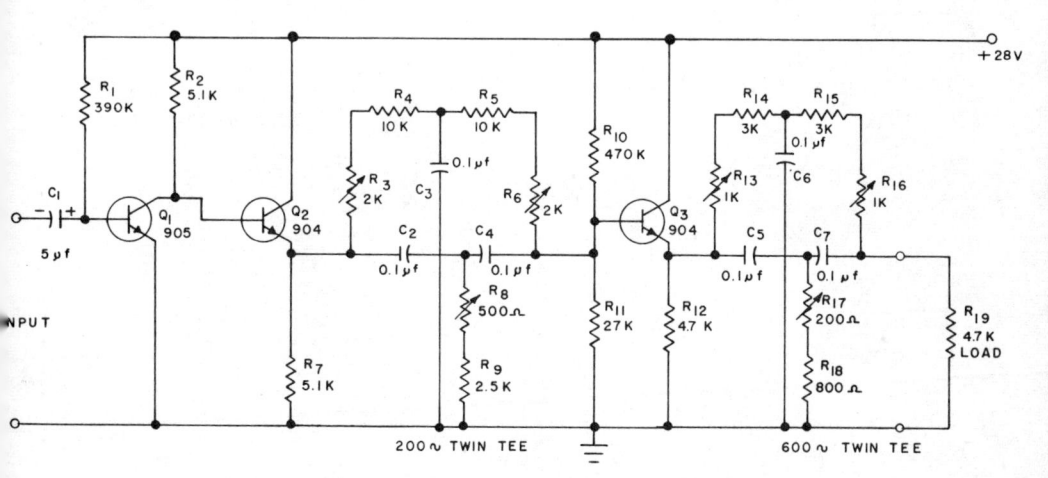

Circuit 3-10 - Twin tee rejection amplifier

Circuit 3-10, Twin Tee Rejection Amplifier (cont'd.)

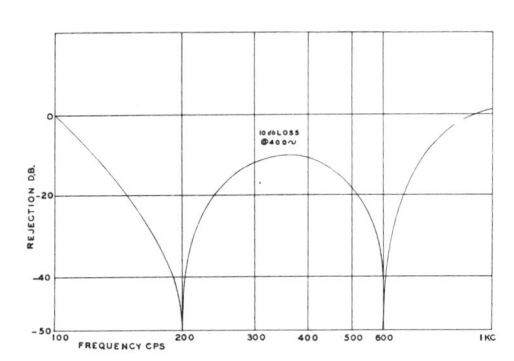

Fig. 3-10.1 - Rejection amplifier response curve

may be designed for transistor circuits.

The circuit is straightforward and needs little comment. Q_1 and Q_2 comprise a preamplifier to raise the input level above the following circuit noise. Q_2 is a common collector stage which drives the 200 cycle twin tee. Q_3 is a buffer amplifier, also a common collector stage.

Performance of the rejection amplifier is shown in Figure 3-10.1.

This amplifier stresses the design of frequency rejection. For wide temperature operation and good interchangeability of transistors, some of the temperature stabilization and negative feedback measures described previously may be employed.

CIRCUIT 3-11

TRANSFORMERLESS QUASI-COMPLEMENTARY AUDIO AMPLIFIER

C.B.S.   Contributed by H. C. Lin*

This circuit features a Class B transformerless output stage using two of the same type power transistors and is driven by a pair of complementary transistors as phase inverters. Due to the lack of matched complementary power transistors available commercially, the use of power transistors of the same type is an advantage in many instances. The output is designed to feed a conventional 16 ohm loudspeaker. While the power output of the amplifier is six watts, greater output can be

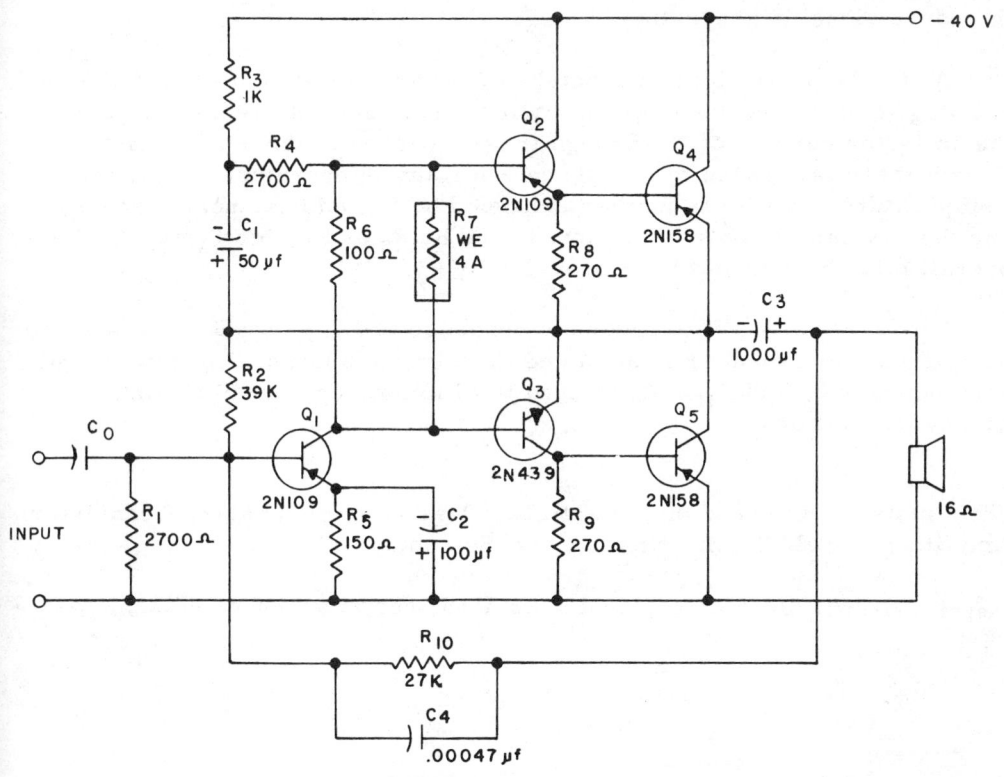

Circuit 3-11 - Transformerless quasi-complementary audio amplifier

---

*Now with Westinghouse Electric Corporation.

Circuit 3-11, Transformerless Quasi-Complementary
        Audio Amplifier (cont'd.)

obtained by using higher supply voltage. The frequency response is flat within 1.5 db from 30 to 15,000 c.p.s. The harmonic distortion is less than 1 percent in the mid-frequencies. Operation is satisfactory from 0° to 50°C.

Besides the driver and output stages, there is a Class A pre-driver. All three stages are direct-coupled. The output is coupled to the load through a capacitor to permit single battery operation. If this capacitor were not used, the loudspeaker would have to return to the center of the power supply.

Capacitor C_1 is connected between the output and the junction between R_3 and R_4. This capacitor serves the same function as C_2 in Circuit 3-6, to reduce signal degeneration.

A small forward bias is furnished the bases of the driver and the output stage to avoid cross-over distortion. This bias is created by passing the collector current of Q_1 through the parallel resistance of R_6 and R_7. R_7 is a thermistor used to compensate for any temperature variations. Temperature stability is further enhanced by the emitter resistance R_5 and the low base resistance R_1 for Q_1 and by R_8 and R_9 for Q_4 and Q_5. An overall d.c. shunt feedback is provided by R_2.

Low distortion, wide frequency response and low output impedance for damping the loudspeaker is achieved through the additional negative feedback network C_4 and R_{10}. C_4 is used to eliminate any possible high frequency oscillations.

(This material is based on the article "Quasi-Complementary Transistor Amplifier" by H. C. Lin, Electronics, September 1956.)

Reprinted from Electronics, September 1956; copyright McGraw-Hill, Inc. 1956

CIRCUIT 3-12

COMPLEMENTARY SYMMETRY AMPLIFIER

National Research Council, Canada   Contributed by D. W. R. McKinley and
R. S. Richards

For reasons of miniaturization and extended frequency response, contemporary design of audio amplifiers and output stages tends to eliminate transformers wherever possible. The complementary symmetry push-pull connections offer single-ended input and output, require no transformer, and permit a low load impedance and frequency response. Circuit 3-12 is an example of this configuration.

The circuit employs three direct-coupled grounded emitter stages, Q_1, Q_2 and Q_3 as preamplifier, feeding the push-pull complementary driver stage, Q_4 and Q_5, which in turn drives the output stage, Q_6 and Q_7.

The complementary stages employ one p-n-p and one n-p-n transistor to achieve Class B operation without the use of a phase inverter. Each transistor conducts on one-half of the cycle.

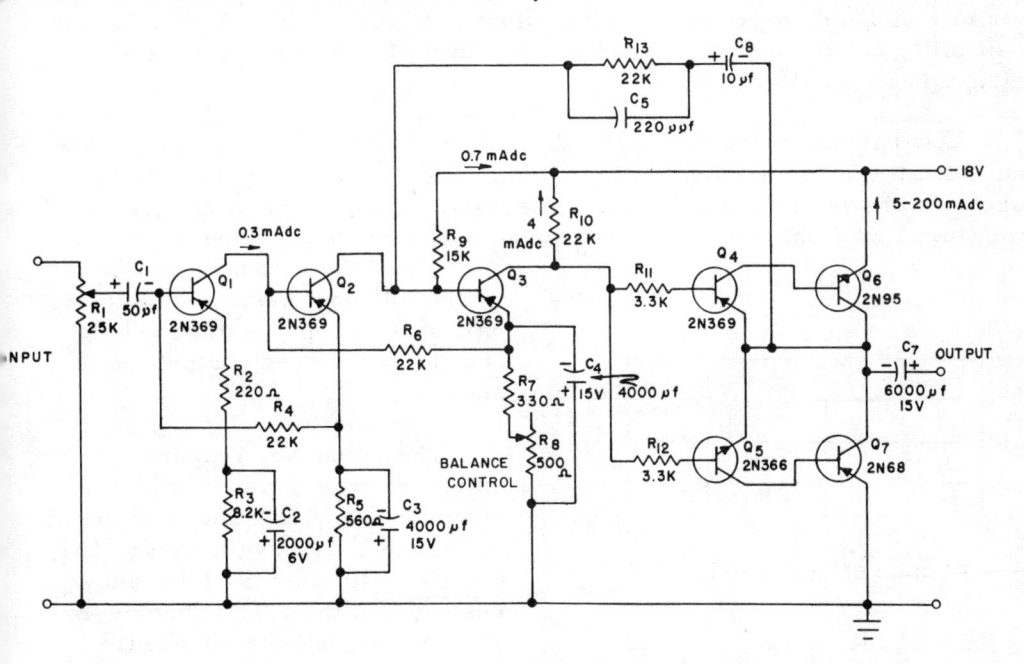

Circuit 3-12 - Complementary symmetry amplifier

Circuit 3-12, Complementary Symmetry Amplifier (cont'd.)

The preamplifier of this circuit contains two d.c. feedback loops provided by R_4 and R_6 which effect a high degree of temperature stabilization. Additional feedback is offered by the use of emitter resistors; this also enhances the possibility of transistor substitution without adversely affecting the operation of the amplifier.

A-C feedback from the output of the amplifier to the base of the last preamplifier transistor Q_3 is introduced by the feedback network R_{13}, C_5 and C_8. This network determines the frequency response of the amplifier. With the parameters shown in the diagram, the response extends from 2 c.p.s. to 9 kc. Modifying this network as shown in Figure 3-12.1 extends the frequency response to 20 kc.

With the lower frequency response, this amplifier is used for driving a pen recorder or similar instrument. The overall voltage gain is 4000; a 1 mVac input signal will produce 4 volts across 10 ohms at the output. The minimum input impedance of the circuit is 10K. For varying load resistances, the efficiency and power output vary. The efficiency is maximum at 20 ohms load resistance, where the power output is over 60% of the d.c. power to the driver and output stages; the power output in the vicinity of 1 watt maximum at this output impedance. The maximum power capability occurs with a 6 ohm load impedance, at which point the efficiency is about 45%.

Low harmonic distortion may be obtained, but the distortion is dependent upon the load resistance which determines the amount of negative feedback, as shown in Figure 3-12.1. The noise figure of the amplifier is measured as 7 db; this can be improved with low noise transistors. The balance of the output stages is set by the 500 ohm balance control R_8; half of the supply voltage should appear across each output transistor.

The frequency response of the amplifier may be modified by the use of the feedback network shown in Figure 3-12.1, with curves (A) and (B) indicating the frequency response of the circuit before and after the modification. Figure 3-12.2 shows the distortion as a function of load resistance. This circuit is useful for a multitude of

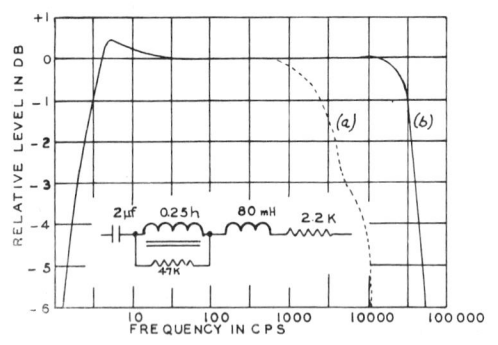

Fig. 3-12.1 - Relative level vs frequency

Circuit 3-12, Complementary Symmetry Amplifier (cont'd.)

audio applications, and may be used to drive a loudspeaker. The power available is 1 to 2 watts with quite low distortion. However, for loudspeaker output some forward bias should be used for Q_4 and Q_6 to eliminate crossover distortion.

The output coupling capacitor may be eliminated by returning the load to a low impedance center-tap point on the power supply. This may be accomplished by using batteries or with the use of two a.c. power supplies of opposite polarities. With direct coupling to the load, the output transistors must be very closely matched or a d.c. current will flow through load.

The supply requirements are 18 volts at 200 mAdc, and regulation of some type should be employed to prevent distortion.

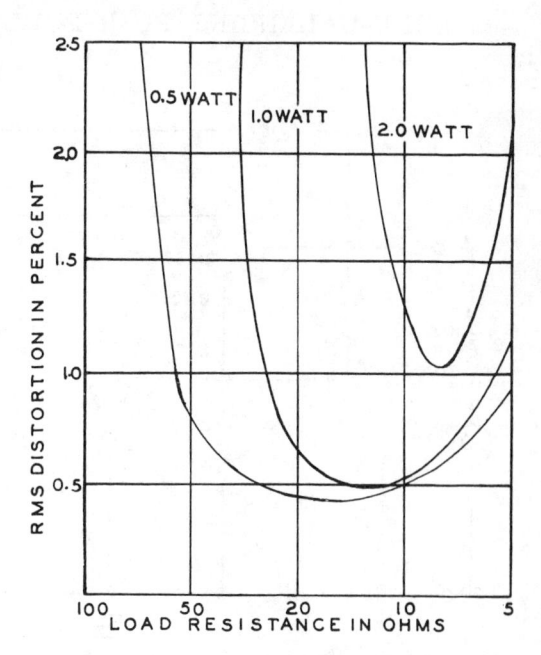

Fig. 3-12.2 - RMS distortion vs load resistance

The amplifier is useful over a reasonable temperature range, but is not designed for extremes of temperature. It is primarily useful in room temperature applications. It is however, typical of the results possible with complementary symmetry. The overall circuit economy is excellent, and the miniaturization possible without the use of transformers is excellent. This amplifier was designed for use as a medical recording instrument, but simple modifications presented suit it for use as a loudspeaker amplifier. With complementary transistors, the complementary symmetry configuration is useful for all low frequency power applications.

Reprinted from Electronics, August 1, 1957; copyright McGraw-Hill, Inc. 1957.

(This material is based on the article "Transistor Amplifiers for Medical Recording," by D. W. R. McKinley and R. S. Richards, Electronics, August 1, 1957.)

CIRCUIT 3-13

SELF-BALANCING PUSH-PULL CLASS A OUTPUT STAGE

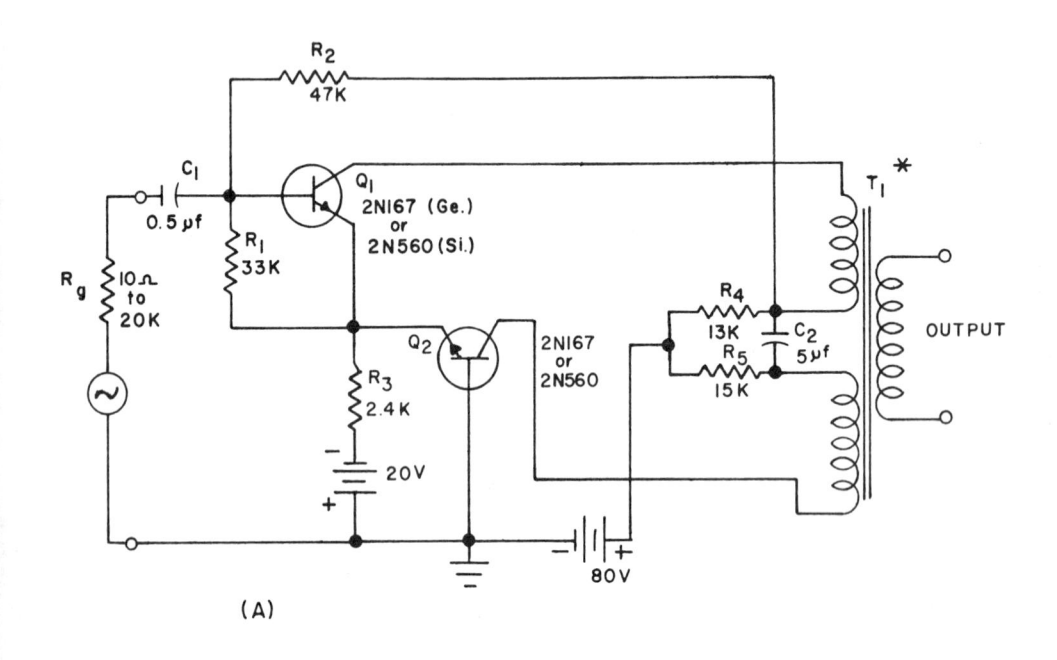

(A)

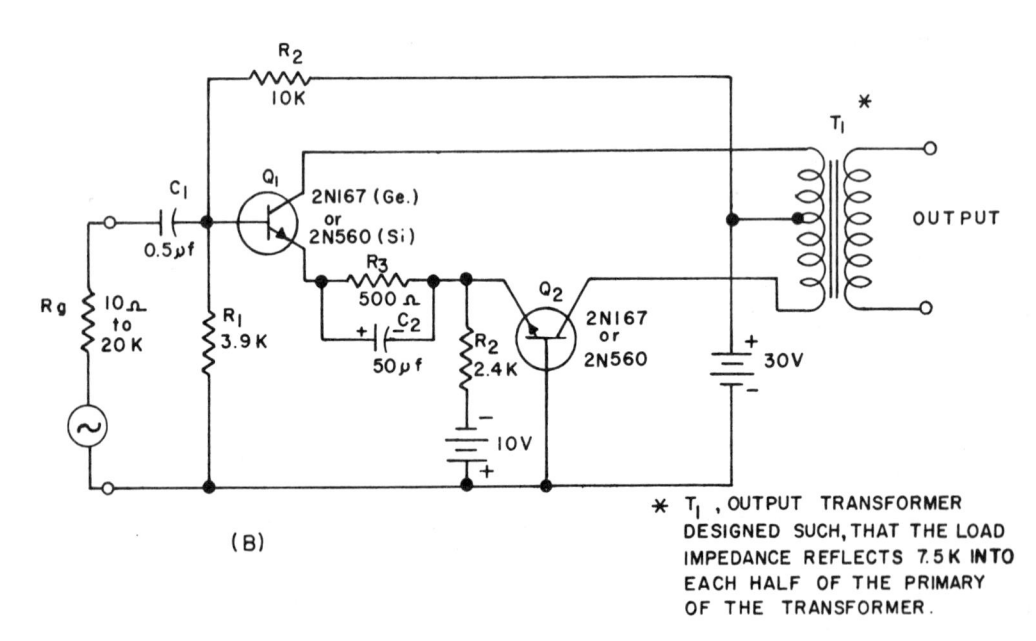

(B)

* T_1, OUTPUT TRANSFORMER DESIGNED SUCH, THAT THE LOAD IMPEDANCE REFLECTS 7.5K INTO EACH HALF OF THE PRIMARY OF THE TRANSFORMER.

Circuit 3-13 - Self-balancing push-pull class A output stage

Circuit 3-13, Self-Balancing Push-Pull Class A Output Stage (cont'd.)

American Bosch Arma Corporation   Contributed by J. Tellerman

Circuit 3-13, shown in two forms, is a push-pull Class A output stage using two medium power transistors of the same type. The transistors need not be matched to achieve balanced operation.

The feature of the circuit is that balanced operation can be achieved without requiring matched transistors. In both (A) and (B), Q_2 is a grounded base stage. The a.c. emitter input impedance of Q_2 is much lower than the 2.4K biasing resistor. Thus the a.c. collector current of Q_1 flows through the emitter of Q_2. Since α_{fb} of any transistor is nearly equal to unity, the collector current of Q_1 is therefore nearly equal but out of phase to that of Q_2, and balanced push-pull operation results.

In (A), Q_1 is stabilized by collector-base shunt degeneration. In (B), Q_1 is stabilized by series emitter degeneration.

The circuit may be used with either germanium or silicon units. Even for transistors of widely differing characteristics, a good balance can be maintained. The circuit can only operate Class A.

The circuit values are for medium or low power transistors. For high power applications this Class A approach is somewhat impractical because of the high power loss in the d.c. bias arrangement.

CIRCUIT 3-14

75 WATT AUDIO AMPLIFIER

Univ. of Cincinnati   Contributed by Alexander B. Bereskin

Circuit 3-14 is outstanding from the standpoint of high power output. The circuit is capable of producing 75 watts over an extended frequency range with a distortion less than 2%.

The output stage is operated in Class B to minimize transistor dissipation, and the minimum possible forward bias is used to avoid crossover distortion. Approximately 25 db of feedback is used to correct the nonlinearities. The margin of stability is greater than 12 db.

A low level n-p-n transistor, Q_1, is direct-coupled to the driver p-n-p transistor, Q_2. The driver transistor is in turn coupled through the driver transformer, T_1, to the output stage transistors, Q_3 and Q_4. An output autotransformer, T_2, is used to couple the load to the output transistors. With this arrangement, the collector of Q_4 is at approximately the same d.c. potential as the emitter of Q_1, and a direct-coupled feedback network

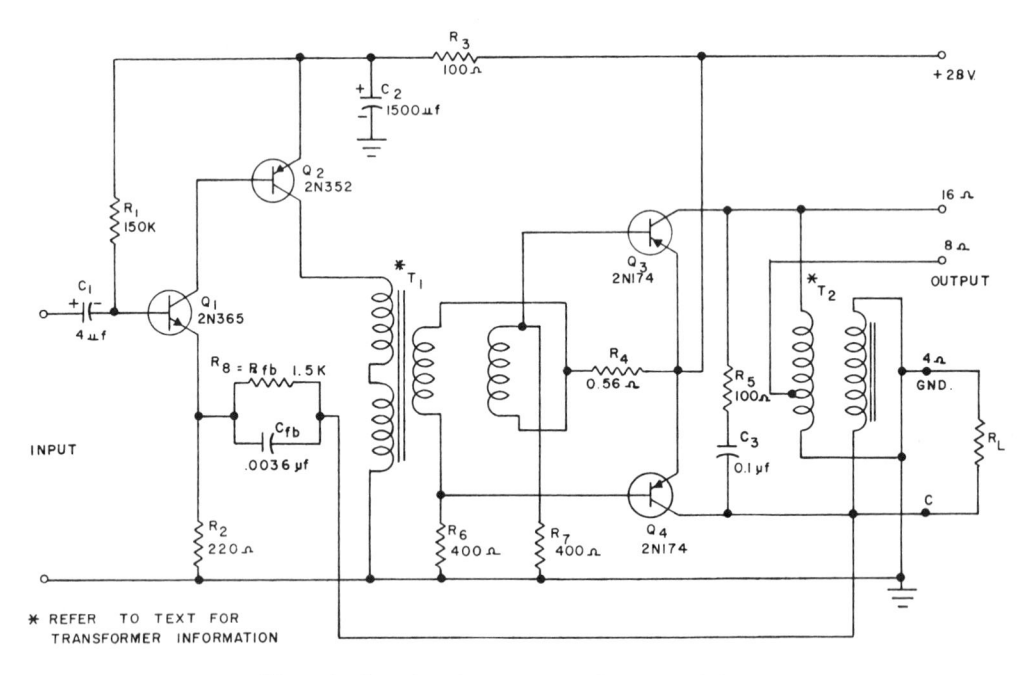

Circuit 3-14 - 75 watt audio amplifier

Circuit 3-14, 75 Watt Audio Amplifier (cont'd.)

is connected between them. The feedback resistor controls the low and middle frequency feedback, while the capacitor controls the high frequency feedback. The two transformers are bifilarly wound, and their construction is briefly described later. The bifilar winding reduces the leakage inductance which causes undesirable transients or crossover distortion when the conducting transistor is cut off.

The output stage is biased by the d.c. current flowing through the 0.56 ohm resistor, the driver transformer secondaries, and the two 400 ohm resistors. Temperature compensation may be had by using a thermistor in place of the 0.56 ohm resistor. To minimize the possibility of thermal runaway, the transistors are mounted on heat sinks as specified by the manufacturer.

The input and output transformers are wound on standard transformer laminations, and the laminations of the output transformer may be fully interleaved. The low leakage inductance of the driver transformer is attained by winding the primary in two equal sections with a single layer bifilar secondary sandwiched between them. The primary of the driver transformer carries d.c. current, and this must be taken into account in its design. The output transformer consists of a relatively small number of turns of parallel wires suitably insulated from each other. The collector terminal of Q_4 is used as a common output terminal, and taps may be brought out to match any impedance up to 16 ohms. The upper frequency is usually limited by the transistors rather than by the transformers.

The amplifier as described provides near perfect linearity and low distortion over a dynamic range of 70 db. If full power is not required, 45 watts can be obtained with an 8 ohm load, and 22 watts can be obtained with a 16 ohm load, connected to the 4 ohm tap; this procedure results in decreased distortion. The output impedance of the amplifier is less than 1 ohm for good damping. A power supply of good regulation is required; 48 volts at 4 amperes is required for full power output. Response to 10,000 c.p.s. should be obtained with the heavy feedback used. The input resistance of the circuit as shown is approximately 50K; there is a slight variation with frequency.

This amplifier is selected to demonstrate the high power capabilities of present circuits. Although special transformers are required, simpler circuits are possible if power and distortion criteria are not as stringent. This amplifier is primarily a room temperature device, but as such it provides a miniaturized circuit usable for applications normally reserved for vacuum tube amplifiers. The savings in heat, size, and weight with the

Circuit 3-14, 75 Watt Audio Amplifier (cont'd.)

use of transistors may be had at high power levels, as well as in low level equipment. This amplifier should find use primarily as a high fidelity public address unit, and in this capacity it employs transistors at a power level well above that normally associated with these devices.

The weight of the amplifier is determined largely by the transformers necessary to satisfy the low frequency requirements of the application. The transformers specified below resulted in a power output, with 2% distortion, of 60 watts at 20 cycles and 75 watts above 40 cycles.

Output Transformer, T_2:

#14 FV Wire - 100 turns of 2 parallel (bifilar) strands, layer wound with one thickness of 0.005" Kraft paper between layers of wire. Core - 2" stack of El-1-1/2"-0.016".

Dynamo grade laminations were used. The laminations were 100% interleaved. A better grade iron may be used to some advantage.

Driver Transformer, T_1:

Primary - 260 turns total of #27 FV wire wound in two sections of 130 turns each, one section below the secondary and the other section above the secondary. Secondary - 40 bifilar turns of #23 FV wire sandwiched between the two primaries. Core - 1-1/2" stack of El-1-1/4" grain oriented laminations. The laminations are butted (not interleaved) without any spacer in the air gap.

(This material is based on the article "A High Power High Quality Transistor Audio Power Amplifier," by Alexander B. Bereskin, 1957, IRE Convention Record, Vol. 5, Part 7.)

CIRCUIT 3-15

HIGH GAIN SERVO AMPLIFIER

Transitron Electronic Corporation   Contributed by G. Wyntjes

    The amplifier shown in Circuit 3-15 was designed for high reliability in low frequency servo applications. Silicon transistors are employed to extend the high temperature operating range. Performance is essentially independent of temperature changes between -65°C and 100°C, supply variations of ±20%, and the interchanging of transistors.

    The amplifier is a 5-stage amplifier, the first three stages giving current gain and the last two stages power gain. The three intermediate stages are coupled by means of forward biased diodes; this reduces the number of capacitors in the circuit to two. The first four stages are in Class A operation, the output stage is in push-pull Class B operation.

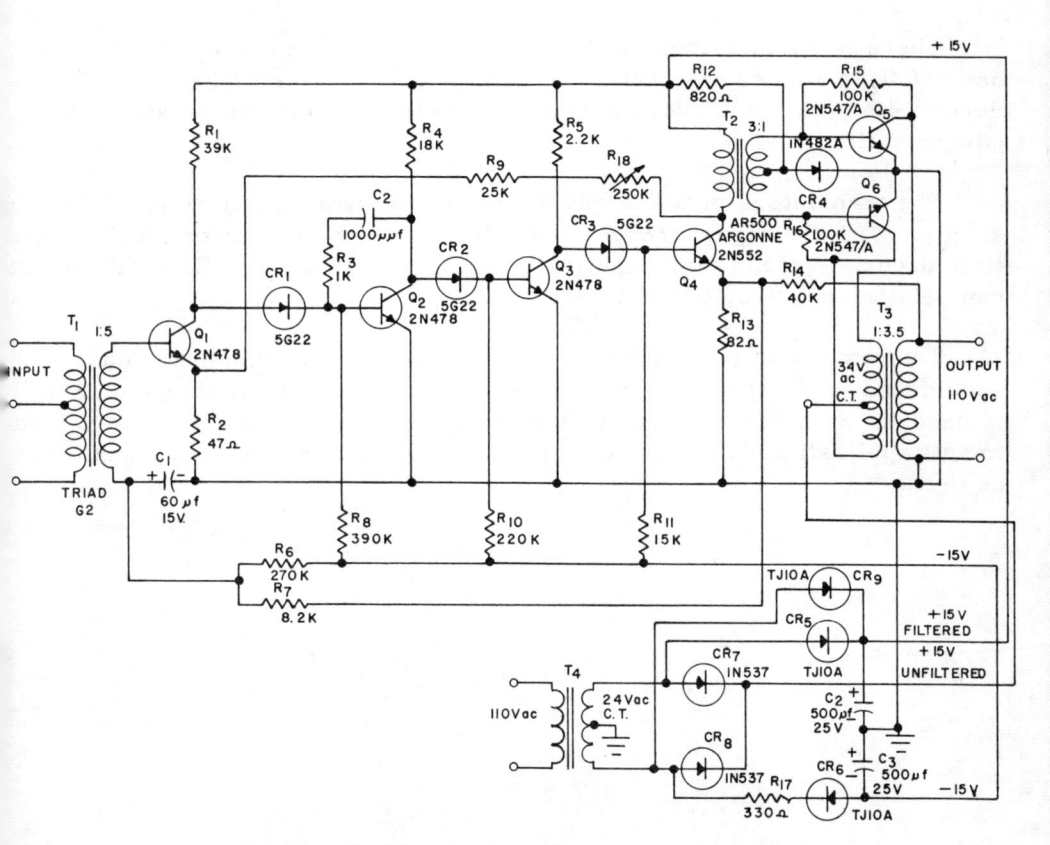

Circuit 3-15 - High gain servo amplifier

Circuit 3-15, High Gain Servo Amplifier (cont'd.)

Stability is achieved with three feedback loops. The d.c. stabilizing loop, bypassed by a 60 μf tantalytic capacitor, is connected from the emitter of the driver stage to the base of the input transistor. A second feedback loop is connected from the collector of the driver transistor to the input emitter; the feedback factor is variable, providing variable gain. This feedback loop increases the input impedance of the first stage, and the amplifier should be driven by a source impedance of less than 500 ohms to make this feedback effective.

By applying the feedback to the base of the first transistor, the amplifier can be fed from high impedance sources. This can be done by omitting the second collector-emitter feedback and by adding a variable resistance of 100 ohms in series with the 60 μf capacitor.

The third feedback loop is from the output to the emitter of the driver. This gives reduced output impedance, less distortion, and compensation for variations among transistors.

There is another local shunt a.c. feedback from the collector to the base of Q_2 to reduce any instability at high frequencies. The output stage derives its bias from a forward-biased diode, CR_4, for temperature compensation.

The transistors of the output stage are fed from a full-wave unfiltered supply. With the signal frequency in phase with the power supply frequency (this occurs in most servo applications), an efficiency of 80% at full output can be attained. (See Circuit 3-16.)

This amplifier is useful over a wide range of servo applications, and should find application where a reliable, wide temperature range circuit is needed. Stability of the circuit is excellent, and the gain may be varied to suit application. Either a 60 c.p.s. or 400 c.p.s. supply and signal may be used.

CIRCUIT 3-16

HIGH EFFICIENCY SERVO OUTPUT STAGE

Boeing Aircraft Company   Contributed by Bruce M. Benton

Circuit 3-16 is a high efficiency servo output stage. Efficiencies approaching 100% are possible, and actual efficiencies of 90 to 95% have been obtained. It is identical to the conventional Class B amplifier, except that the collector power for the transistors is derived from full-wave rectification of the servo power supply.

The transistors with unfiltered full-wave rectified power operate with maximum full-load efficiency because the wave shapes of the load voltage and the supply voltage differ only by the saturation voltage of the transistor. This is illustrated in Figure 3-16.2(A), where the voltage drop v_c across the transistor at full load is shown for the one-half cycle of a given transistor. Figure 3-16.2(B) shows the same transistor voltage drop for pure d.c. collector supply power as found in conventional Class B amplifiers.

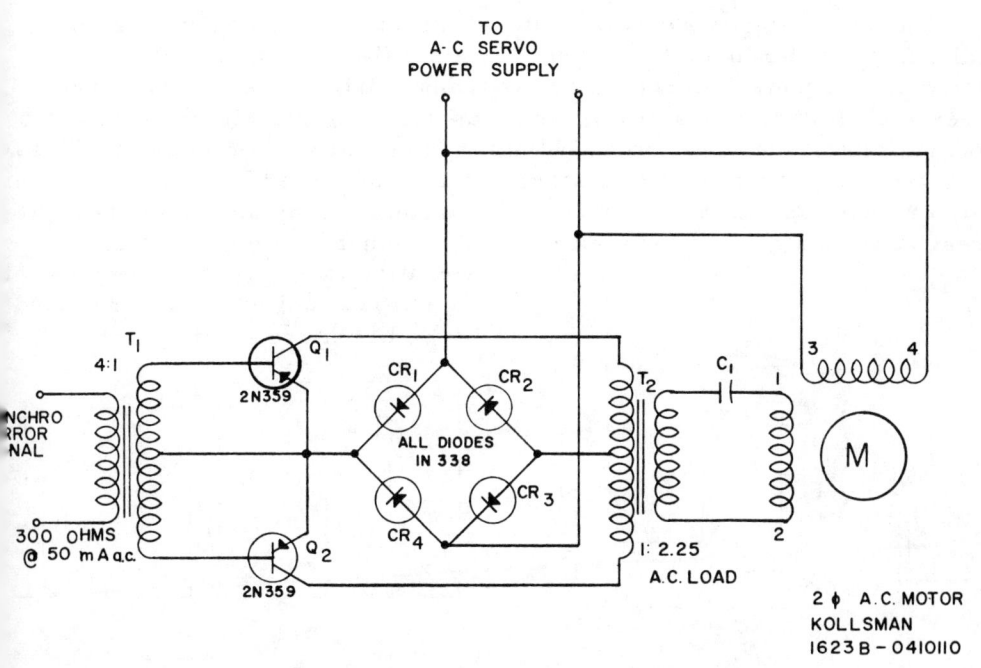

Circuit 3-16 - High efficiency servo output stage

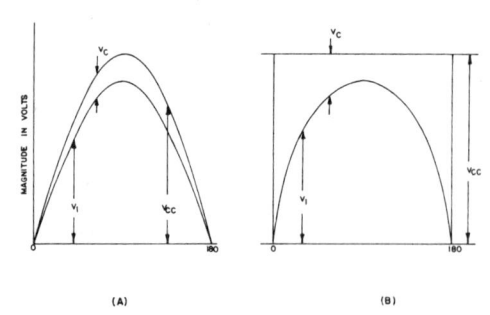

Circuit 3-16, High Efficiency Servo Output Stage (cont'd.)

The area between the two curves in each case gives an indication of the power dissipated in the transistor, and it is clearly much smaller in the case of Figure 3-16.2(A).

Maximum efficiencies of 95% at full load can be obtained. The maximum collector dissipation occurs at 1/4 power, and is 25% of full load power as compared to 40% for a Class B stage. Thus, the high efficiency stage is capable of 1-1/2 times the power output of the same stage operated in Class B.

A method of applying this amplifier to d-c loads is shown in Figure 3-16.1. The circuit amplifies and rectifies the servo error signal to supply a d-c voltage across controller loads 1-2 or 3-4, depending upon the phase relationship of the power source and the error signal. This circuit has the advantage of low drift associated with a.c. systems; straight d.c. control amplifiers are subject to the drift found in d.c. amplifiers. For maximum efficiency the loads should be resistive. If necessary, smoothing capacitors may be used as shown. This will, however, result in decreased efficiency. Complete smoothing will result in Class B operation.

The output stages shown operate well over the frequency range normally found in servo applications. The amplifier operates well with distorted a.c. power sources, such as transistorized square-wave power inverters. A 20 watt amplifier can be made, using transistors with rated dissipation of 5 watts or more; reliability is greatly improved over Class B stages using the same transistors at the same power levels. The circui is useful with any of the readily-available power transistors, and is of great value in any servo applications. This scheme of high efficiency operation may be applied to most of the Class B servo amplifiers found in the literature and in practice,

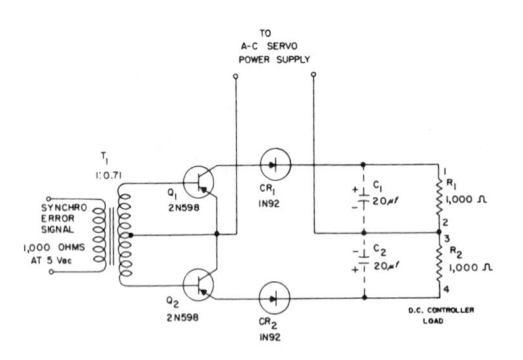

Fig. 3-16.1

Fig. 3-16.2 - Supply voltage phase in deg.

Circuit 3-16, High Efficiency Servo Output Stage (cont'd.)

and in many other single frequency control applications such as voltage regulators and (transistor equivalent) magnetic amplifiers.

(This material is based on the article "Servo Amplifiers Use Power Transistors," by Bruce M. Benton, Electronics, September 1956.)

Reprinted from Electronics, September 1956; copyright McGraw-Hill, Inc. 1956.

PART 4

HIGH-FREQUENCY AMPLIFIERS

Section 1. Design Philosophy

A. Introduction

This section deals with topics of importance to the application of transistors to high-frequency band-pass and video amplifiers.

B. Equivalent Circuits

There are a variety of equivalent circuit representations of transistors (1), ranging from the extremely simple in which accuracy may suffer, to the accurate but complex and quite likely useless. The choice of equivalent circuit is one of compromise between accuracy and reasonable utility. One would like the equivalent circuit to be as simple as possible within the desired range of accuracy.

Equivalent circuits, based on theoretical models of the junction transistor, have the advantage of relating the elements of the equivalent circuit with the physical properties of the transistor. These equivalent circuits generally have a common origin in a model such as shown in Figure 4-0.1. An ideal one-dimensional transistor, called the intrinsic transistor, is first considered. To the intrinsic transistor are added the extrinsic element r_e', r_c', r_b', C_{te}, and C_{tc} and possibly parasitic elements such as stray capacitances.

In turn, r_e' and r_c' represent ohmic resistances of the emitter and collector regions. These usually are not important in alloy-type transistors, but may be important in diffused-base type transistors. The base spreading resistance, r_b', may be complex in grown junction type transistors (2). Both C_{te} and C_{tc} are the emitter and collector depletion layer or transition capacitances. In ordinary alloy transistors, C_{te} is usually negligible, but may be important in diffused-base type transistors, due to the low resistivity of the base region near the emitter side.

Exact equivalent circuits (4,5) require r_c transmission line to

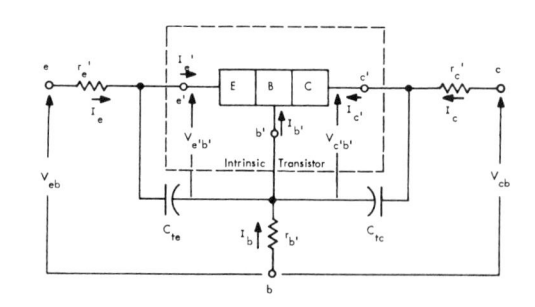

Fig. 4-0.1 - Model of transistor

represent the diffusion of minority carriers across the base region and are therefore too complicated. The differences in equivalent circuits then, based on the model of Figure 4-0.1, come about in the approximations used to derive a simple equivalent circuit.

Figure 4-0.2 shows a high-frequency, common-base equivalent circuit derived by Scarlett (6), and the hybrid-π common-emitter equivalent circuit of Giacoletto (7). Figure 4-0.3 shows an equivalent circuit for a diffused base transistor having a mesa-type structure (8). Capacitors C_{be}, C_{cb}, and C_{ce} are parasitic capacitors. $C_{c'}$ is the collector-depletion capacitance of that part of the collector that is under the emitter stripe and $r_{c'}$ is the collector body resistance for that part that is under the emitter stripe, while $C_{c''}$ and $c_{c''}$ are the collector depletion capacitance in collector body resistance for the portion of the collector that lies outside of the emitter stripe, and $r_{e'}$ is the emitter body resistance; not to be confused with $r_{e'}$ of Figure 4-0.2(A).

Thomas and Moll (9) have shown that the common-base short-circuit current gain alpha may be accurately expressed as

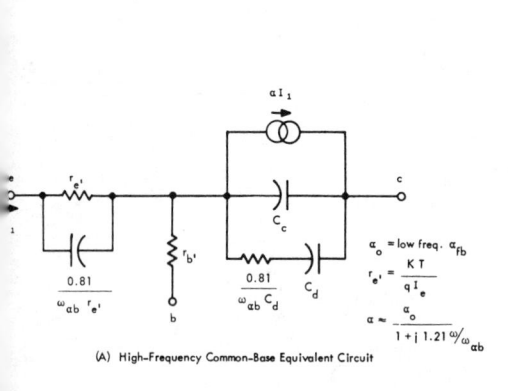

$$a_b = \frac{a_{bo} e^{j\left(\frac{K-1}{K}\right) f/f_{ab}}}{1 + j \frac{f}{f_{ab}}}, \text{ for } f < f_{ab},$$

4-0.1

where

$$K = \frac{f_{ae}}{(1 - a_{bo}) f_{ab}}$$

Fig. 4-0.2 - Small signal equivalent circuits for alloy and surface barrier type transistors

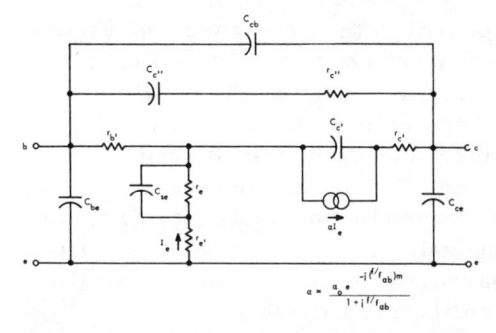

Fig. 4-0.3 - Equivalent circuit for a mesa type diffused base transistor

a_{bo} is the low-frequency short-circuit common base current gain

f_{ab} is the frequency at which the magnitude of a_b is 3 db below a_{bo}, and

f_{ae} is the frequency at which the magnitude of the common emitter short circuit current gain, a_e, is 3 db below its low frequency value, a_{eo}.

K in equation 4-0.1 depends on the base layer purity distribution. For uniform base layer distribution, K is 0.82 and decreases for built in fields corresponding to the decrease in purity concentration from emitter to collector. In Figure 4-0.3, m in the expression for alpha corresponds to - (K-1/K) of equation 4-0.1.

The use of equivalent circuits may be avoided in some instances and accurate results obtained by using measured terminal characteristics.

C. Power Gain and Stability

The non-unilateral nature of the transistor, i.e., the internal feedback, complicates gain relationships, makes the input and output impedances functions of load and source impedances respectively, and may lead to instability. (This is not intended to imply that unilateralized transistor circuits may not be unstable.)

Gain and instability will be discussed with reference to Figure 4-0.4 which represents the transistor in terms of the small signal Y parameters, Y_L is the load admittance and Y_G is the source admittance. The Y parameters are those appropriate to the transistor configuration, i.e., common emitter parameters for common emitter configurations, etc.

In Figure 4-0.4 (B), expressions showing the dependence of input and

(A) Small Signal Representation of Transistor Amplifier

$I_1 = y_{11} V_1 + y_{12} V_2$
$I_2 = y_{21} V_1 + y_{22} V_2$
$Y_G = G_g + j B_G$
$Y_L = G_L + j B_L$

$y_{in} = y_{11} - \dfrac{y_{12} y_{21}}{y_{22} + Y_L}$

$y_{out} = y_{22} - \dfrac{y_{12} y_{21}}{y_{11} + Y_g}$

$P_g = \dfrac{P_{out}}{P_{in}} = \dfrac{|y_{21}|^2 G_L}{|y_{22} + Y_L|^2 \operatorname{Re}(y_{in})}$

(B) General Relationships

$y_{in} = g_{11} [N_1 - j N_2] + j b_{11}$

$y_{out} = g_{22} [N_1 - j N_2] + j b_{22}$

$P_{g\,(max)} = \dfrac{|y_{21}|^2}{g_{22} g_{11} [(1+N_1)^2 + N_2^2]}$

where

$N_1 = \sqrt{1 - A - \left(\dfrac{B}{2}\right)^2}, \quad N_2 = \dfrac{B}{2}$

$A = \dfrac{g_{12} g_{21} - b_{12} b_{21}}{g_{11} g_{22}} \qquad B = \dfrac{g_{12} b_{21} + g_{21} b_{12}}{g_{11} g_{22}}$

g_{ij} = real part of y_{ij}, and b_{ij} = imaginary part of y_{ij}

(C) Relationships for Conjugate Matching at Input and Output to Yield Maximum Gain ($Y_G = y_{in}*$ and $Y_L = y_{out}*$)

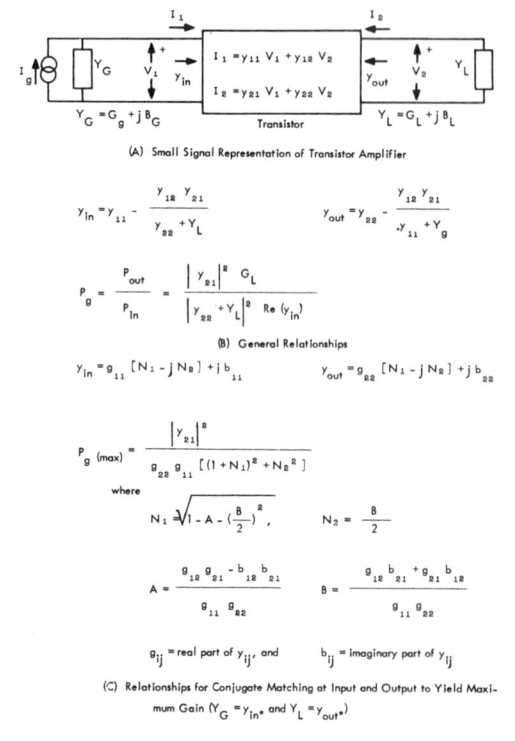

Fig. 4-0.4 - Small signal model of transistor amplifier and some applicable relationships

utput admittances on load and source admittances are given. This dependence arises from the non-unilateral nature of the transistor and would disappear if y_{12} were zero (or if y_{21} were zero in which case the gain would e zero also). An expression for power gain is also given.

The power gain is maximum when the source admittance is the conjugate of the input admittance and the load admittance is the conjugate of he output admittance. Under these conditions, the maximum power available from the signal source goes into the transistor input and the maximum ower available from the transistor is applied to the load. The relationships for conjugate matching at input and output (10) are given in Figure -0.4(C). The numerators in the expressions for A and B may be recognized as the real and imaginary parts respectively of $y_{12} \, y_{21}$.

If the transistor is unilateralized, then N_1 becomes unity, N_2 becomes ero, and the expressions of Figure 4-0.4(C) may be written

$$y_{in} = y_{11}' \qquad y_{out} = y_{22}'$$

$$P_{g\,(max)} = \frac{|y_{21}'|^2}{4 g_{22}' g_{11}'} \qquad 4\text{-}0.2$$

where the primes signify that the parameters apply to the unilateralized ransistor circuit arrangement and are not those of the transistor.

Another useful gain relationship is the transducer gain, defined as the atio of power delivered to the load to the power available from the input ignal source. In terms of the Y parameters, the transducer gain is given y

$$G_T = \frac{|y_{21}|^2 \, 4 G_G \, G_L}{|(y_{22} + Y_L)(y_{11} + Y_G) - y_{12} \, y_{21}|^2}. \qquad 4\text{-}0.3$$

he conjugate matching at input and output, G_T becomes identical with the ower gain expression given in Figure 4-0.4(C).

Mason (11) has derived an expression for power gain with loss-less nilateralization, which is much used as a standard for comparison in valuating neutralizing arrangements in transistor amplifiers in general. 'his gain function

$$U = \frac{|y_{21} - y_{12}|^2}{4 \, (g_{11} g_{22} - g_{12} g_{21})}, \qquad 4\text{-}0.4$$

where

g_{ij} is the real part of y_{ij}

is the same, regardless of the external loss-less passive network used for unilateralization and is independent of the set of transistor parameters used (common base, common emitter, common collector).

An amplifier is potentially unstable if passive terminations can be found to cause it to oscillate. From the gain expression of Figure 4-0.4(C) it is seen that unless g_{11} and g_{22} are positive, the gain may become infinitely large, implying oscillation. So two of the requirements for unconditional stability are

$$g_{11} > 0$$
$$g_{22} > 0 .$$
4-0.5

With g_{11} and g_{22} positive, instability may still be possible due to internal feedback. In Figure 4-0.5, Figure 4-0.4(A) is redrawn to show the input admittance of the amplifier loaded at both ends. If the real part of y'_{in} can be made negative by adjusting the external loads having positive real parts, the amplifier is potentially unstable, for then the imaginary part of Y_G may be adjusted to make the amplifier oscillate, as far as the input side is concerned, as a two-terminal negative conductance oscillator. Similar statements could be made regarding Y'_{out}, but it is only necessary to consider Y'_{in} or Y'_{out} as the conditions for stability in terms of the transistor parameters come out the same in both cases. For the amplifier to be unconditionally stable, then, in addition, the requirements of inequalities of equation 4-0.5, there is the further requirement that for any load and source admittances having positive real parts, the real parts of Y'_{in} shall be positive.

The requirement that the real part of Y'_{in} be positive for unconditional stability is expressed as

$$g_{11}'(g_{22}')^2 + g_{11}'(b_{22}')^2$$
$$- g_{22}' \operatorname{Re}(y_{12} y_{21})$$
$$- b_{22}' [\operatorname{Im}(y_{12} y_{21})]^2 > 0,$$
4-0.6

where

Re $(y_{12} y_{21})$ means the real part of $(y_{12} y_{21})$ and

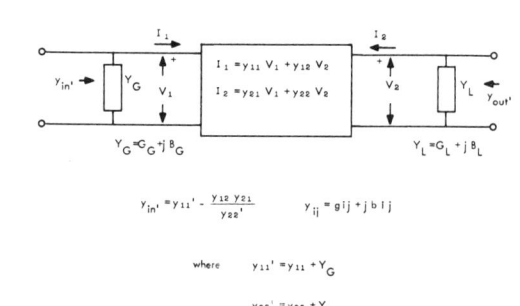

Fig. 4-0.5 - Model used to explain potential instability

$\text{Im}(y_{12} y_{21})$ means the imaginary part of $(y_{12} y_{21})$.

By adjustment of B_L, b'_{22} (equals $b_{22} + B_L$) may take on any real value. When

$$b_{22}' = \frac{\text{Im}(y_{12} y_{21})}{2 g_{11}'} \qquad 4\text{-}0.7$$

Equation 4-0.6 is minimized to give

$$4 g_{11}' g_{22}' [g_{11}' g_{22}' - \text{Re}(y_{12} y_{21})] > [\text{Im}(y_{12} y_{21})]^2 \qquad 4\text{-}0.8$$

as the requirement for unconditional stability to the amplifier loaded with conductances G_G and G_L. The requirements for unconditional stability are most severe when G_G and G_L are zero, for which equation 4-0.8 becomes

$$4 g_{11} g_{22} [g_{11} g_{22} - \text{Re}(y_{12} y_{21})] > [\text{Im}(y_{12} y_{21})]^2 \qquad 4\text{-}0.9$$

The relationship of equation 4-0.9 can be expressed in a number of equivalent forms listed below

$$4 g_{11} g_{22} [g_{11} g_{22} - \text{Re}(y_{12} y_{21})] > [\text{Im}(y_{12} y_{21})]^2 \qquad 4\text{-}0.9 \text{ (A)}$$

$$g_{11} g_{22} > |y_{12} y_{21}| + \text{Re}(y_{12} y_{21}) \qquad 4\text{-}0.9 \text{ (B)}$$

$$\sqrt{g_{11} g_{22}} > \text{Re} \sqrt{y_{12} y_{21}} \qquad 4\text{-}0.9 \text{ (C)}$$

$$g_{11} g_{22} > |y_{12} y_{21}| (1 + \cos \theta) \qquad 4\text{-}0.9 \text{ (D)}$$

where

$$\theta = \tan^{-1} \frac{\text{Im}(y_{12} y_{21})}{\text{Re}(y_{12} y_{21})}$$

$$\text{Re}(y_{in})\big|_{min} > 0 \qquad 4\text{-}0.9 \text{ (E)}$$

$$\text{Re}(y_{out})\big|_{min} > 0. \qquad 4\text{-}0.9 \text{ (F)}$$

The stability problem has been discussed by many authors (11-18).

If the inequalities of equations 4-0.9 and 4-0.5 are not met, then the amplifier is said to be potentially unstable. This does not mean that it will be unstable, but that source and load admittances may be found for which it is unstable. For narrow band tuned amplifiers, it is safe to assume that if the amplifier circuit is potentially unstable, it is going to oscillate.

By inserting transistor parameter values into equation 4-0.9, the frequency ranges in which the transistor is potentially unstable for a different transistor configuration, this may be determined. Scarlett (6) has done this, using approximate high frequency parameters and finds that the common base, the common collector configurations are potentially unstable over the range of frequencies for which the high frequency parameters are a good approximation.

The common emitter configuration may be unconditionally stable above a critical frequency given by Pritchard (2,19) as

$$\omega_{crit} = \frac{0.4 r_e'}{r_b'} \omega_{ab} . \qquad 4\text{-}0.10$$

Thus, narrow band RF amplifiers using transistors in the common emitter configuration will be stable without neutralizing when operated at frequencies greater than ω_{crit}. There are some cautions to be observed, particularly in case the amplifier stage is AGC controlled, for if the emitter current is controlled by the AGC signal, as is frequently the case, then ω_{crit}, being proportional to r_e' and therefore inversely proportional to the emitter current, will be a function of the AGC signal. If the AGC signal places ω_{crit} above the operating frequency, the circuit may oscillate. Also, in such an amplifier, relying on operation above ω_{crit} for stability, it would not do to substitute a transistor having a higher alpha cutoff frequency for the design type without considering what the new critical frequency would be. The common emitter transistor is also unconditionally stable at low frequencies, up to something in the neighborhood of $(1-a_{bo})(\omega_{ab})$. There is then, a frequency band where the common emitter configuration may be potentially unstable.

If equation 4-0.8 is written as

$$(g_{11} + G_G)(g_{22} + G_L) = k \, [\,|y_{12} y_{21}| + \text{Re}\,(y_{12} y_{21})] , \qquad 4\text{-}0.11$$

where $k > 1$, then the amplifier is stable, but the likelihood of instability decreases as k is made larger. It is seen that the amplifier can be stabilized to any degree desired by making G_G and G_L sufficiently large.

Studies have been made (15) to determine the optimum values of G_G and G_L, yielding the maximum power gain for a given value of stability factor k. Values of k up to 10 may be desirable, taking into account tolerances of transistor and circuit parameters, effects of temperature,

AGC, etc. The analysis is quite complicated for the single stage, and much more so for multiple stages. Increasing stability by increasing k results in a decrease in the maximum gain obtainable, but it does not follow that a gain reduction obtained by arbitrarily floating the amplifier results in increased stability.

The high frequency gain of ordinary alloy type transistors has been given by Pritchard (2), and Giacoletto (7) as

$$G_p \approx \frac{a_o \, f_{ab}}{25 \, f^2 \, C_c \, r_{b'}} = \frac{g_m}{4\omega^2 \, C_{b'e} \, C_{b'c} \, r_{bb'}} \qquad 4\text{-}0.12$$

where $0.05 - 0.1 < \omega/\omega_{ab} < 2$.

The maximum frequency of oscillation is given as the frequency at which the power gain is unity from equation 4-0.12, then

$$f_{max} = \left[\frac{a_o \, f_{ab}}{25 \, C_c \, r_{b'}}\right]^{1/2} = \frac{1}{4\pi}\left(\frac{g_m}{r_{bb'} \, C_{b'c} \, C_{b'e}}\right)^{1/2} \qquad 4\text{-}0.13$$

For high frequency operation, f_{max} is a useful figure of merit.

D. Unilateralization and Neutralization

All three transistor configurations are potentially unstable in some part of the useful frequency range. In the preceding section, it was shown that stability may be achieved by introducing losses. This, however, does not remove the dependence of input and output impedances on load source impedances. This dependence on load and source can lead to difficulties in alignment of tuned circuits, or distort the shape of the bandpass characteristics. Neutralization or unilateralization circuits are often used to stabilize the transistor and should reduce or eliminate the dependence of input and output impedances on source and load.

In unilateralization, the feedback parameter of the unilateralized circuit (y_{12}, h_{12}, etc.) is made zero. Neutralization is a similar technique carried to the extent necessary to achieve a desired result. For example, making an amplifier stable while allowing some dependence of source and load on input and output impedances. Under this sort of a definition, neutralization would include unilateralization.

Angell (20), Chu (21), and Stern (13) have discussed neutralization and have given a number of circuit arrangements. The approach used in reference 13 to explain unilateralization is outlined below.

The transistor in a given configuration is represented by

$$\begin{bmatrix} D_1 \\ D_2 \end{bmatrix} = \begin{bmatrix} k_{11} & k_{12} \\ k_{21} & k_{22} \end{bmatrix} \times \begin{bmatrix} J_1 \\ J_2 \end{bmatrix} \qquad 4\text{-}0.14$$

(For example, in terms of the common emitter parameters, $k_{ij} = y_{ije}$. $D_1 = i_b$, $D_2 = i_c$, $J_1 = V_{be}$, $J_2 = V_{ce}$.)

A network (usually passive but not necessarily so) described by

$$\begin{bmatrix} D_1' \\ D_2' \end{bmatrix} = \begin{bmatrix} k_{11}' & k_{12}' \\ k_{21}' & k_{22}' \end{bmatrix} \times \begin{bmatrix} J_1' \\ J_2' \end{bmatrix} \qquad 4\text{-}0.15$$

is connected to the transistor in such a way that J_1' and J_2' equal J_1 and J_2 respectively.

The composite network represented by

$$\begin{bmatrix} D_1'' \\ D_2'' \end{bmatrix} = \begin{bmatrix} k_{11}+k_{11}' & k_{12}+k_{12}' \\ k_{21}+k_{21}' & k_{22}+k_{22}' \end{bmatrix} \times \begin{bmatrix} J_1 \\ J_2 \end{bmatrix}$$
$$4\text{-}0.16$$

is unilateral if it satisfies the following

$$k_{12} + k_{12}' = 0. \qquad 4\text{-}0.17$$

The requirements for unconditional stability for the unilateralized composite network reduced to

$$\text{Re } (k_{11} + k_{11}') > 0$$
$$4\text{-}0.18$$
$$\text{Re } (k_{22} + k_{22}') > 0.$$

Some methods of interconnection of the transistor in neutralizing network are shown in Figure 4-0.6. Some practical neutralizing arrangements are shown in Figure 4-0.7.

The unilateralized transistor may be unstable if inequalities of

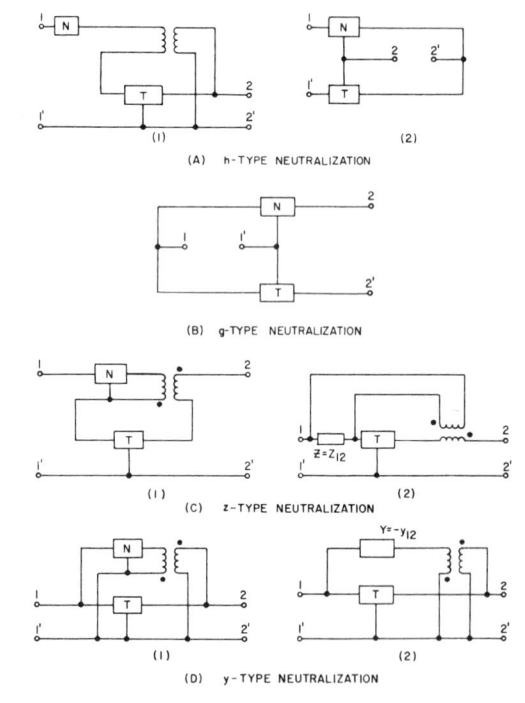

Fig. 4-0.6 - Some methods of interconnection of transistor [T] and neutralizing network [N] (13)

(A) h-TYPE (OR BRIDGE) NEUTRALIZED COMMON-BASE CIRCUITS[20]

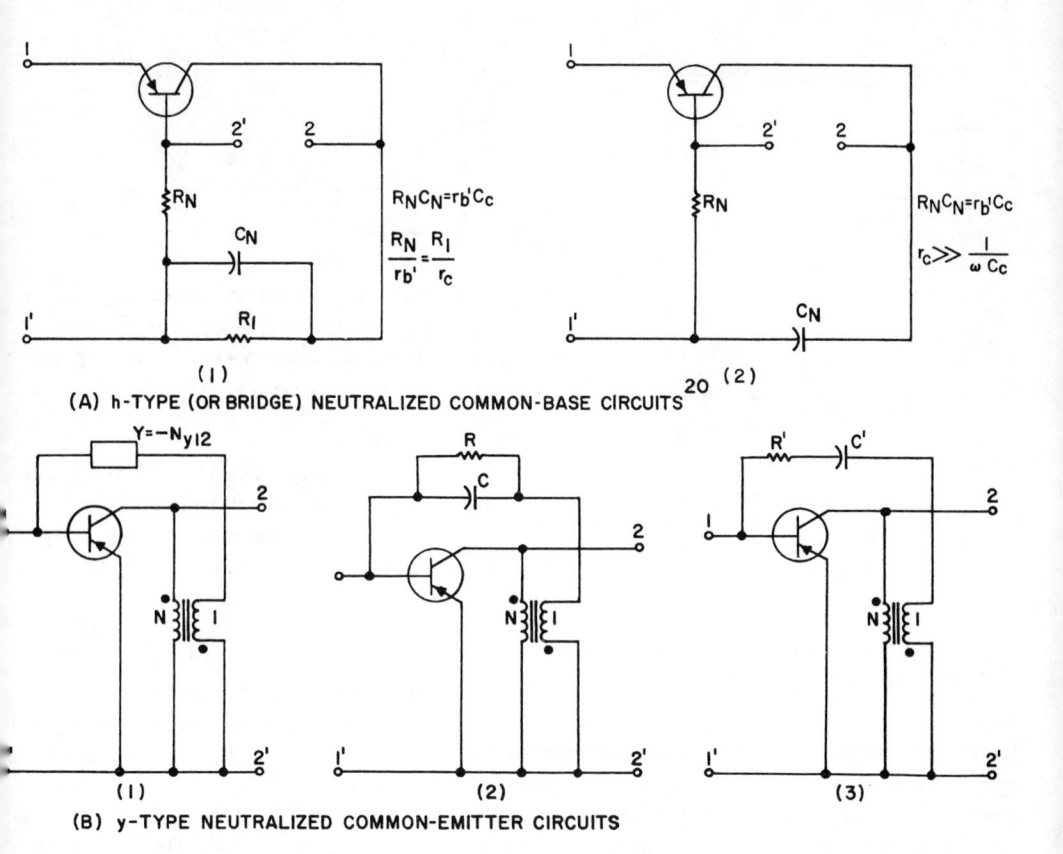

(B) y-TYPE NEUTRALIZED COMMON-EMITTER CIRCUITS

Fig. 4-0.7 - Some practical neutralization circuits

equation 4-0.18 are not met. Cote (22) has considered this problem and notes that the common base h type neutralization of Figure 4-0.7(A) may be unstable at high frequencies as h_{22} (or z_{22}) of the common base transistor configuration frequently has a negative real part. Cote has also taken into account practical transformer effects and calculated optimum parameter values for various unilateralization networks to yield maximum gain.

E. High Frequency Noise (23 to 31)

Neilson's (25) simplified noise equivalent circuit of Figure 4-0.8 and the noise figure expression which it yields, namely

$$F_b = F_e = 1 + \frac{r_b'}{R_g} + \frac{r_e}{2R_g} + \frac{[(1-\alpha_{ofb})]\left[1+\left(\frac{f}{\sqrt{1-\alpha_{ofb}}\, f_{ab}}\right)^2\right][R_g + r_b' + r_e]^2}{2\alpha_{ofb}\, r_e\, R_g} \quad 4-0.19$$

$$\overline{e_g^2} = 4kTR_g\,df$$

$$\overline{e_b^2} = 4kTr_{b'}\,df$$

$$\overline{e_e^2} = 2kTr_e\,df$$

$$\overline{e_c^2} = \frac{2kT\alpha_o(1-\alpha_o)\,|Z_c^2|\,\left[1 + (f/\sqrt{1-\alpha_o}\;f_{\alpha b})^2\right]}{r_e\left[1+(f/f_{\alpha b})^2\right]}\,df$$

$$r_e = kT/qI_E$$

$$I_E = \text{d-c emitter current}$$

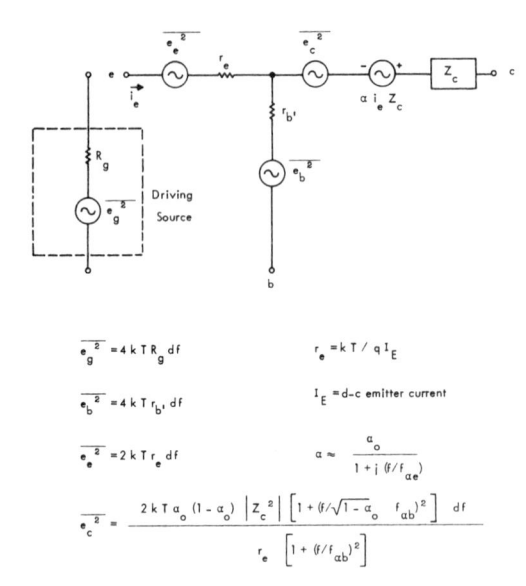

$$\alpha \approx \frac{\alpha_o}{1 + j(f/f_{\alpha e})}$$

Fig. 4-0.8 - Simplified noise equivalent circuit (25)

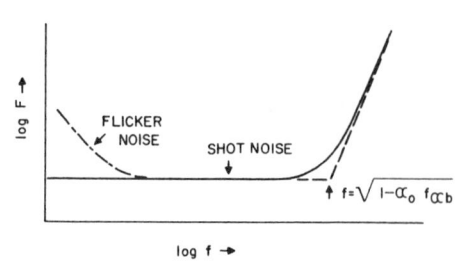

Fig. 4-0.9 - Sketch showing noise fig., F, as function of frequency, f for common-emitter or common-base configuration

enables an easy visualization of the noise behavior of transistors at frequencies where shot noise prevails over the low frequency flicker noise, i.e., above a few kilocycles. Neilson's simplifications relative to the results of van der Ziel (23) and Guggenbuehl and Strutt (24) for example, are mostly the neglect of the correlation between emitter and collector noise sources. The diffusion capacitance is shunting r_e and frequency dependence of the emitter noise source. Despite the simplifications, Figure 4-0.8 and equation 4-0.19 represent quite well the noise behavior of ordinary junction transitions up to the alpha cutoff frequency.

From equation 4-0.19, the noise figure (ratio of the total output noise power to that part of the output noise power due to the driving source thermal noise) is seen to be constant from low frequencies up to a frequency comparable to $\sqrt{1-\alpha_o}\;f_{\alpha b}$ where the noise figure commences to increase eventually in proportion to f^2 (6 db per octave) at high frequencies. The frequency behavior of the noise figure is sketched in Figure 4-0.9. The noise figure for the common collector configuration is similar to that given by equation 4-0.19, except that it becomes flat again above the alpha cutoff frequency.

It is apparent from equation 4-0.19 that for a given source resistance, and at a given frequency, there is a corresponding optimum value of emitte resistance (or emitter current) that will minimize the noise figure. If both emitter current and source resistance are optimized, and if alpha were independent of emitter current, equation 4-0.19 would indicate that I_E should go to zero, but of course this cannot be because alpha does decrease at small emitter currents and also effects of collector and emitter saturation currents are not included in equation 4-0.19. For ordinary alloy type transistors, the minimum noise figure does call for small emitter current (fraction of a milliampere typically).

The transistor requirements for low noise operation are the same as those for good high frequency power gain; i.e., a_o close to unity, r_b, small, and f_{ab} high.

The bias conditions for graded base type transistors in which the emitter transition capacitance is appreciable will be somewhat different than indicated above. In these types, the cutoff frequency will be dependent on emitter current to a greater extent than was the case of ordinary uniform base types in which the transition capacitance is small. The cutoff frequency of the graded base type increases with emitter current until the transition capacitance is swamped out by the diffusion capacitance (proportional to I_E) or high level effects set in. Taking this dependence of frequency response into account, and in considering equation 4-0.19, it is seen that graded base type transistors will then require a larger emitter current for a minimum noise figure. Both types call for a high collector voltage to the extent that this increases the cutoff frequency.

The common emitter configuration has the advantage that the source resistance required for minimizing noise is nearly that for which the gain is a maximum for frequencies in the flat portion of Figure 4-0.9.

F. Automatic Gain Control

Automatic gain control (AGC) may be accomplished in transistor amplifiers by: 1) controlling bias conditions with the gain control signal (32); 2) elements external to the transistor acting as an attenuator controlled by the AGC signal (33); and 3) combinations of 1) and 2) (34).

The most common method of AGC is that in which the bias conditions of the transistor are controlled. The gain may be reduced by reducing the emitter current or the collector voltage; the latter finds little application, however, for in order to reduce the gain, the collector voltage must be very low, which results in an increase in collector capacitance and a reduction in cutoff frequency. The input and output impedances change considerably with change in bias conditions, while the matched power gain remains fairly insensitive so gain control is achieved primarily by mismatch. The AGC control stages should be designed for optimum performance for the condition of minimum AGC signal when the greatest gain is desired.

In order to limit distortion, AGC control stages must be those having low signal levels, since their signal handling capabilities may be much reduced. Careful consideration must be given to the power which is necessary to drive the control stage (the better the bias stability the harder it is to drive). A considerable amount of power may be required from the AGC source to control emitter current directly. Therefore, it is common practice to apply the AGC voltage to the base of the transistor being controlled and use the transistor as a d.c. amplifier of the gain concontrol signal. A separate d.c. amplifier may be used to augment the power available from the AGC source.

When the AGC signal acts to alter the bias conditions in controlling transistor gain, this is reflected as a shift in bandwidth in center frequency due to changes in input and output impedances. For example, in a common emitter stage decreasing I_E to reduce the gain results in an increase in input resistance and capacitance, to reduce the bandwidth of a parallel tuned circuit (due to increased Q) and shift the center frequency upward.

Hurtig (34) has developed a circuit to limit changes in bandwidth by placing a diode in parallel (to a.c.) with the emitter diode of the transistor and dividing the signal current between the diode and the emitter in accordance with the AGC signal in such a way that the conductance of the combination remains constant. This circuit is quite successful in maintaining bandwidth although the frequency shift problem remains.

Chow and Lazar (33) have devised an AGC circuit which does not involve controlling transistor bias, and which may control a signal exceeding the dynamic range capabilities of the first stage, reduce the next stage gain below unity and requires little control power. Their circuit consists of a bridge utilizing reverse biased silicon diodes as variable capacitances controlled by the AGC signal. The transistor in the controlled stage acts as the detector of the bridge. One of the bridge diodes is reverse-biased at the maximum anticipated AGC voltage and the reverse bias on the second varies with the AGC signal; being very small at minimum AGC and approaching the reverse bias applied to the first at maximum AGC signal. Thus, with minimum AGC signal, the bridge is greatly unbalanced, with most of the power being coupled to the transistor amplifier. Then, as the AGC signal increases, the bridge output decreases as balance is approached.

G. Tuned Amplifier Interstages

The function of an interstage network is to match input and output circuit impedances, provide a desired bandpass, or both. Some typical interstages are shown in Figure 4-0.10.

The interstages of most tuned amplifiers, except those with exacting specifications, can be designed with conventional techniques (35, 36, 37). The commonly used circuits are the single and double tuned transformers. Attached primary and/or secondary is often used to limit the effect of

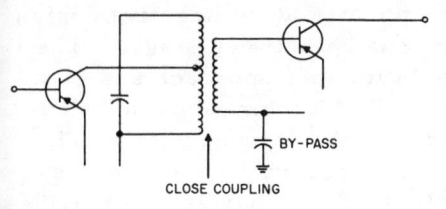

(A) SINGLE TUNED INTERSTAGE

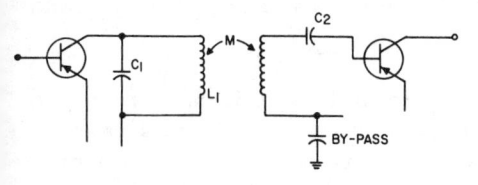

(B) DOUBLE TUNED INTERSTAGE (INDUCTIVE COUPLING)

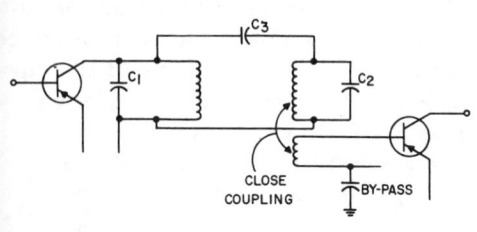

(C) DOUBLE TUNED INTERSTAGE (INDUCTIVE COUPLING)

Fig. 4-0.10 - Examples of typical interstages

transistor capacitance changes and to afford a means of impedance match. This can also be done with a capacitive rather than an inductive tap. Specially prepared design charts may often be of great help in interstage design (38). Very narrow bandwidths can be achieved using crystal or ceramic interstages.

H. Video Amplifiers

Let us now consider the special problems inherent in the design of the broadband, or video, amplifier. In this type of amplifier the transistor plays a determining role in the limitations on bandwidth. The maximum power gain decreases with the frequency of operation, reaching unity at a frequency designated f_{max}. This is a fundamental upper limit on the operating frequencies usable.

For a large number of transistors, the maximum power gain of the transistor rises with decreasing frequency at a rate approximately 6 db per octave until it reaches a limiting value at low frequencies. This value of slope may be slightly different with some transistors; for example, it is approximately 4-1/2 db per octave in grown-junction units and may exceed 6 db per octave at the high end of the frequency range of a graded-base unit.

If the upper cutoff frequency of the video amplifier to be realized falls in the region where the slope of the maximum power gain is 6 db per octave and other considerations are not limiting, it is valid to use an amplification-bandwidth product as a figure of merit and look for ways to trade gain for bandwidth in a reciprocal manner.

In general, this is a difficult thing to achieve in practice as it requires impedance matching devices such as transformers and the use of transformers over wide fractional bandwidths is usually impractical. It should be noted however that matching is required only at the amplifier cutoff frequency as there will be excess gain available below this frequency.

Most design approaches tend to use an approximately iterative design procedure where there is a considerable mismatch between stages. The common emitter connection is most desirable for this approach and gain is achieved because of the current gain, h_{fe}. The fundamental limits on cutoff frequency in this case are designated f_{ae}, the frequency at which h_{fe} drops by 3 db, and f_T, the product of high frequency h_{fe} and the frequency of measurement; f_T is approximately the frequency at which h_{fe} is unity. Below f_T, the current amplification will increase to its low frequency value at approximately 6 db per octave. In this case too, it is valid to use the gain bandwidth product and attempt to trade gain for bandwidth.

If high load impedances are used, the terminal capacitances may provide a limitation on the bandwidth which may be obtained. In this case there will be an RC cutoff frequency which prevents the achievement of the desired bandwidth. The incorporation of terminal capacitance into interstage networks may then be necessary. The use of peaking inductances is an aid to reduce the effect of the shunt capacitance across a load impedance due to the output capacitance of a stage and the input capacitance of the following one. In most cases where stages are repeated without impedance transformations, the low input resistance of the following stage keeps this relatively low shunt capacitance from being a major limitation on bandwidth.

With the gain and bandwidth determined by the above transistor limitations, the main design problem is that of obtaining the desired gain vs. frequency characteristic which may be required to be flat or even to rise with frequency. The problem, then, is to reduce the low frequency gain to a value approximating the high frequency gain. There are basically two methods available. One method is to incorporate the response correction in the interstage coupling networks (39, 40, 41). The other is to reduce the low frequency gain by the use of inverse feedback at low frequencies and reducing this feedback as the higher frequencies are approached (42, 43, 44).

The first method used alone has the disadvantage that the low frequency gain and f_{ae} of the higher frequency transistors is likely to vary more from transistor to transistor than the high frequency parameters. However, some use of this method may have considerable advantages.

The second method has the advantage that the low frequency gain tends to become independent of the transistors, but considerable care may be necessary to prevent regenerative effects and instability at the higher frequencies. Partially bypassed emitter bias resistors are commonly used for this purpose where the high input impedance at low frequency may be tolerated. Most authors on this subject recommend against loops including more than two transistors because of the large phase shift when the cutoff frequency is high.

It should be noted that if the amplifier application can tolerate some regeneration and if adequate stability can be obtained, the high frequency gain can be increased by this means.

If the video amplifier is used to drive a cathode ray tube, rather high output voltage may be required. In general, it may be observed that the same collector voltage swings may be obtained in video amplifier circuits as in lower frequency circuits, subject only to power dissipation and collector breakdown voltage limitations. When higher output swings are required, it is necessary to have recourse to such methods as the push-pull amplifier and the voltage doubler amplifier.

REFERENCES AND BIBLIOGRAPHY

1. R. L. Pritchard, "Electric Network Representation of Transistors - A Survey," IRE Trans. on Circuit Theory, Vol. CT-3, March 1956, pp 5-21.

2. R. L. Pritchard, "High-Frequency Power Gain of Junction Transistors," Proc. IRE, Vol. 43, September 1955, pp 1075-1085.

3. R. L. Pritchard, "Frequency Variation of Junction Transistor Parameters," Proc. IRE, Vol. 42, May 1954, pp 786-799.

4. G. Y. Chu, "A New Equivalent Circuit for Junction Transistors," 1954 IRE Convention Record, Part 2 - Circuit Theory, pp 135-140.

5. R. M. Scarlett, "Some New High-Frequency Equivalent Circuits for Junction Transistors," Tech. Report No. 103, March 20, 1956, Stanford Electronics Laboratories, Stanford University.

6. L. J. Giacolletto, "Study of P-N-P Alloy Junction Transistors from D-C Through Medium Frequencies," RCA Review, Vol. 15, December 1954, pp 506-562.

7. W. E. Ballentine and F. H. Blecher, "Broadband Video Amplifiers," Digest of Technical Papers, 1959 Solid-State Conference, pp 42-43.

8. D. E. Thomas and J. L. Moll, "Junction Transistor Short-Circuit Current Gain and Phase Determination," Proc. IRE, Vol. 46, June 1958, pp 1177-1184.

9. Lo, Endres, Zawels, Waldhauer, and Cheng, <u>Transistor Electronics</u>, Prentice-Hall, Englewood Cliffs, 1955.

10. S. J. Mason, "Power Gain in Feedback Amplifiers," Research Laboratory for Electronics, M.I.T., Technical Report No. 257, August 25, 1953.

11. J. G. Linvill and L. G. Shimpf, "The Design of Tetrode Transistor Amplifiers," Bell System Technical Journal, Vol. 35, July 1956, pp 813-840, and Bell Telephone System Monograph 2657.

12. A. P. Stern, C. A. Aldridge and W. F. Chow, "Internal Feedback and Neutralization of Transistor Amplifiers," Proc. IRE, Vol. 43, July 1955, pp 838-847.

13. A. P. Stern, "Considerations on the Stability of Active Elements and Applications to Transistors," 1956 IRE Convention Record, Part 2 - Circuit Theory, pp 46-52.

14. A. P. Stern, "Stability and Power Gain of Tuned Transistor Amplifiers," Proc. IRE, Vol. 45, March 1957, pp 335-343.

15. E. F. Bolinder, "Survey of Some Properties of Linear Networks," IRE Trans. on Circuit Theory, Vol. CT-4, September 1957, pp 70-77.

16. G. S. Bahrs, "Stable Amplifiers Employing Potentially Unstable Transistors," 1957 IRE Wescon Convention Record, Part 2, pp 185-189.

17. M. A. Karp, "Power Gain and Stability," IRE Trans. on Circuit Theory, Vol. CT-4, December 1957, pp 339-340.

18. R. L. Pritchard, "Measurement Considerations in High-Frequency Power Gain of Junction Transistors," Proc. IRE, Vol. 44, August 1956, pp 1050-1051.

19. J. B. Angell and F. P. Keiper, Jr., "Circuit Applications of Surface-Barrier Transistors, Part III of the Surface Barrier Transistor," Proc. IRE, Vol. 41, December 1953, pp 1709-1712.

20. G. Y. Chu, "Unilateralization of Junction Transistor Amplifiers at High Frequencies," Proc. IRE, Vol. 43, August 1955, pp 1001-1006.

21. A. J. Cote, Jr., "Evaluation of Transistor Neutralization Networks," IRE Trans. on Circuit Theory, Vol. CT-5, June 1958, pp 95-103.

22. A. van der Ziel, "Theory of Shot Noise in Junction Diodes and Junction Transistors," Proc. IRE, Vol. 43, pp 1639-1646.

23. W. Guggenbuehl and M. J. O. Strutt, "Theory and Experiments on Shot Noise in Semiconductor Diodes and Transistors," Proc. IRE, Vol. 45, June 1957, pp 839-854.

24. E. G. Neilson, "Behavior of Noise Figure in Junction Transistors," Proc. IRE, Vol. 45, July 1957, pp 957-963.

25. G. H. Hansen and A. van der Ziel, "Shot Noise in Transistors," Proc. IRE, Vol. 45, November 1957, pp 1538-1542.

26. W. N. Coffey, "Behavior of Noise Figure in Junction Transistors," Proc. IRE, Vol. 46, February 1958, pp 495-496.

27. A. van der Ziel and A. G. T. Becking, "Theory of Junction Diode and Junction Transistor Noise," Proc. IRE, Vol. 46, March 1958, pp 589-594.

28. A. van der Ziel, "Noise in Junction Transistors," Proc. IRE, Vol. 46, June 1958, pp 1019-1038.

29. R. D. Middlebrook, "Optimum Noise Performance of Transistor Input Circuits," Semiconductor Products, July/August 1958, pp 14-20.

30. B. Schneider and M. J. O. Strutt, "Theory and Experiments on Shot Noise in Silicon P-N Junction Diodes and Transistors," Proc. IRE, Vol. 47, April 1959, pp 546-554.

31. W. F. Chow and A. P. Stern, "Automatic Gain Control of Transistor Amplifiers," Proc. IRE, Vol. 43, Sept. 1955, pp 1119-1127.

32. W. F. Chow and H. Lazar, "A New Method of Automatic Gain Control For H.F. and V.H.F. Transistor Amplifiers," 1959 Solid-State Circuits Conference Digest of Technical Papers, pp 40-41.

33. C. R. Hurtig, "Constant-Resistance AGC Attenuator for Transistor Amplifiers," IRE Transactions on Circuit Theory, Vol. CT-2, June 1955, pp 191-195.

34. S. H. Colodny, "Single Tuned Transformers for Transistor Amplifiers," 1958 National IRE Convention Record, Part 7, pp 118-124.

35. R. C. Rand and J. B. Oakes, "Single and Double Tuned Transistor IF Amplifiers," Reprint Series Report No. 382, April 1956, Johns Hopkins University Applied Physics Laboratory.

36. R. B. Hurley, "Designing Transistor Circuits - Tuned Amplifiers, Parts 1 and 2," Electronic Equipment, Vol. 5, July 1957, pp 14-17, and August 1957, pp 20-23.

37. A. E. Hayes, "High Frequency Transistor Circuit Design," Ampex Corporation Publication, 934 Charter Street, Redwood City, California.

38. G. Bruun, "Common Emitter Transistor Video Amplifiers," Proc. IRE, Vol. 44, November 1956, pp 1561-1572.

39. J. J. Spilker, "A Multistage Video Amplifier Design Method," 1957 IRE Wescon Convention Record, Part 2, pp 54-59.

40. C. R. Zimmer, "Transistor Video Amplifiers," Syracuse University Research Institute, E.E. Dept. Report No. EE466-5711F.

41. C. A. Steggerda, "A Study of Gain and Bandwidth in Transistor Video Amplifiers," Philco Research Report No. 242.

42. F. H. Blecher, "Design Principles for Single Loop Transistor Feedback Amplifiers," IRE Transactions on Circuit Theory, Vol. CT-4, September 1957, pp 51-55.

43. R. P. Abraham, "A Wide-Band Transistor Feedback Amplifier," IRE Solid State Convention Record, February 1959.

CIRCUIT 4-1

ONE STAGE NEUTRALIZED 455 KC I.F. AMPLIFIER

Radio Corporation of America Contributed by Lova Plus

Circuit 4-1 represents a commercial radio-receiver i.f. amplifier with center frequency at 455 kc, embodying standard good practice of design.

Circuit 4-1 is a single stage amplifier with a double tuned input transformer and a single tuned output transformer. The double tuned input transformer, T_1, is designed to provide the proper load (100K) for the mixer or converter. The secondary of T_1 is tapped to provide the proper mismatch to the input of the i.f. transistor 2N373 for stability considerations.

The single tuned transformer T_2 has a tapped primary, such that the proper load is established for the output of the i.f. transistor and in turn

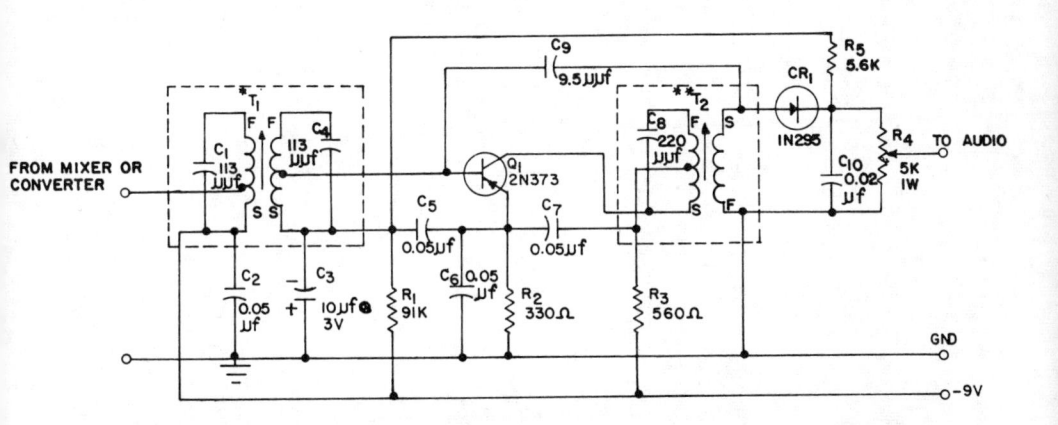

*T_1=DOUBLED-TUNED I.F. TRANSFORMER, UNLOADED Q OF PRIMARY AND SECONDARY=100, LOADED Q OF PRIMARY=50, TUNED RESISTANCE OF PRIMARY=300,000 OHMS, LOADED Q OF SECONDARY=50, TUNED RESISTANCE OF SECONDARY=300,000 OHMS, REFLECTED RESISTANCE OF SECONDARY=300,000 OHMS, SECONDARY LOAD RESISTANCE = 1550 OHMS, CO-EFFICIENT OF COUPLING = CRITICAL.

**T_2= SINGLE-TUNED I.F. TRANSFORMER, UNLOADED Q=105, LOADED Q=50, REFLECTED RESISTANCE OF SECONDARY AT TAP=21,500 OHMS, SECONDARY LOAD RESISTANCE=1000 OHMS, TUNED RESISTANCE AT TAP 21,500 OHMS.

TURNS RATIO	T_1	T_2
TOTAL SECONDARY / TAP	13.9	—
TOTAL PRIMARY / TAP	—	2.78
TAP / SECONDARY	—	4.63

Circuit 4-1 - One stage neutralized 455 kc i.f. amplifier R.C.A.

Circuit 4-1, One Stage Neutralized 455 kc i.f. Amplifier (cont'd.)

provides the mismatch necessary for stability. The total mismatch (16.5 db) in both the input and output of the transistor combined provides a maximum usable gain of 37.5 db.

The turns ratio, tap to secondary, is established so that the reflected impedance of the detector in parallel with the tuned resistors of the primary at the tap provides the proper load for the i.f. transistor. A 180° phase shift is provided in the transformer T_2 by a reversal of the secondary. This is necessary to provide a means of neutralization by means of capacitor C_9, which couples a signal equal and opposite in phase to the signal which is inherently coupled from output to input through the internal feedback capacity of the transistor, thus providing neutralization.

A simple straightforward AGC system is included. With no signal applied to the amplifier, the base of the transistor, and the junction of R_4 and R_5, are negative and the emitter current is at its quiescent value of about 1 mAdc. As the input signal to the amplifier increases, the detected signal makes the junction point of R_4 and R_5 go positive. Thus the base bias voltage, the emitter current, and the gain of the stage will decrease.

CIRCUIT 4-2

TWO STAGE 455 KC I.F. STRIP

General Electric Company   Contributed by E. Gottlieb

Circuit 4-2 represents a two-stage amplifier with a basic structure similar to Circuit 4-1. The stages are not neutralized because the collector to base capacitance of the transistors is very low. The stabilization of the d.c. operating point and the basic AGC system used are the same as with Circuit 4-1, except that voltage polarities are reversed because n-p-n transistors are used. The diode CR_1 serves to extend the range of the AGC control beyond the limitation imposed by the minimum gain of the controlled stage. The d.c. load resistors of Q_1 and Q_2 are chosen such that under quiescent operating conditions the diode CR_1 is back biased. This diode is placed (with respect to a.c. signal) across the primary tank circuit of Q_1. As the AGC action reduces the emitter current of Q_2, the

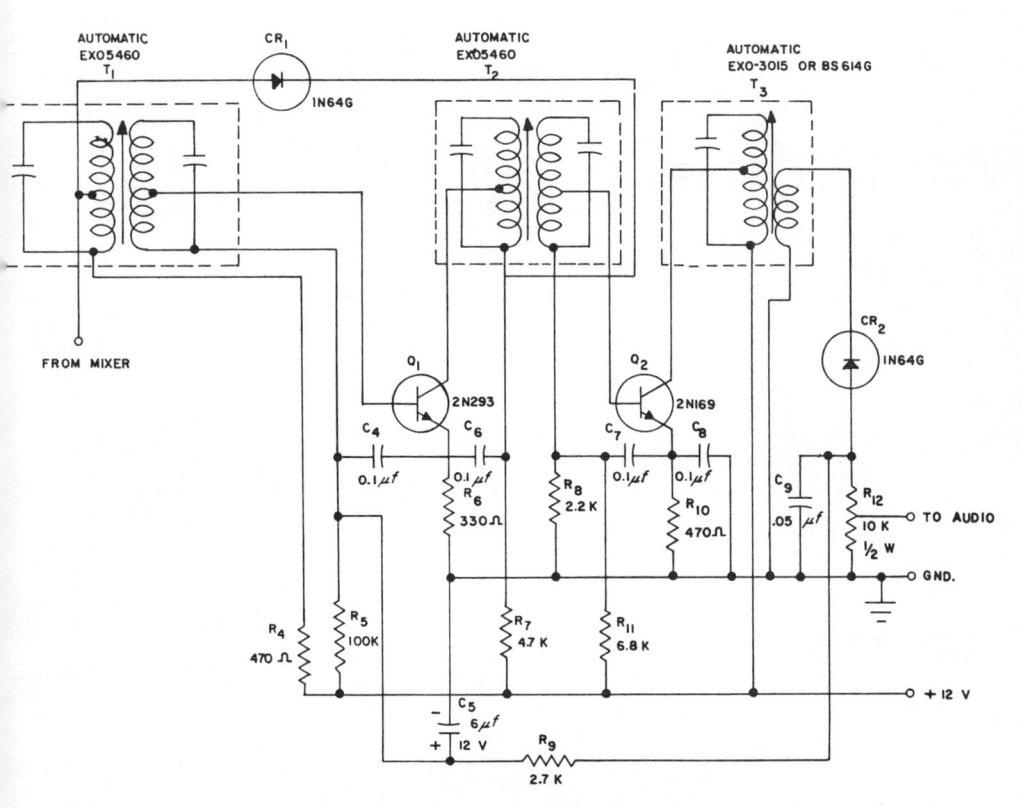

Circuit 4-2 - Two stage-455 kc i.f. strip

Circuit 4-2, Two Stage 455 kc i.f. Strip (cont'd.)

collector voltage will rise until CR_1 becomes forward biased and partially shunts the i.f. signal to ground through C_6, C_4, and C_5. The exact point at which CR_1 becomes effective is adjusted by R_4 and R_7. R_9 and C_5 serve to filter the audio signal level before it is applied to the base of the transistor for AGC control.

CIRCUIT 4-3

STABLE WIDE BAND AMPLIFIER

Sanders Associates, Inc.　　　　　　　　Contributed by Ralph O. Goodwin

This design was necessitated by the need for an extremely stable amplifier in gain and phase, to operate at 10 mc with reasonable gain, over a temperature range of -30°C to 65°C. The amplifier had to meet definite design requirements which are tabulated in Table 1 below:

TABLE 1

R_{in}	50 Ω ± 5 Ω
R_L	50 Ω ± 5 Ω
φ error	1° ⎫ at 10 mc
G error	0.2 db ⎭
G_p	30 db ± 1 db
Temp.	-30°C to 65°C

Circuit 4-3 - Stable wide band amplifier

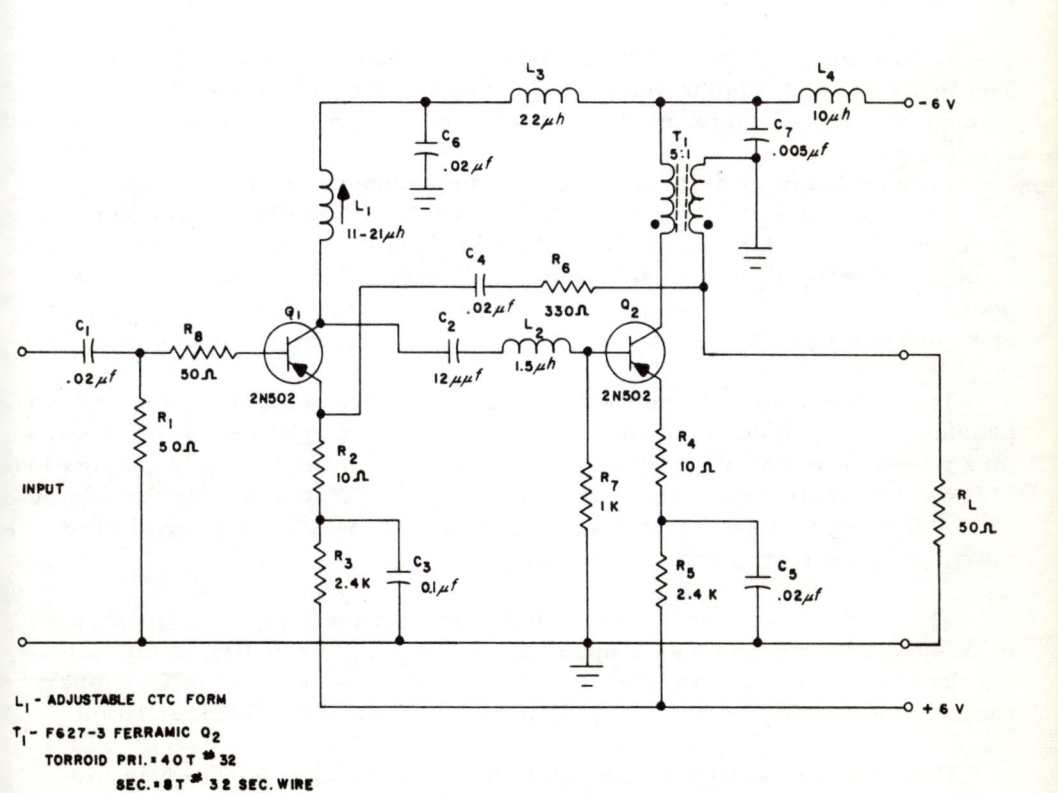

Circuit 4-3, Stable Wide Band Amplifier (cont'd.)

The main requirements were that the amplifier was not to drift or deviate more than 0.2 db or 1° relative to another similar amplifier, and that the amplifier input should terminate a 50 Ω coaxial cable. Also it is desirable to make the amplifier as versatile as possible so that it may be used at other frequencies besides 10 mc.

To obtain the most gain from the least number of transistors, it was decided to transformer-couple the transistors. The first stage uses an LC design where L_1 C_2 and the input and output capacity of Q_1 and Q_2 tune the open loop response to the frequency of interest. C_2 and Q_2 input capacity are chosen so that the series input resistance of Q_2 appears as a high parallel resistance across the collector load of Q_1. This provides a near match between transistors. L_2 is used as series peaking to improve the high frequency cutoff point. L_2 may be adjusted for best high frequency response if desired.

Resistor R_7 is used along with R_4 to maintain an essentially constant input impedance to Q_2.

T_1 is a toroidal transformer wound on Ferramic Q-2 material, #627-3. The turns ratio is 40-8 to feed a 50 Ω load and the primary 40 turns can also be slightly adjusted in number to affect the low frequency response.

Voltage feedback is provided from the secondary of T_1 through R_6 to R_2. The ratio of R_2 to $R_2 + R_6$ is the feedback ratio set to obtain 30 db of gain. Feeding back to the first emitter at Q_1 will increase the input impedance allowing R_1 to be used, which will determine the 50 Ω input impedance. C_3 must be large so as not to become part of the feedback loop at R_2 and it must also remain a small impedance at high frequencies.

Two battery bias is used with small base impedance to ground, which provides a very small stability factor, S, for the amplifier. The transistors are operated at 2.5 ma and -6 volts, which yields a 15 mW dissipation. This is the maximum dissipation that can be used and still be operable at 65°C. The highest voltage and current available was used to obtain the maximum gain from each transistor.

The feedback loop is opened and L_1 tuned to the frequency of interest, in this case 10 mc. The gain open loop at 10 mc is typically 52 db. Closing the loop yields an amplifier that is flat within ±1 db over a frequency range from 3 to 30 mc at 30 db. This gives 22 db of feedback at 10 mc.

Temperature variations are applied from -30°C to 65°C while the output of two units are differenced against each other. One unit is in the oven or cold box while the other is kept at room temperature. The relative

Circuit 4-3, Stable Wide Band Amplifier (cont'd.)

gain and phase difference variations measured over this temperature range were equivalent to a maximum of 0.6° or 0.095 db at 65° and 0.81° or 0.12 db at -30°C. With more feedback applied at the frequency of interest, greater stability can be realized.

The unit may also be operated at other frequencies by varying L_1. If the frequency at interest is changed radically beyond the adjustment of L_1, then a new L_1 can be wound and the value of C_2 checked to be sure that match is maintained. Tuning the open loop below 10 mc more gain will be realized, tuning above 10 mc will yield less open loop gain. Thus the feedback margin will change at different frequencies, dependent upon where the loop is tuned.

With this circuit a wide band amplifier is obtained with excellent phase and gain stability at the center frequency of interest. The units can be cascaded with bandwidth shrinkage the same as double tuned circuits. When cascading it is best to place in series a 50 Ω resistor between R_1 and the base of Q_1. This reduces the gain less than 0.5 db and provides the second amplifier a 50 Ω impedance to look back upon.

The noise figure of 12 db is realized with 50 Ω terminations present on the amplifier as source and load impedances.

CIRCUIT 4-4

R.F. STAGE FOR 3-BAND RECEIVER

Radio Corporation of America   Contributed by J. W. Englund

Circuit 4-4 represents the r.f. stage of a typical commercial-type short wave receiver, covering the broadcast band and the short wave band up to 23 megacycles. The amplifier uses drift transistor and embodies the best standard design techniques.

The frequency is covered in three bands: Broadcast, 4.75-11 mc, 11-23 mc. Each band is associated with a separate single tuned antenna and output tank circuit. Tapped autotransformer coils are used on all tank circuits except on the broadcast band antenna coil which is transformer wound on a ferrite rod. The taps on the antenna and r.f. transformers are chosen such that the necessary gain is sacrificed for stability. The mismatch is obtained by proper selection of these taps at the high end of each frequency band where the transistor has inherently the lowest gain. The usable unneutralized gain of the transistor is about 14 db at 23 mc and increases at a rate of 6 db per octave with decreasing frequency.

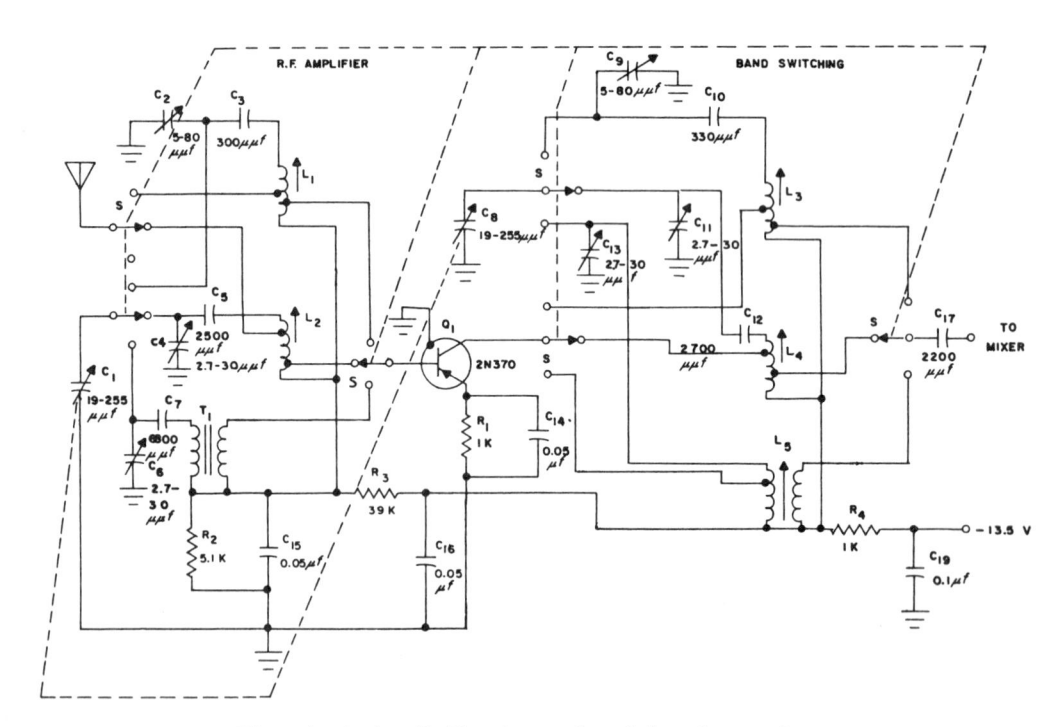

Circuit 4-4 - R.F. stage for 3 band receiver

Circuit 4-4, R.F. Stage for 3-Band Receiver (cont'd.)

However, the input and output impedances of the transistor increase with decreasing frequency, which tends to equalize the effective gain variations of the stage over each band. The tank circuits are tuned with gang condensers and provisions are made for individual slug tuning of the coils as well as individual trimmers on each band.

In circuit 4-4, the r.f. transistor (2N370) is biased in a conventional manner with the voltage divider in the base circuit and a bypassed emitter resistor. In the case of an actual receiver, means would be provided for the application of AGC in the base network, in which case the selection of the emitter resistor would be a compromise between d.c. stability and AGC.

The mixer and oscillator circuitry is straightforward, and uniform performance is possible over the entire frequency range.

The coil construction is specified in Figure 4-4.1.

COIL	ANTENNA			INTERSTAGE		
	T1 (Standard Broadcast Band)	L2 (4.5 to 11.5 Mc)	L1 (10.5 to 23 Mc)	L5 (Standard Broadcast Band)	L4 (4.5 to 11.5 Mc)	L3 (10.5 to 23 Mc)
Total primary turns	127	21	13	225	21	12
1st primary tap - turns from bottom	-	2	1	13	2	1
2nd primary tap - turns from bottom	-	7	4	-	6	6
Secondary turns	7	-	-	12	-	-
Wire size	10/38 Litz	#22	#24	7/41 Litz	#22	#24
Coil diameter (inside) - inches	.33	5/8	3/8	1/4	5/8	3/8
Turns per inch	24	24	24	Universal	24	24

NOTE: Standard Broadcast Band Antenna Coil (T1) should be wound on an 8" × .33" ferrite rod.

Fig. 4-4.1 - Radio-frequency tuner circuit coil data

CIRCUIT 4-5

60 MC I.F. AMPLIFIER USING SILICON TETRODES

Texas Instruments, Inc.

Contributed by Glenn E. Peniston
and Donald B. Hall

Circuit 4-5 has been selected to illustrate the use of silicon tetrode transistors in the design of high-frequency i.f. strips.

The main features of this eight-stage i.f. amplifier were a maximum gain of 105 db with a bandwidth of 20 mc at a center frequency of 60 mc, with a maximum gain variation from room temperature of only -8 db at 85°C.

No neutralization of the amplifier was required. To insure stability, mismatch in the interstage coupling networks was provided. A mismatch ratio of 5:1 proved appropriate over the range of transistor parameter differences. This resultant loss in gain was only 2.5 db. The coupling method chosen was the transitionally-coupled double-tuned interstages designed to provide the required 5:1 mismatch required for stability.

The proper biasing of the stage is important to insure optimum gain and interchangeability of units. Measurements made on large quantities

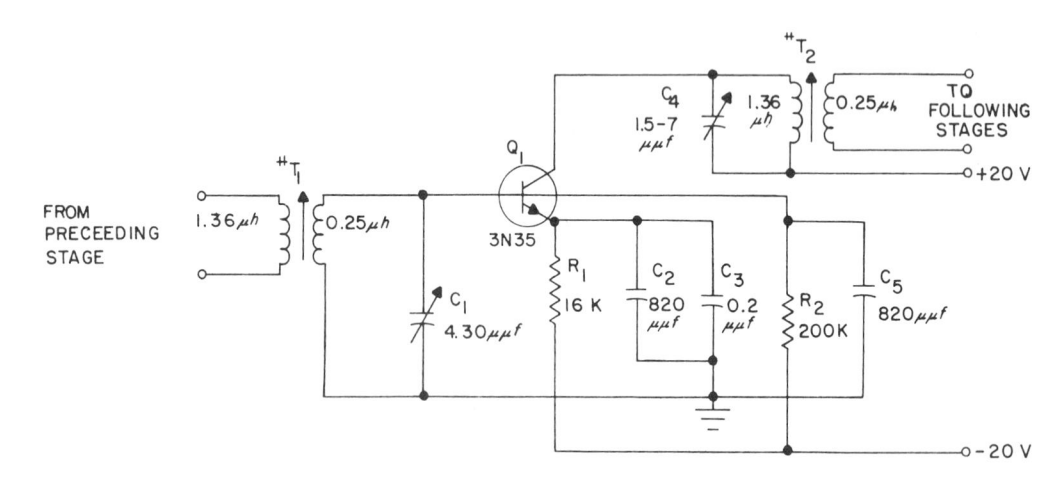

++ COEFFICIENT OF COUPLING = 0.43
I.F. STRIP CONTAINS 8 STAGES

Circuit 4-5 - 60 Mc i.f. amplifier using silicon tetrodes

Circuit 4-5, 60 Mc i.f. Amplifier Using Silicon Tetrodes (cont'd.)

of 3N35 units show that this device has optimum gain characteristics at an operating point of V_{CE} = 20 volts, I_E = -1.3 mAdc, and I_{B_2} = -0.1 mAdc. To insure adequate bias circuit performance from unit to unit under conditions of large ambient temperature variations, a two-battery circuit was employed. The negative supply was made 20 volts thereby providing a symmetrical arrangement. Each transistor is biased common base even though the RF circuitry is common emitter. The large resistors in the emitter and base-two leads assure that the currents in these elements will remain constant. These resistors are bypassed for signal frequencies by appropriate capacitors. An additional 0.2 μf capacitor helps insure that the bias point will remain constant with a pulsed input signal.

Circuit 4-5 shows a single stage of the eight-stage amplifier. All stages are identical except for the input and output stages whose transformers may be designed for the appropriate driving and load resistances, respectively.

CIRCUIT 4-6

NARROW-BAND SELECTIVE AMPLIFIERS

Hermes Electronic Co.   Contributed by Carl R. Hurtig

In many modern communication and navigation systems, amplifiers with extremely narrow bandwidths are required. If the ratio of center frequency to overall bandwidth is large or the operating temperature range is large, the use of crystal filters is indicated. Furthermore, a marked performance improvement in the presence of interfering signals results from the use of a passive, linear filter at the antenna of a receiver or the first stage of an intermediate frequency amplifier.

In general, the use of a crystal filter requires a reasonable accurate termination of the filter to preserve the desired attenuation characteristics in the pass-band region. The stop-band attenuation is not influenced by the termination impedances but by the filter network itself. Since stop-band attenuations greater than 80 db can be achieved in a relatively small size, adequate precautions in mechanical layout and shielding of the amplifier are required to preserve the stop-band attenuation of the filter.

Transistors can be readily used with crystal filters. The major requirement is reasonably accurate control of impedance level. The Circuit 4-6 shows a two-stage amplifier with a standard crystal filter employed in the interstage network.

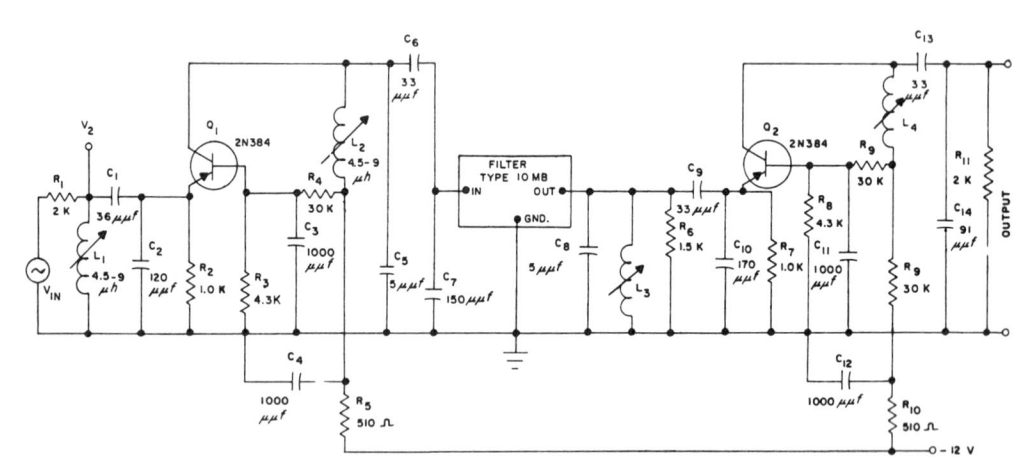

Circuit 4-6 - Narrow-band selective amplifiers

Circuit 4-6, Narrow-Band Selective Amplifiers (cont'd.)

The filter has a center frequency of 10.7 mc, a 6 db bandwidth of 15 kc, and a 60 db bandwidth of 30 kc. The attenuation characteristic of the circuit is shown in Figure 4-6.1. The maximum insertion loss of the filter is 3 db. The terminating impedances of the filter are 1000 ohms at both input and output terminals. Both of the transistors are operated in the grounded-base connection without neutralization for signals and a stabilized grounded emitter connection for bias. The effective generator resistance is, to a major part, controlled by the losses in the tuning coil L_2. A capacitance divider is employed to transform the equivalent loss of the coil (in parallel with the output impedance of the transistor) to 1000 ohms. The load resistance is a parallel combination of the 1.5K resistor and the input impedance of a transistor suitably transformed to a high value by an LC network. Although there is some loss in gain by the use of the padding network, the stability of the overall characteristic with temperature and the interchangeability of transistors are both excellent.

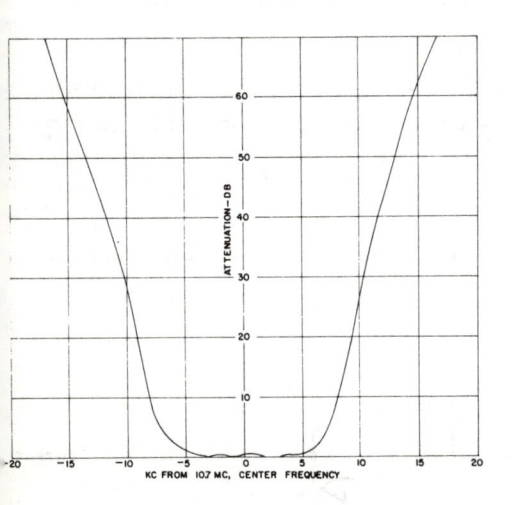

Fig. 4-6.1 - Circuit attenuation characteristic

The overall power gain from the input point labelled e_2 to the output across the 2.0K load resistor is approximately 28 db. Changes of transistors of any of the types listed result in less than a 6 db variation (with coils retuned) and negligible change in overall attenuation characteristic. The value of generator and load resistances can be easily varied by means of the capacitance coupling networks.

This circuit has been used to illustrate the design techniques associated with one particular filter. It should be kept in mind that a very large variety of crystal filters are available covering the frequency range from 20 kc to 60 mc.

CIRCUIT 4-7

EMITTER-TUNED I.F. AMPLIFIER

General Electric Company　　　　　　　　Contributed by W. F. Chow
and D. A. Paynter

    The emitter-tuned amplifier shown in Circuit 4-7 represents an interesting and useful deviation from the usual tuned amplifier. The selectivity is achieved through the use of series resonant networks which are placed in the emitter circuits of each transistor stage, and the amplifier is directly-coupled without the use of impedance matching interstage transformers. This arrangement results in several desirable features: 1) units can be designed in which the operating frequency is almost independent of transistor property variations and supply voltage changes, 2) circuit complexity is reduced, 3) alignment procedure is simplified. Stability of the individual stages is accomplished by mismatch so that neutralization is unnecessary; however, the cascading of common emitter stages results in an amplifier with lower overall gain than a similar unneutralized amplifier with interstage matching networks.

    Several of the circuit parameters effect the gain and the bandwidth of the amplifier. The bandwidth is directly proportional to the load resistor

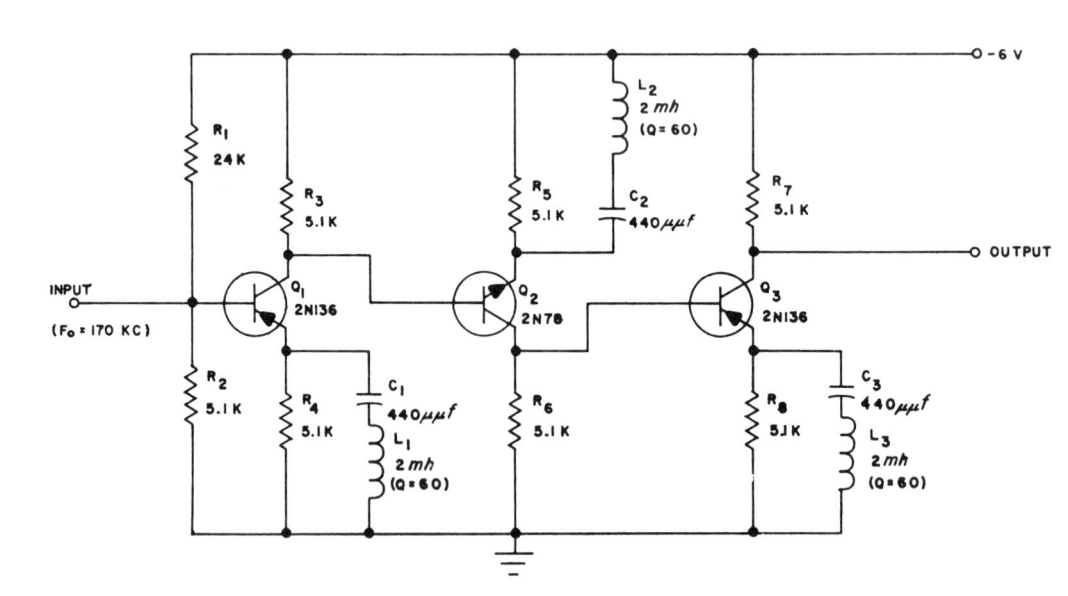

Circuit 4-7 - Emitter-tuned i.f. amplifier

Circuit 4-7, Emitter-Tuned i.f. Amplifier (cont'd.)

R_3 until the load resistor is several times greater than the input impedance. The emitter resistor R_4 is used to provide d.c. biasing of the transistor. This resistor should be as large as the available power supply voltage allows. The effect of R_4 is to raise the selectivity skirts. The LC ratio has a large effect on the bandwidth of the amplifier; the bandwidth decreases as the LC ratio is increased. The Q of the coil effects the gain of the stage more than it effects the bandwidth. It is thus better to use a coil of lower value of L and better Q than a coil of high value of L and low Q. The frequency dependence of the transistor current amplification may result in a shift of the center frequency towards a lower value and in some skewing of the passband shape, but with most narrow band amplifiers this effect can be neglected.

With careful design the emitter degenerative amplifier can be made into a very reliable circuit with excellent transistor interchangeability. The circuit is particularly attractive for the use with crystal, ceramic or mechanical filters, which can be used in place of the series-tuned circuits for added selectivity.

CIRCUIT 4-8

30 MC I.F. AMPLIFIER

General Electric Company   Contributed by L. D. Wechsler

This 4-stage 30 mc i.f. amplifier is an example of the simplicity possible with the use of transistors with very high cutoff frequencies. The circuit as shown utilizes synchronous, single-tuned transformers for interstage coupling. The base of each transistor is at d.c. ground and separate supplies are used for collector and emitter; this allows excellent bias stability with simple circuitry.

Resistive loading of the tank circuit increases the bandwidth while simultaneously reducing the possibility of oscillation. This is, of course, accomplished at the expense of gain; it has the advantage of being non-critical while eliminating the necessity for neutralization. Furthermore, resistive loading reduces bandwidth variation as a function of temperature and transistor operating points.

Capacitive loading of the tank circuit is employed to further reduce the sensitivity to temperature and bias drifts. If the tanks were to be tuned with the transistor and stray capacities alone, the variation of the center frequency with varying environment would be excessive.

The amplifier as shown is capable of operating over the temperature range of -40°C to +65°C. The gain is greater than 82 db, and 84 db is generally obtained. The center frequency is 30 mc (±2%), and the bandwidth is 3.6 mc (±6%). This circuit is an excellent example of simple, yet effective, i.f. circuitry using presently available germanium transistors.

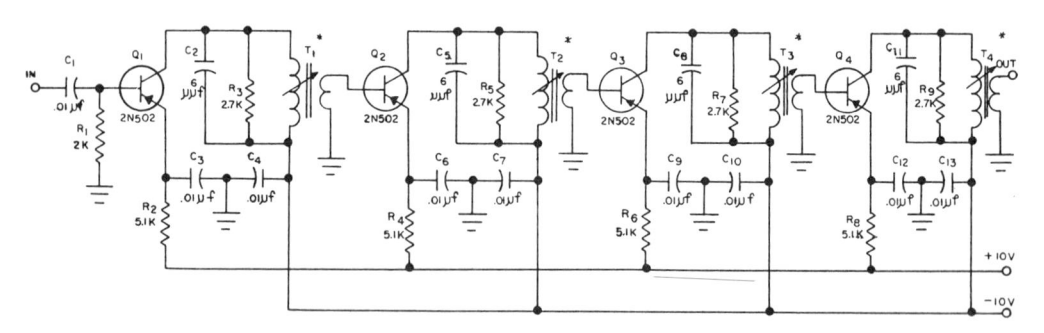

T_1, T_2, T_3 & T_4; PRIMARY 16 TURNS,
SECONDARY 4 TURNS, AWG #34, CAMBRIDGE
THERMONIC COIL FORM TYPE LS9 WITH
GREEN (20-50MC) POWDERED IRON SLUG.

Circuit 4-8 - 30 Mc i.f. amplifier

CIRCUIT 4-9

WIDE BAND TRANSISTOR FEEDBACK AMPLIFIER

Texas Instruments, Inc.   Contributed by R. P. Abraham

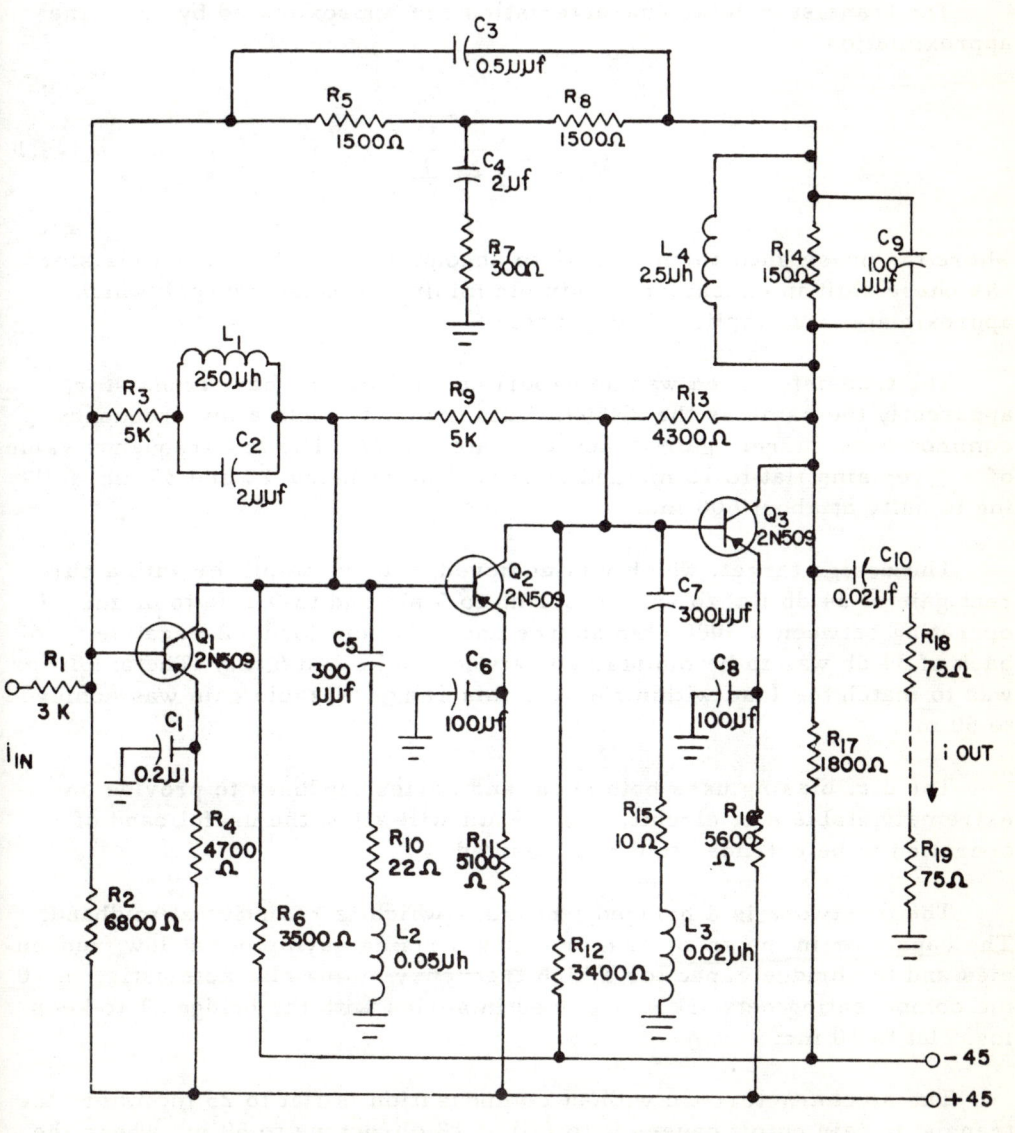

Circuit 4-9 - Wide band transistor feedback amplifier

Circuit 4-9, Wide Band Transistor Feedback Amplifier (cont'd.)

As mentioned in the design philosophy section, it is difficult to apply feedback over more than two stages in a wide-band amplifier, because of the possible instability at high frequencies. Circuit 4-9 is a good example of the maximum which can be achieved in the way of feedback with a good high frequency transistor when careful attention is paid to standard stability criteria for wide band amplifiers.

The transistor delay characteristics are approximated by using the approximation

$$\alpha_{fb} = \alpha_o \frac{e^{-jm\frac{f}{f\alpha}}}{1 + j\frac{f}{f\alpha}} \qquad 4\text{-}9.1$$

where the phase shift term takes into account the fact that the transistor has phase shift in excess of the simple minimum phase network which approximates its amplitude response.

The transistor used was an experimental diffused base transistor, apparently the same as the 2N509. This transistor has a low frequency common base current gain of approximately 0.97. The low frequency value of α_{fe} remains flat to 10 mc and is down 3 db at below 20 and 30 mc, falling to unity at about 600 mc.

The design target, which was achieved, was an amplifier with a current gain of 34 db flat to within 0.01 db to 5 mc and to 0.1 db to 10 mc operating between a 3000 ohm source and a 75 ohm load. A negative feedback of 34 db was to be maintained between 50 kc and 5 mc. The amplifier was to match the load within 5% over this range. Usable gain was achieved to 50 mc.

The d.c. biasing uses both shunt and series feedback to provide an extremely stable d.c. circuit, this circuit will allow the useful band of operation to be extended to d.c. if desired.

The β network is a bridged T network which is resistive at midband. The capacitor in the shunt arm provides a rising asymptote at low frequencies and the bridge capacitor a high frequency rising characteristic. A 10 mc compensating network was placed in series with the bridged T to keep loss flat to 10 mc.

The $A\beta$ characteristic without compensation is flat to 25 mc where the transistor gain cutoff causes it to fall at 18 db/octave to 50 mc where the β characteristic reduces this to 12 db/octave up to over 500 mc.

Circuit 4-9, Wide Band Transistor Feedback Amplifier (cont'd.)

The series interstage networks are used to shape the gain between mc and 100 mc to approximate an ideal Bode cutoff. A fillet is added in the feedback circuit around the first stage so that $A\beta$ is down 3 db at 7 mc.

The low frequency characteristics are controlled by the emitter by-asses. The first stage bypass provides a roll off frequency of 50 kc while the other two extend to under a half kc to provide stability.

Considerable care was exercised in the physical layout because of the need to control the loop gain to over 200 mc.

This work was done at Bell Telephone Laboratories while Mr. Abraham was employed there.)

CIRCUIT 4-10

VIDEO AMPLIFIER

Sanders Associates, Inc.   Contributed by Ralph O. Goodwin

Circuit 4-10 was selected to illustrate a video amplifier design for pulse application. This design was necessitated by the need for a pulse amplifier with gain and phase stability of 0.1 db and 1° at 2.5 megacycles over a dynamic range of 66 db.

The circuit of the open loop amplifier was designed to have a peak in its response, in this case approximately 1 mc. For most cases, it would be better to peak the open loop at the frequency of interest, however, the low frequencies are also required in the bandpass thus the need for significant impedance in the loop. The loop was closed from second emitter to first base (Steggerda's design (42)). This controls the output current and drives the input impedance of the transistor down making the series 100 ohm resistor at the input the effective input impedance. At 2.5 mc 16 db of feedback is realized, at 1 mc 28 db of feedback is obtained and at 400 kc 16 db is again seen, so that at least over the band from 400 kc to 2.5 mc the minimum amount of feedback is 16 db. The 3 db cutoff of the

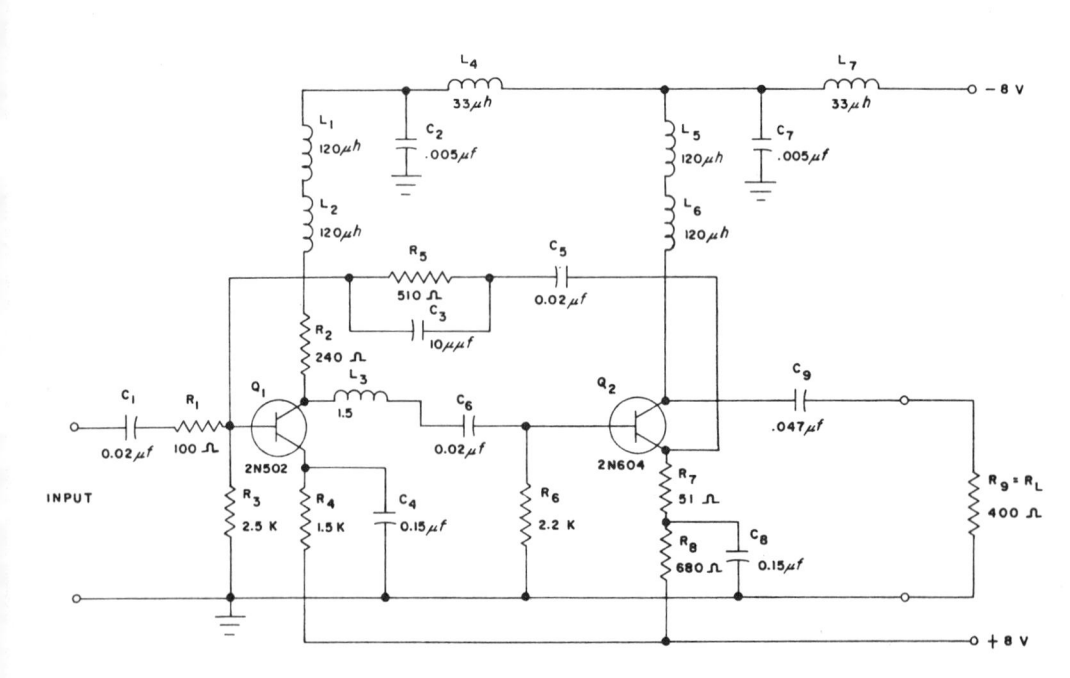

Circuit 4-10 - Video amplifier

Circuit 4-10, Video Amplifier (cont'd.)

amplifier, closed loop, varies from 15 mc to 17.5 mc with first stage beta variations from 28 to 140 and second stage variations of 55 to 280.

Measurements of phase and gain were made by the differencing method over a dynamic range of 66 db where the upper end of signal level was 2 V rms across a 400 ohm load. At 2.5 mc the output vs. input phase and amplitude varies less than 1° and 0.1 db over the dynamic range.

Design of the amplifier was straightforward. A simple RLC coupled amplifier was constructed with the shunt coils in the collector leads so as not to limit the available collector voltage from the -8 volt supply. The interstage coil is used for series peaking so as to keep the open loop bandwidth wide enough with adequate gain to make it possible to close the feedback loop with stability. The emitter resistance in Q_2 is chosen to achieve adequate feedback but at the same time not to increase the input impedance to a point where the open loop gain suffers appreciably. Q_2 was a 2N604 transistor which has adequate frequency response and a high power capability. This was necessary because of the power limitations of the 2N502, which was used in the input stage. In order to provide a 2 volt rms swing across the 400 ohm load, the output transistor has to be biased heavily. The 2N604 is capable of 120 mW of collector dissipation. The operating point was set at 10 mAdc, -8 volts.

CIRCUIT 4-11

WIDE BAND VIDEO AMPLIFIER

Philco Corporation   Contributed by P. G. Thomas and C. A. Steggerda

Circuit 4-11 is a two-transistor feedback pair designed in accordance with the principles outlined by C. A. Steggerda (42). It is an excellent example of what can be achieved by this means.

The feedback which is used to exchange gain for bandwidth is inserted from the second emitter to the first base to avoid the serious loading effec which accompany other forms of 2-stage feedback. The circuit is in many respects similar to a grounded collector stage as regards its terminal impedances, but incorporates voltage gain. A further advantage of this form of feedback is that the load resistance has negligible effect on the feedback loops providing that $R_L \ll r_c(1-\alpha)$, which is generally desired for wide band amplifiers. This circuit may be designed to operate for any value of source and load resistance.

An efficient trade of gain for bandwidth can be made only for small values of source and load resistance. When using source and load resistances up to 300 ohms, bandwidth on the order of 20 mc were obtainable with degradation in gain-bandwidth of approximately 20%. The use of source and load impedance greater than 1 K led to degradation in gain-bandwidth of approximately 50% for bandwidths of approximately 10 mc.

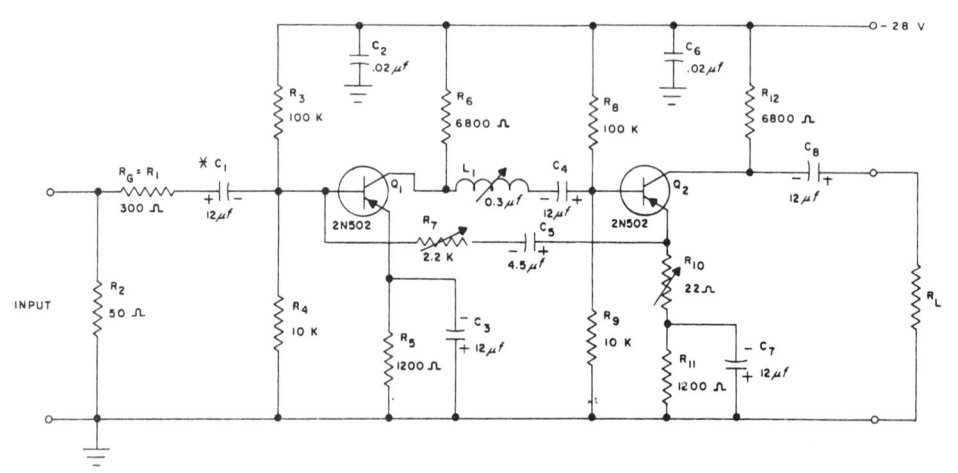

NOTE *
ALL POLARIZED CAPACITORS ARE "TANTALYTIC".

Circuit 4-11 - Wide band video amplifier

Circuit 4-11, Wide Band Video Amplifier (cont'd.)

The interstage peaking coil improves the gain bandwidth in two ways. It provides some measure of impedance matching at high frequencies and increases the amplifier phase shift so that positive feedback exists at high frequencies. Both of these effects increase the gain bandwidth as discussed in the design philosophy section. Steggerda's analysis indicates that the value of the coil should be approximately $L = [\sqrt{\omega}\,(1 - \alpha_o)]/C_c$. It is also pointed out that the coils used should be adjustable and first adjustments should be on the low inductance side as too large an inductance can lead to instability of the amplifier. Despite this, tests made on a properly adjusted amplifier shows no ringing when fast rise transients were applied to the input. The transistors used were 2N502's. The obtainable bandwidth may be varied to any desired value of varying R_7 and R_{10}. As for an example, a bandwidth of 18 mc is obtained when $R_7 = 1300\,\Omega$, and $R_{10} = 10\,\Omega$, with a resultant gain bandwidth of 1330 (where $R_G = 300\,\Omega$, $R_L = 300\,\Omega$).

CIRCUIT 4-12

WIDE-BAND HIGH-FREQUENCY I.F. AMPLIFIER

Bell Telephone Laboratories, Inc.   Contributed by V. R. Saari

Circuit 4-12 is a 15 mc wide i.f. amplifier strip centered at 70 mc using seven diffused-base germanium transistors. This was one of the first high frequency i.f. amplifiers designed when diffused-base transistors were first introduced.

This circuit was selected to illustrate use of a digital computer as a tool in the circuit design.

A rough first approximation to the desired design was made on the basis of the measured transistor parameters neglecting feedback and frequency effects. The changes necessary were then determined experimentally, and the computer was used to verify the goodness of the design and to determine the sensitivity of the adjustable parameters.

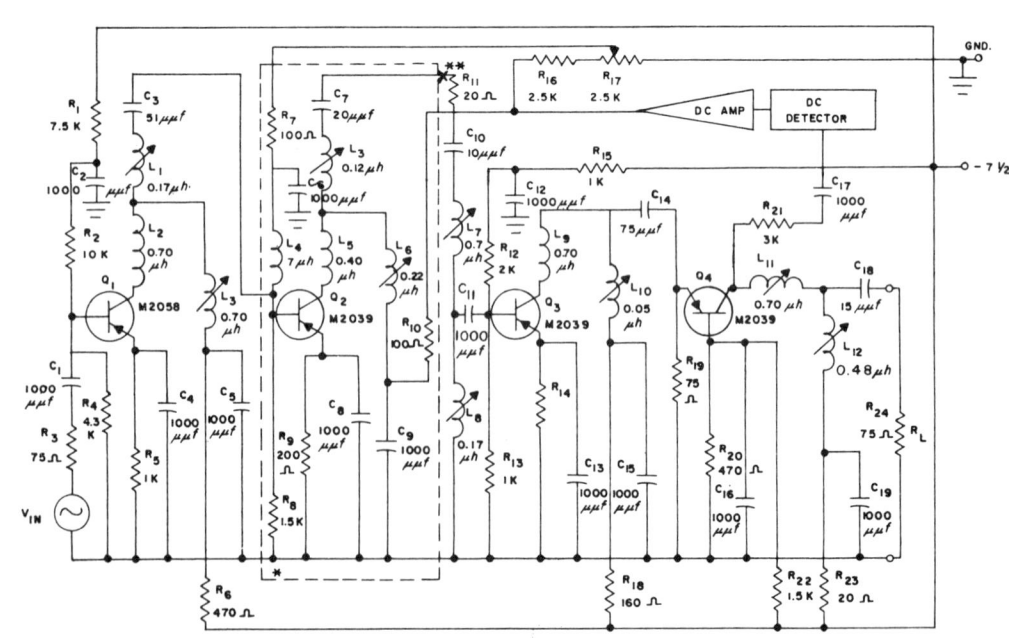

* THE SECTION WITHIN THE BROKEN LINES CONTAINS FOUR STAGES IDENTICAL EXCEPT IN THAT THE LAST ONE HAS FIXED BIASES.

** COAXIAL LINE CONNECTORS (75 OHM) WERE INSERTED HERE (OUTER CONDUCTER GROUNDED) SO THAT A PHASE EQALIZER COULD BE INSERTED.

Circuit 4-12 - Wide-band high frequency i.f. amplifier

Circuit 4-12, Wide-Band High-Frequency I.F. Amplifier (cont'd.)

The T equivalent of a low-Q double-tuned circuit is used as the basic interstage network. Two adjustments are provided in each circuit so that the drop in transistor gain with frequency can be compensated and the passband can be adjusted to be flat to 0.3 db.

All transistors are operated in common emitter configuration except the last stage, which is common base. A current step-down transformer matches the low impedance load to the collector of this stage so that high current can be supplied without excessive signal compression. The transistors are stabilized by mismatch because the use of neutralization becomes difficult with the wide band covered. This results in a loss of gain from the 19 db available from the transistor to 13 db actually realized per stage. The d.c. stabilization is conventional.

The stages of Q_2 to Q_4 are provided with AGC. The control voltage is obtained by detecting the output voltage with a high impedance detector circuit. The d.c. is compared with a fixed reference in a two-transistor difference amplifier so that the controlled output level can be adjusted by changing of the reference voltage level. A final emitter follower driver is used to supply the bias voltage to the three controlled stages.

The 2.5K potentiometer is used to adjust the balance between the emitter current and the collector voltage so that the flatness of the passband is not affected excessively by AGC over the control range.

SPECIFICATIONS

Center frequency:	70 mc
Bandwidth:	15 mc
Gain:	90 db ±0.3 db
Output Level:	12 dbm
Output Impedance:	75 ohms
AGC range:	35 db
Noise Figure:	5 db
Power Consumption:	0.4 watt

(This material is based on the technical paper, "Transistor 70-MC I.F. Amplifier," given by Mr. Saari at the 1958 Transistor and Solid-State Circuits Conference.)

PART 5

OSCILLATORS

Section 1. Design Philosophy

A. Introduction

An oscillator will be defined here as an electronic device for converting a direct current into an alternating current. This portion of the handbook will be concerned only with junction transistor harmonic oscillators, whose current or voltage waveforms are nearly sinusoidal. The frequency range of interest is that covered by the junction transistors presently available; it extends from audio to the lower ultra-high frequencies. The discussion is concerned with low-level oscillators in which frequency, amplitude stability, and waveform are of primary importance.

B. Frequency Stability

Considering frequency stability, it is convenient to study the oscillator outlined in Figure 5-0.1. Separation of the oscillator into an amplifier of gain A and a feedback network or resonator with transmission β will permit the evaluation of the effects of changing amplifier parameters upon the operating frequency. If the produce $A\beta$ is greater than unity, and if it has zero phase angle at a particular frequency, the amplifier is capable of supplying its own input and oscillation will normally occur at that frequency. The amplitude of the oscillation will increase until limiting in the amplifier or in some external amplitude-control circuit adjusts the gain so that $A\beta = 1$ (Barkhausen's condition for oscillation). If there is a phase shift associated with A, there must be an equal but opposite phase shift in β in order to continue to satisfy this condition. The feedback network or resonator can only satisfy this condition by experiencing a change in frequency. If the resonator has a quality factor Q, the fractional frequency change $\Delta\omega/\omega$ associated with a phase angle change $\Delta\phi$ is

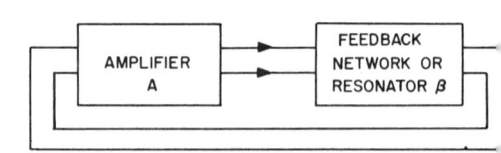

Fig. 5-0.1 - Basic oscillator containing amplifier and feedback network

$$\frac{\Delta\omega}{\omega} = \frac{\Delta\phi}{2Q} \qquad 5-0.1$$

provided that the resonator is operating close to its resonant frequency, both initially and after $\Delta \varphi$ has been applied.

In simple oscillators, one important source of phase instability is associated with the intermodulation of harmonic frequencies in the nonlinear input or transfer characteristics of the amplifier (1). Such harmonic frequencies are a natural consequence of the limiting that must take place in the amplifier unless some form of AVC circuit is employed. The phase instability results from the fact that the harmonic-frequency voltages or currents are shifted in phase with respect to the fundamental by the various reactances associated with the amplifier and resonator. If intermodulation occurs between two or more of these frequencies to produce a component at the fundamental, the component at the fundamental frequency will have been shifted in phase. If the degree of limiting changes as a result of the modification of some circuit parameter, it can easily be seen that the amplitude of the harmonic-produced fundamental component can change, producing a resultant phase shift at the operating frequency.

Unfortunately such nonlinear effects, which are of great importance in simple oscillators, are very difficult to evaluate quantitatively. The effect of limiting is normally to decrease the frequency of oscillation. Other circuit effects may be in the opposite direction so that partial compensation may be obtained under some conditions. Circuits 5-1.1 and 5-1.2 have been chosen and studied to illustrate these various effects. In a simple self-limiting oscillator having a low-Q resonator, the frequency stability may be largely determined by the nonlinear action of the circuit. As the resonator Q is increased, however, thermal and secular effects in the resonator may predominate. It is of interest, therefore, to assume linear operation and examine the other types of phase shifts that may occur in a junction transistor amplifier.

Thus there is a transit time t associated with the diffusion or drift of minority carriers through the base region of the transistor. This time is a function of the operating point, and therefore it is reasonable to assume a change Δt associated with a modification of the operating point due, for instance, to a decrease in collector supply voltage. Then, $\Delta \varphi = -\Delta t \, \omega$ and therefore

$$\frac{\Delta \omega}{\omega} = - \frac{\omega \Delta t}{2Q}. \qquad 5\text{-}0.2$$

It would appear that the transit time fractional frequency instability is directly proportional to frequency.

Other phase shifts may be encountered at the amplifier terminals and also, unfortunately, within the transistor itself. Consider that a resistive impedance Z may be measured across one of the couplings between the amplifier and the resonator. It is inevitable that a capacitance C will appear across Z, but it may be tuned out by some suitable means. It is also inevitable that some instability ΔC will be encountered. Such a ΔC may result, for instance, from a change in average collector potential, which will affect the width of the depletion region in the collector-base junction and produce a change in collector capacitance. Considering the phase instability due to ΔC, $\Delta\phi = -Z\omega\Delta C$ for small angles, and therefore

$$\frac{\Delta\omega}{\omega} = -\frac{Z\omega\,\Delta C}{2Q}. \qquad 5\text{-}0.3$$

It is apparent that the frequency stability will be improved by using low-impedance couplings between the amplifier and the resonator. In particular it is desirable to maintain low impedances at the transistor input and output terminals. There is, however, a limit to the benefit to be derived from decreasing these impedances. Examine the base-input hybrid-π transistor equivalent circuit shown in Figure 5-0.2. The base resistance $r_{bb'}$ and the emitter-junction capacitance $c_{be'}$ will function as a phase shifter whose effect cannot be eliminated by the use of any positive impedance at the base terminals.

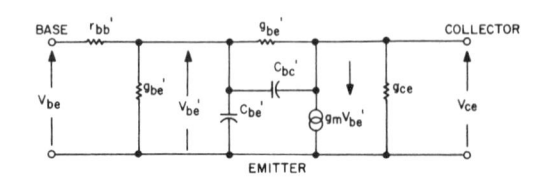

Fig. 5-0.2 - Base-input hybrid-π transistor equivalent circuit

As mentioned above, it is assumed that the resonator is operated at or near resonance, where its phase slope is maximum. As the frequency of oscillation is increased, however, the fixed phase shifts due to transit time and base impedance may become so large that this condition is violated, and $\Delta\omega/\omega$ may become much greater than is indicated by equations 5-0.2 and 5-0.3. Thus, if the phase angle of α is 90 degrees, which will occur at approximately $2f_{ab}$ in a diffusion transistor and 0.9 f_{ab} in a drift transistor (2), the resonator will exhibit little control over the frequency of oscillation unless the large phase angle is compensated by some device other than the resonator.

It would appear, then, that when frequency stability is the primary concern, the oscillator designer should choose the most phase-stable amplifier and the most stable high-Q resonator, and then should connect them in the optimum manner while satisfying other requirements such as output voltage or power, and waveform. Reactive elements - normally capacitors or transformers - are best used for impedance transformation between the amplifier and resonator. The frequency stability will always be degraded when power output is required because the resistance of the load must become associated with the amplifier or resonator in such a way as to degrade the Q of the resonator or increase the coupling impedance. The same can be said for the resistors that are sometimes used for coupling. Although some improvement might result from the connection of a resistor across a coupling that exhibits poor phase stability, it would be better to use the proper impedance transformation and derive the low coupling impedance from the losses in the resonator itself.

The choice of a suitable resonator depends upon the application of the oscillator. If moderately poor waveform and frequency stability can be tolerated in an audio-frequency oscillator, a simple RC network such as that used in Circuit 5-11 may apply. If the waveform must be excellent, but the frequency-stability requirement is still moderate, the resonator can consist of an RC bridge (3), which may contain a provision for amplitude control. The RC networks are particularly valuable, of course, where a wide frequency range must be covered. For more critical frequency-stability requirements an LC tuned circuit can be used at audio or radio frequencies. Circuits 5-1 and 5-10 employ simple tuned circuits. For the highest frequency stability crystal units are necessary, with tuning forks filling some intermediate requirements.

The amplitude stability of an oscillator is determined by the functioning of the amplitude-control mechanism, which must exist in any practical oscillator. In a simple, self-limiting oscillator the output voltage or current may be directly related to the supply voltage, particularly if bottoming occurs. This fact is employed in a simple oscillator having excellent amplitude stability (4). However, the analysis of nonlinear oscillators for both amplitude and frequency stability is difficult, and will not be considered here. The amplitude-stability theory for oscillators employing some type of AVC system is more readily understood, however. Edson has discussed this in detail (5), and his results apply equally to vacuum-tube and transistor oscillators.

At this point it should be mentioned that there are transistor temperature effects that will affect both the amplitude and the frequency of oscillation. First, the variation of I_{co} with temperature must be recognized. If the initial bias is supplied by cutoff current alone, the oscillator may not start at the extremes of a temperature range. The use of some form of d.c. stabilization (6) is indicated. If the operating point is not stabilized, the variation of emitter and collector voltage and current will change practically all of the transistor parameters. Even if the operating point is stabilized, most of the transistor parameters will change with temperature, although not as rapidly as they might otherwise. Thus the base-emitter capacitance is inversely proportional to its absolute temperature, while the base resistance and base-collector capacitances are functions of resistivity, which depends upon temperature. The phase angle and magnitude of h_{fb} also vary with temperature. Compensation of these effects is possible to some extent but, in critical applications, the temperature of the transistor must be controlled.

C. Types of Oscillators

In surveying the various types of transistor oscillators it is convenient to discuss the LC oscillator, which is useful at both ratio and audio frequencies, the RC oscillator, which is used at audio and ultrasonic frequencies, and the crystal-controlled oscillator, which is encountered from the middle of the audio range upward. The following simplified oscillators have been chosen to illustrate various problems in circuit design, and will be considered before the selected circuits are presented.

A simple Hartley-type oscillator is shown in Figure 5-0.3. The collector circuit is tuned, and feedback is applied to the emitter from a portion of the inductor, which is used as a transformer. The circuit is easy to employ, and will be discussed in detail as a design example in Circuit 5-1. The Colpitts-type circuit of Figure 5-0.4 is similar, but is often used at high frequencies where it may be difficult to obtain tight coupling between the tapped portions of the transformer of Figure 5-0.3. Conversely the capacitor C_1 in Figure 5-0.4 may become inconveniently large at low audio frequencies, suggesting the use of Figure 5-0.3. One of the many possible phase-shift RC oscillators is shown in Figure 5-0.5. This RC network, which is of the current-derived type, is used in Circuit 5-11.

Figure 5-0.6 is a simple crystal oscillator based on the Colpitts circuit, in which the crystal unit is used as a series inductor and resistor. Circuits 5-5, 5-6 and 5-7 are of this type. Noting that the feedback is from collector to base, it might at first be supposed that the relatively low cut-off frequency (and large phase angle at a particular frequency) associated with h_{fe}, the base-to-collector current gain, might degrade the frequency

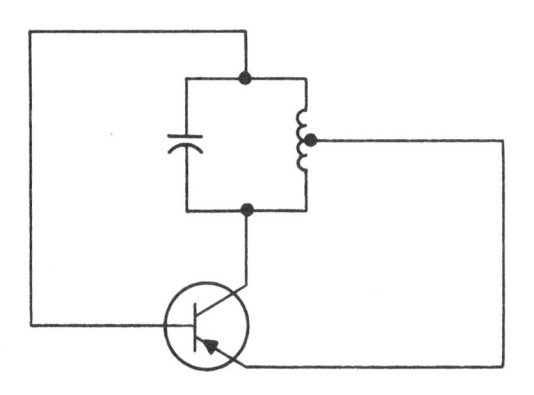

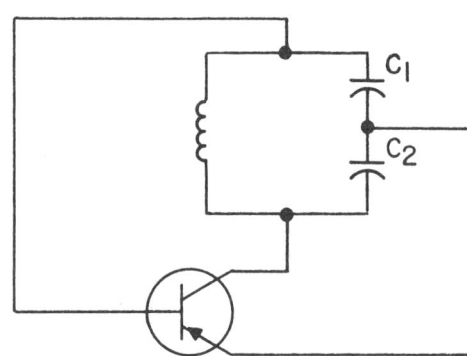

Fig. 5-0.3 - Hartley oscillator   Fig. 5-0.4 - Colpitts oscillator

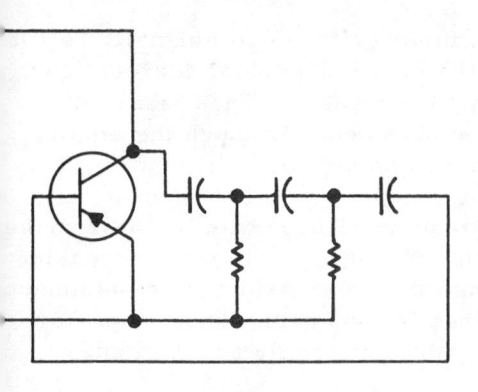

Fig. 5-0.5 - Phase shift RC oscillator

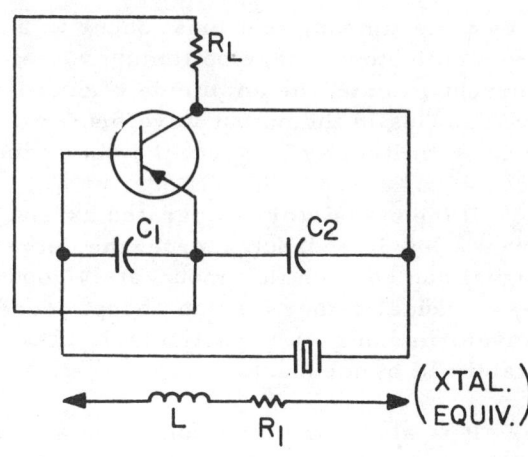

Fig. 5-0.6 - Colpitts type crystal oscillator

ability of this circuit in comparison with an emitter-driven type such as Figure 5-0.7, in which controls. However, C_1 of Figure 5-0.6 can be chosen so that the base is driven by a low impedance generator. In this event it matters little whether the base or the emitter is driven. Thus, if a voltage V_b appears across C_1,

$$I_b \approx \frac{V_b}{r_b + h_{fe} r_e} \qquad 5\text{-}0.4$$

$$I_c \approx h_{fe} I_b = \frac{h_{fe} V}{r_b + h_{fe} r_e} \qquad 5\text{-}0.5$$

$h_{fe} r_e \gg r_b$, as is normally true, $I_c \approx V/r_e = g_m V$. The phase angle associated with g_m is that due to transit time. It should be mentioned that the low cutoff frequency associated with h_{fe} is of importance in the circuit of Figure 5-0.5, where the base is driven by a constant-current generator.

It should be noted here that the current gain of a transistor, particularly if it is of the small-signal type, will increase as the average emitter current is increased. If an oscillator is designed to use a transistor as a current amplifier, which implies that the base is driven from a high-impedance source, the

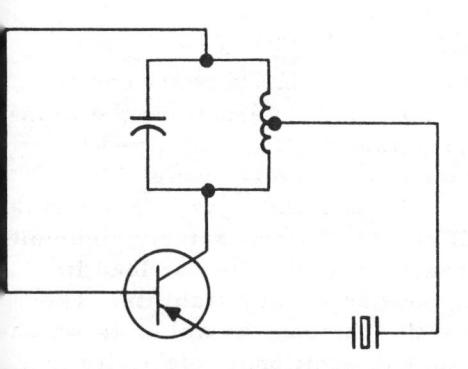

Fig. 5-0.7 - Simple oscillator for use of overtone crystal

157

inevitable limiting that must occur in a simple system will normally result from collector voltage bottoming rather than from a gradual decrease in current gain, as the amplitude of oscillation increases. Thus, sharp discontinuities in the output waveform will be obtained. Although the amplitude stability may be good, the distortion will be high.

If the transistor is operated as a voltage amplifier, with the base drive from a low impedance source, the decrease in voltage gain with increasing signal can be such that moderately good amplitude control can be obtained by a gradual rather than an abrupt decrease in loop gain. Thus, superior waveform can result, particularly if the gain is carefully adjusted and stabilized by feedback.

It is also worth mentioning, however, that collector-voltage bottoming will cause relatively little distortion in a tuned collector oscillator if the operating Q of the tuned circuit is high.

The Colpitts-type circuit of Figure 5-0.6 is of particular interest because it readily permits a design for high frequency stability. Here C_1 and C_2 can be chosen in ratio and magnitude to develop the optimum impedances at the transistor terminals. The source impedance for the base is equal to $(X_{C_1})^2/R_1$, if $X_{C_1} \gg R_1$, while the load impedance for the collector is equal to $(X_{C_2})^2/R_1$, assuming R_L to be large. The collector barrier capacitance is small, relatively stable with collector current, and has a low temperature coefficient, while the base-to-emitter capacitance is large and unstable. Thus, C_1 should be greater than C_2, and each should be as large as possible while satisfying the condition $\mu\beta > 1$. It should be noted here that the relatively high transistor terminal impedances connected across C_1 and C_2 will normally transform to low values of resistance in series with R_1, and therefore the resonator Q will be degraded but little.

If the crystal unit is replaced with an inductor and capacitor, the Clapp circuit (7) is obtained.

The simple crystal oscillator shown in Figure 5-0.7 permits control of the impedance at the collector but not at the emitter, assuming the crystal resistance to be fixed. If the crystal resistance is low compared to the emitter resistance, its Q may be degraded excessively, while if its resistance is very high, the attenuation associated with the ratio of crystal resistance to emitter input impedance will require an excessively high voltage gain from emitter to collector. The required high collector load impedance will then result in a design having poor frequency stability. The circuit has found its principle application with overtone crystal units which will operate at the fundamental frequency unless additional selectivity is provided. Circuits 5-4 and 5-8 are of this type.

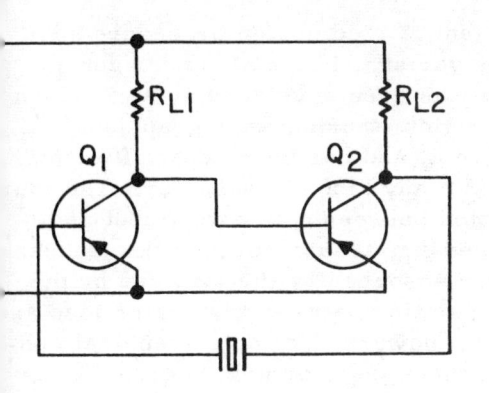

Fig. 5-0.8 - Example of less desirable oscillator

The two-transistor circuit shown in Figure 5-0.8 is included in the discussion because it illustrates the undesirable conditions that can readily be obtained. The oscillator has merit at audio frequencies where the crystal resistance is very high, but here the circuit may suffer from a parasitic oscillation through the crystal shunt capacitance. At radio frequencies, where the crystal resistance may be between 10 and 1000 ohms, the crystal Q will be degraded excessively by R_{L_2}. If R_{L_2} is decreased, R_{L_1} must be increased to obtain sufficient gain, increasing the phase instability at the interstage. Furthermore, the complex impedance at the base of Q_1 will have an undesirable effect. If a very low resistance is connected from base to emitter here, the gain of the rest of the system must be increased. Even if the system could be designed to minimize the trouble at the various interstages to a reasonable extent, the built-in transistor phase shifts are additive, and their effects will be doubled as ambient or operating conditions are varied.

Although the circuit can be modified by connecting the crystal from emitter to emitter - two low-impedance points - a high-impedance coupling is required from the collector of one transistor to the base of the other.

0. Summary

In summary, it can be stated that, when frequency stability is of paramount importance, a single-transistor oscillator coupled to a resonator by networks of negligible loss will give the best results. Unfortunately the stability of the system can, at best, only approach that of the resonator alone.

The selected circuits to follow have been chosen to represent currently used designs over the frequency range of interest. Circuit 5-1.1 and 5-1.2 have been tested carefully, and the rather lengthy accompanying discussion is presented to give a reasonably complete picture of transistor effects. Radio-frequency oscillators are considered first, and then audio-frequency oscillators.

The discussion so far has utilized Barkhausen's criterion as a starting point, and has indicated the effects of various circuit parameters upon

the performance of the oscillator. A generalized design procedure has
not been presented because of the wide variety of circuits under consideration. Recently such a procedure has been developed by McSpadden
and Eberhard (12, 13). Their method, which is principally graphical,
makes use of the loop-power-gain concept, and has been kept sufficiently
general to permit its use with practically any kind of oscillator. The four
quadrant graph used contains the relation between transistor input and
output impedance, the effect of the impedance transformer in the feedback
network, the relation between the power absorbed by the load and by the
losses in the feedback method and the relation between transistor load and
the load resistance which absorbs useful power. A simple graphical construction permits the determination of the transistor gain required to assure oscillation. Unfortunately the method gives no direct information on
frequency stability. It should also be mentioned that reference 12 contains
a useful start in the standardization of oscillator-circuit nomenclature.

Figure 5-0.9 contains an excellent chart summarizing the characteristics of military type crystal units.

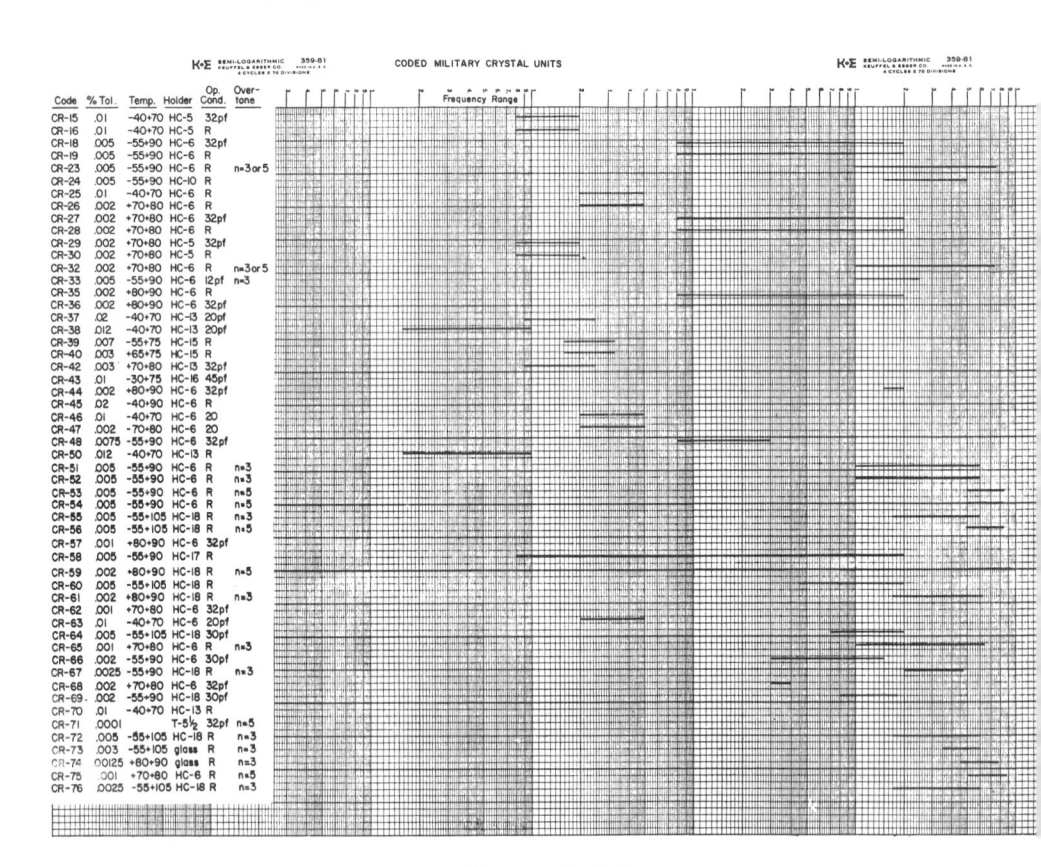

Figure 5-0.9

REFERENCES

1. "Vacuum-Tube Oscillators" (Chapter 4) - W. A. Edson, McGraw-Hill, 1953

2. "A Design Basis for Junction-Transistor Circuits" - D. F. Page, June 1958, PROC. I.R.E.

3. "Low-Distortion Transistor A-F Oscillator" - P. G. Sulzer, Sept. 1953, ELECTRONICS

4. "Transistor Oscillator Supplies Stable Signal" - Leon H. Dulberger, January 31, 1958, ELECTRONICS

5. Edson, Chapter 7

6. "Transistor Operation: Stabilization of Operation Points" - Richard F. Shea, November 1952, PROC. I.R.E.

7. "An Inductance-Capacity Oscillator of Unusual Frequency Stability" - J. K. Clapp, March 1948, PROC. I.R.E.

8. "Analysis of Junction Transistor Audio Oscillator Circuits" - J. B. Oakes, August 1954, PROC. I.R.E.

9. "Junction-Transistor Oscillators" - Mahmoud A. Melehy and Myrie B. Reed, Vol. XIII, Proceedings of the National Electronics Conference, Oct. 1957

10. "Current Derived Resistance--Capacitance Oscillators Using Junction Transistors" - D. E. Hooper and A. E. Jackets, August 1956, ELECTRONIC ENGINEERING

11. "Twin Tee Networks," White Instrument Laboratories, Bulletin No. 500

12. Final Progress Report #P-2208-4, "Transistor Crystal Oscillator Circuitry," Motorola, for Signal Corps Contract DA-36-039-SC-72837

13. W. R. McSpadden and E. Eberhard, "Graphical Design of Transistor Oscillators," Electronics, Vol. 31, Dec. 5, 1958, pp 90-93

14. Coded Military Crystal Units, S. Schodowski, Frequency Control Division, USASRDL, Ft. Monmouth, N.J.

BIBLIOGRAPHY

Dasher, B. J. and Witt, S. N., Jr., "Transistor Oscillators of Extended Frequency Range," Final Report, Proj. 236-206, contract DA-36-039-sc-42712, June 30, 1955.

Dewitt, David and Rossoff, Arthur L., "Transistor Electronics," McGraw Hill, New York, 1957.

Dulberger, Leon H., "Transistor Oscillator Supplies Stable Signal," Electronics, January 31, 1958.

Edson, W., "Vacuum Tube Oscillators," John Wiley & Sons, Inc., New York, 1953.

Foster, William H., "Strain Gage Oscillator," Electronics, January 31, 1958.

Hooper, D. E., and Jackets, A. E., "Current-Derived Resistance-Capacitance Oscillators Using Junction Transistors," Electronic Engineering, August 1956.

Keonjian, E., "Stable Transistor Oscillator," I.R.E. Trans. CT3 (1956).

Kretzmer, E. R., "An Amplitude-Stabilized Transistor Oscillator," Proc. I.R.E., February 1954.

McSpadden, W., "Transistor Crystal Oscillator Circuitry," Final Report, Contract DA-36-039-sc-72837, August 23, 1957.

Melehy, M. A. and Reed, M. B., "Junction Transistor Oscillators," Proc. National Electronics Conference, Vol. XIII, October 7, 1957.

Oakes, J. B., "Analysis of Junction Transistor Audio Oscillators," Proc. I.R.E., August 1954.

Page, D. F., "A Design Basis for Junction Transistor Oscillator Circuits Proc. I.R.E., July 1958.

Shea, Richard F., "Transistor Circuit Engineering," John Wiley & Sons, Inc., New York, 1957.

Wolfendale, E., "The Junction Transistor and Its Applications," The Macmillan Company, New York, 1958.

CIRCUIT 5-1

HIGH-STABILITY SELF-EXCITED OSCILLATOR

Sulzer Laboratories, Inc.   Contributed by Peter G. Sulzer

Circuits 5-1.1 and 5-1.2 will serve to illustrate some of the problems discussed in the design philosophy, and will also indicate the magnitude of the frequency stability obtained by using a simple circuit - first with self-limiting, and second with an effective amplitude-control system producing Class A operation.

Circuit 5-1.1 is designed to operate in the region of 1 mc with inductive tuning. The power supply required is a nominal 20 volts at 1.5 mAdc. The power output is only sufficient to drive a Class A transistor buffer amplifier. In this design, a type 2N384 drift transistor was chosen to minimize the phase shift due to transit time, as the alpha cutoff frequency of this unit is 100 mc. The collector-base capacitance of the transistor is

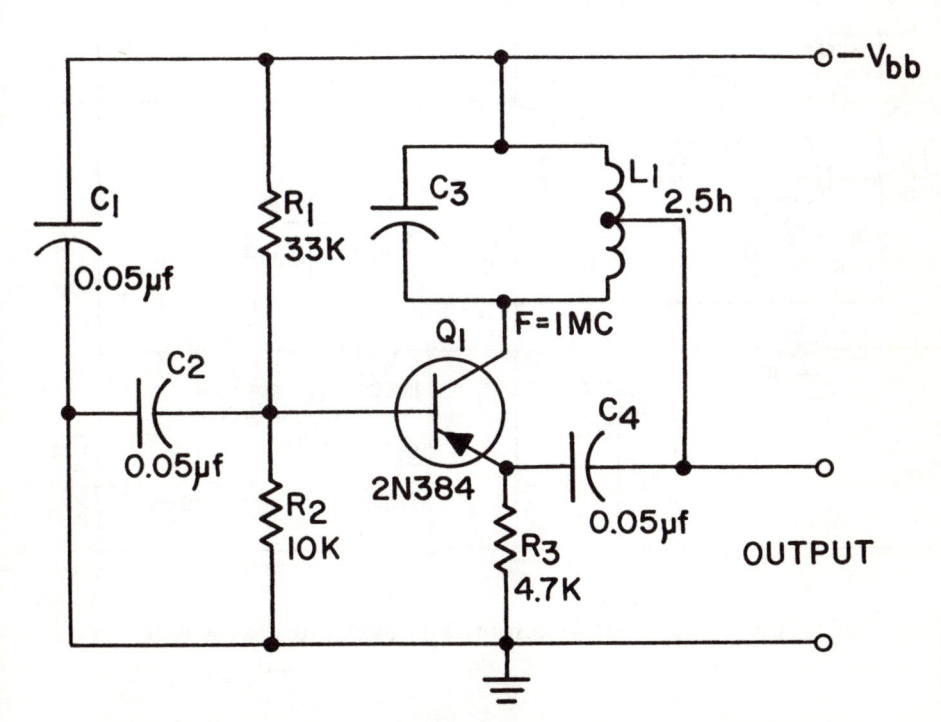

Circuit 5-1.1 - Self-limiting, self-excited 1MC oscillator

Circuit 5-1, High Stability Self-Excited Oscillator (cont'd.)

approximately 2 $\mu\mu$f, while the emitter-base capacitance is approximately 100 $\mu\mu$f. It can be assumed, for want of better information, that the instabilities associated with these capacitances will have twice the ratio of the capacitances themselves. Therefore, to obtain equal frequency instability from these two sources, the collector-load impedance should be 100 times the driving point impedance for the emitter. This is easily accomplished with the Hartley circuit, by the proper choice of the emitter tap on the inductor. It is seen that a voltage gain in excess of 10 from emitter to collector will be required to produce oscillation. The transconductance of the transistor is almost equal to the emitter conductance, which is 40 I_E. With I_E = 1 mAdc, $g_m \approx 0.04$ mho, and therefore the collector-load impedance should be greater than 250 ohms. The operating point, V_C = -15 volts, I_C = -1 mAdc, is established by a conventional means of d.c. stabilization.

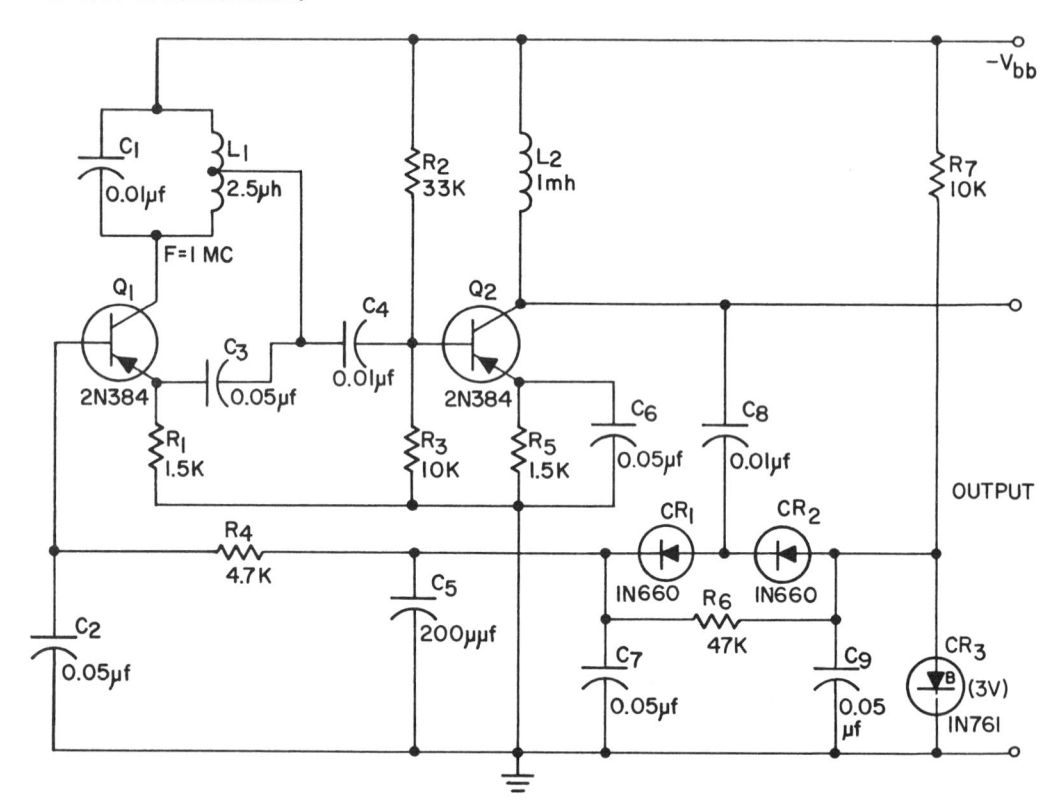

Circuit 5-1.2 - Self-excited IMC oscillator with AVC

Circuit 5-1, High Stability Self-Excited Oscillator (cont'd.)

The tuned circuit selected had a resonant impedance of 1600 ohms, and Q of 100. This high a collector load impedance will ensure rapid and reliable starting, but will result in some degradation of the frequency stability. Normally a safety factor of two times would be used, or, conversely, the design would be worked out for unity loop gain using a transconductance of one half the rated value. It should be noted here that the Clapp type of circuit (7) could have been used to effect a more advantageous impedance transformation, however, the stable capacitors required would have been of excessive physical size. A similar transformation could be obtained by tapping the collector on the inductor, but parasitic oscillations would almost certainly have been obtained.

Frequency and output voltage vs. supply voltage are shown on the Figure 5-1.1. It will be observed that the frequency change with a supply voltage change $\pm 5\%$ is approximately $\mp 25 \times 10^{-6}$. The instability results from a combination of the effects discussed, with nonlinearity predominating.

In Circuit 5-1.2 the same oscillator has been made linear by the addition of an AVC circuit containing transistor amplifier Q_2 and rectifier-doubler $CR_1 - CR_2$. In the absence of oscillation on operating point is established for Q_1 by a regulated voltage from CR_3 which is applied to the transistor base through R_4 and R_6. The resulting constant potential across R_1 determines the emitter current. When oscillation starts, CR_1 and CR_2 rectify, decreasing the magnitude of the base bias applied to Q_1, therefore decreasing the emitter current and transconductance until an equilibrium is reached. The frequency stability with variable supply voltage is improved fifteen (15) times which, in some applications would be sufficient to justify the increased circuit

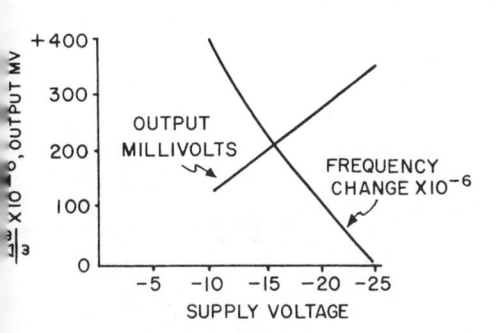

Fig. 5-1.1 - Output and frequency vs. supply voltage for circuit 5-1.1

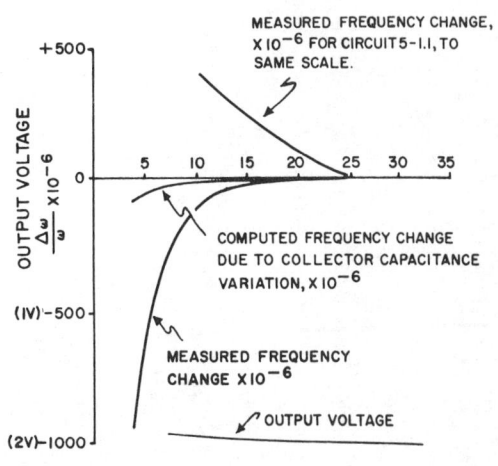

Fig. 5-1.2 - Frequency vs. supply voltage for circuit 5-1.2

Circuit 5-1, High Stability Self-Excited Oscillator (cont'd.)

complication. In a high-performance oven a frequency stability of 1×10
per hour was obtained. The output is approximately 1 volt to a high-
impedance load.

It is interesting to speculate on the origin of the frequency instability
remaining after nonlinearity is eliminated. Class A operation is assured
because the emitter signal is only 0.02 volt. The AVC circuit tends to
keep the emitter current of Q_1 constant, and therefore the base emitter
capacitance should remain constant as the collector voltage is changed.
The collector-base capacitance will vary with supply voltage, however,
and the computed frequency change due to collector-capacitance variation
is shown in the Figure 5-1.2. At the higher collector voltages this would
appear to be of importance. As the collector voltage is lowered, the increase in transit time is probably responsible for the large decrease in
frequency, which far exceeds that computed from collector-capacitance
variation.

In summary, at a 20-volt operating point the frequency of the self-
limiting oscillator decreases 25×10^{-6} per volt. At the same point, the
frequency of the linear oscillator increases 1.5×10^{-6} per volt, which is
very nearly the rate computed for collector-capacitance variation alone.

CIRCUIT 5-2

WIDE-RANGE CONSTANT-AMPLITUDE OSCILLATOR

M.I.T., Lincoln Laboratory   Contributed by Baker, Chatterton and Parker

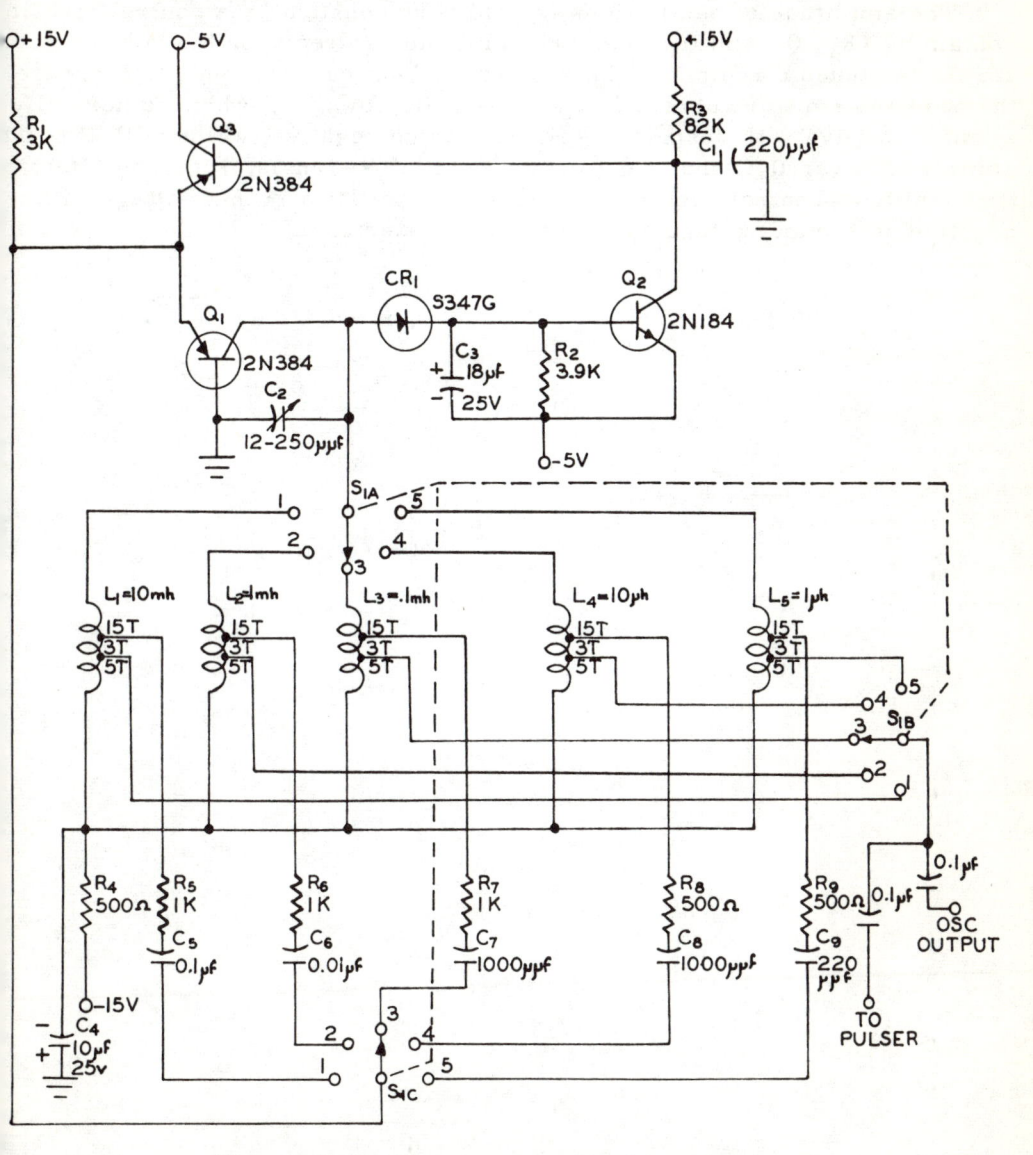

Circuit 5-2 - Constant amplitude oscillator

Circuit 5-2, Wide-Range Constant-Amplitude Oscillator (cont'd.)

Circuit 5-2 is a three-transistor device that is capable of being tuned over the frequency range from 100 kc to 20 mc in five bands with an amplitude stability of $\pm 2\%$. The output is 4 volts peak to peak, to a 500 ohm load. The circuit is included here because it should be of great value in signal generators and other wide-range devices. As in Circuits 5-1.1 and 5-1.2, a tuned collector is used, with feedback to the emitter.

The amplitude of oscillations is held very constant by a control circuit containing CR_1, Q_2, and Q_3. The crystal diode CR_1 rectifies the voltage across the tuned circuit, and the resulting direct current tends to increase the base and emitter current of Q_2. The collector of Q_2, which is normally slightly positive with respect to ground, can go negative, which will, through emitter follower Q_3, tend to cut off the oscillator transistor Q_1. In effect, the amplitude-control system compares the oscillator output voltage with a negative reference voltage applied to the emitter of Q_2.

CIRCUIT 5-3

VHF OSCILLATOR

Bell Telephone Laboratories, Inc.   Contributed by R. L. Lowell

Circuit 5-3 is a tuned-collector oscillator designed for use in the VHF and low UHF ranges. It should find application as a high-frequency receiver local oscillator and also as a component of test equipment.

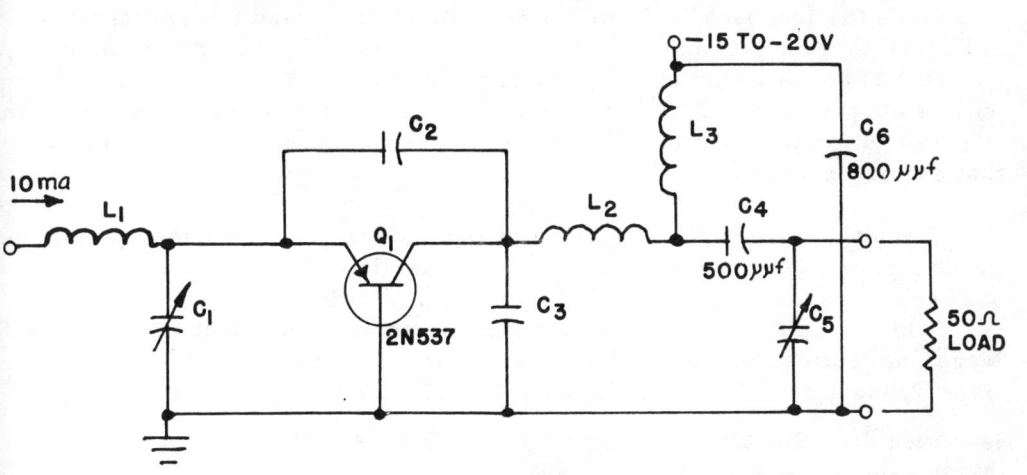

	100 MC	250 MC	450 MC
C_1	3-12 µµf	2-7 µµf	<1 µµf (STRAY)
C_2	2 µµf	stray	stray
C_3	stray	stray	stray
C_4	500 µµf (50VDC) CERAMIC CAPACITOR		
C_5	15-90 µµf	7-45 µµf	8-13 µµf
L_2 - DIA.	1/2	3/8	1/4
LENGTH	1 1/2	1	1/2
WIRE SIZE	16	14	16
TURNS	8	6	2

L_1 & L_3 ARE RADIO FREQUENCY CHOKES WITH VALUES GIVEN BELOW.
   FOR 100 MC   JEFFERS-10102-21 (3.3 µh)
   FOR 250 MC   JEFFERS-10100-31 (1.0 µh)
   FOR 450 MC   JEFFERS-10100-29 WITH TURNS REMOVED TO SELF RESONATE AT APPROX. 500 MC.

| OUTPUT | 90 mW | 60 mW | 35 mW |

Circuit 5-3 - V.H.F. oscillator

Circuit 5-3, VHF Oscillator (cont'd.)

The type 2N537 diffused-base transistor is used which may have an alpha cutoff frequency as high as 1000 mc. It will be observed that feedback takes place from collector to emitter through capacitors C_1 and C_2. The collector circuit is tuned by L_2, and C_3, and impedance matching is accomplished by L_2 and C_5.

Since the feedback is from collector to emitter, and the emitter-to-collector current gain must be slightly less than unity, there may be a question of how a loop current gain of greater than unity is obtained. It will be observed that C_2 is smaller than C_1, so that an impedance stepdown is obtained for the emitter, and thus the current into the emitter can exceed that out of the base.

Circuit constants for operation of these different frequencies are given in the figure. The maximum power output is also stated.

(This circuit is reproduced with the permission of Electronic Design Magazine from an article entitled "1000 mc Range Reached With Diffused Base Transistors" by C. H. Knowles and E. A. Temple, July 9, 1959.)

Reprinted from Electronic Design Magazine, July 9, 1959; copyright Hayden Publishing Company, Inc. 1959.

CIRCUIT 5-4

100 KC REFERENCE OSCILLATOR

James Knights Company   Contributed by John Silver

Circuit 5-4 is a relatively simple crystal oscillator designed for operation at 100 kilocycles with a silicon transistor. The use of a silicon transistor permits operation of the oscillator at the high ambient temperature encountered in military equipment. The crystal unit is placed in series with the feedback connection from the capacitively-tapped tuned collector circuit to the emitter. The use of a tuned collector circuit prevents oscillation at spurious crystal frequencies. Stable components must be used in the tuned circuit, particularly if wide-temperature-range operation is indicated. Silvered-mica capacitors and a carbonyl-iron toroidal inductor were used in one model of the oscillator. If the highest frequency stability is required, the complete oscillator, including the crystal unit, should be contained in a temperature-controlled oven. The

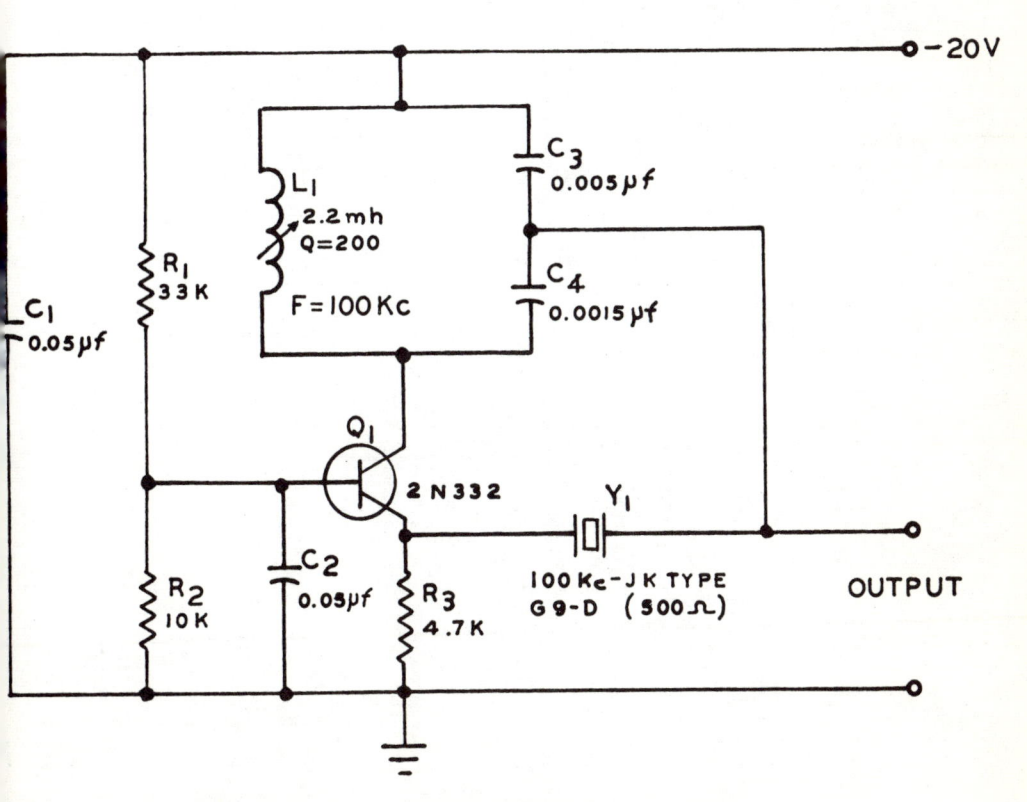

Circuit 5-4 - 100 kc reference oscillator

Circuit 5-4, 100 Kc Reference Oscillator (cont'd.)

impedance at the transistor emitter is such that the crystal Q is degraded but slightly.

The frequency is stable to 2×10^{-7} for supply voltage variations of $\pm 10\%$. The output is 1 volt rms at an impedance of 100 ohms. Changing the load from an open circuit to 300 ohms will produce a frequency shift of 1×10^{-6}. Changing the load from an open circuit to the matched value (100 ohms) will provide a frequency shift of 5×10^{-6}.

CIRCUIT 5-5

450 KC CRYSTAL OSCILLATOR

Motorola, Inc.   Contributed by R. L. Steele

Circuit 5-5 is an excellent 450 kc oscillator designed for use with military type CR-26-U crystal unit. In the design, which is of the modified Colpitts type, it will be noted that the capacitance connected from base to emitter is several times that connected from emitter to ground, thus effecting an impedance transformation. The frequency is adjusted over a range of $\pm 37 \times 10^{-6}$ by means of a variable inductor connected in series with the crystal unit.

An interesting feature of the circuit is that the output is taken from a load connected in series with the collector. If the collector source

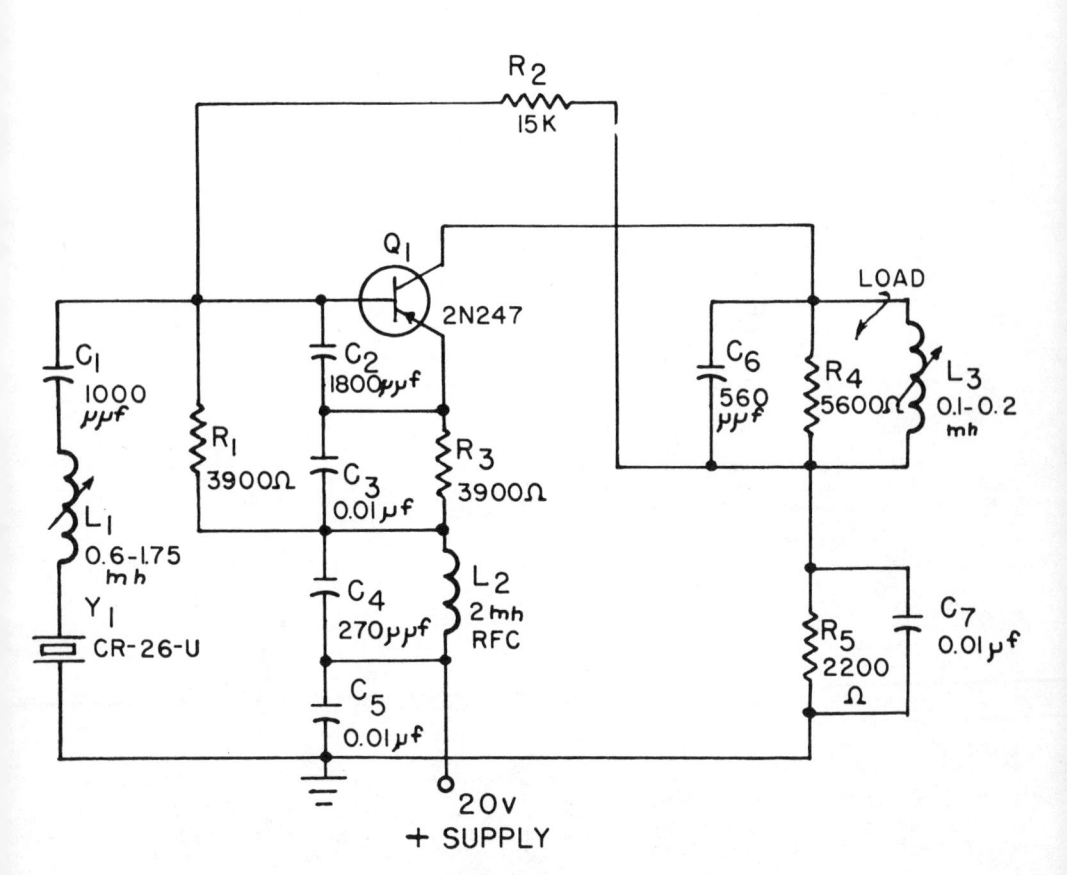

Circuit 5-5 - 450 kc crystal oscillator

Circuit 5-5, 450 Kc Crystal Oscillator (cont'd.)

impedance is high, and if the collector-base capacitance is low, useful isolation of the oscillating circuit from the load can be obtained in this manner. The effect is similar to that of the electron-coupled vacuum-tube oscillator and, indeed, some of the same precautions must be observed to assure good isolation. In particular, the instantaneous collector potential must not be permitted to fall below the knee of the collector voltage current curve corresponding to the maximum instantaneous emitter current.

The frequency is stable to 1×10^{-6} over the supply-voltage range from 5 to 30 volts. The maximum output is 7.5 milliwatts. The maximum frequency change caused by the substitution of several different type 2N247 transistors was 2×10^{-6}. Varying the transistor temperature from -40 to +140°F caused a frequency change of less than 1×10^{-6}.

CIRCUIT 5-6

1 MC CRYSTAL STANDARD OSCILLATOR

James Knights Company · Contributed by John Silver

Circuit 5-6 is a high-quality 1 mc crystal oscillator designed for use as a frequency reference in an airborne single-sideband system. Like Circuit 5-5, it employs the Colpitts-derived crystal oscillator, with a low driving point impedance for the transistor base. The use of silicon transistors permits operation at high ambient temperatures. The entire assembly is maintained at a constant temperature between 85°C and 125°C. A common-emitter buffer amplifier is used to isolate the oscillator from a variable load. Note that the buffer is driven from the lowest impedance point of the oscillator through a series isolating resistor, which is also used to adjust the output level. A silicon-diode voltage regulator is used to provide a relatively constant supply voltage for both transistors. The use of a simple base-collector bias resistor is satisfactory to establish the transistor operating point because of the very low I_{co} that is characteristic of silicon transistors. These particular transistors are selected to have a narrow range of h_{fe}, which further facilitates the use of such a simple biasing scheme. Note that in Circuit 5-5, a base voltage divider and emitter bias resistor are used with the germanium transistor.

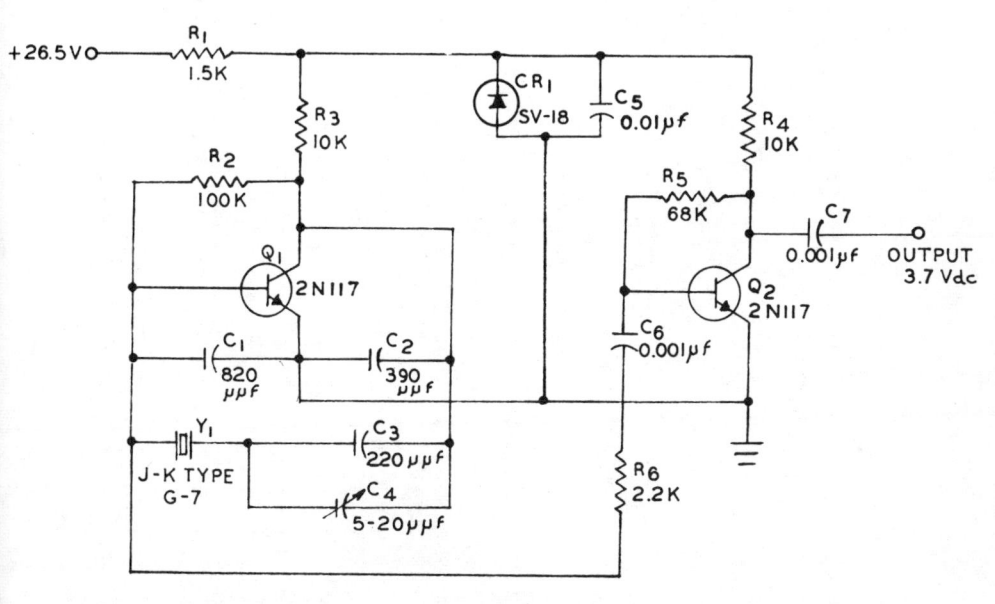

Circuit 5-6 - 1 mc crystal standard oscillator

Circuit 5-6, 1 Mc Crystal Standard Oscillator (cont'd.)

Performance of the oscillator is good with respect to supply voltage change; a frequency deviation of less than $\pm 3 \times 10^{-8}$ is held over the supply voltage range 24-29 volts. The regulator diode is largely responsible for such frequency stability. The frequency changes less than 5×10^{-9} as the load is changed from an open circuit to a short circuit. However, the substitution of different transistors will cause a frequency spread of $\pm 3 \times 10^{-6}$.

CIRCUIT 5-7

9 MC CRYSTAL STANDARD OSCILLATOR

Motorola, Inc. Contributed by R. L. Steele

Circuit 5-7, which is similar to Circuit 5-5, has been designed for stable operation at 9 mc. It has been included here because it illustrates the necessity of using a higher-frequency transistor as the oscillator frequency is increased, if frequency stability is to be preserved. The alpha cutoff frequency of the 2N384 transistor used is approximately ten (10) times the operating frequency. The frequency stability is 2×10^{-6} over the input range from 6-30 volts. The frequency is adjustable over a range of 200×10^{-6} when the military type CR-19-U crystal unit is used. Transistor temperature variation from -40°F to +149°F produced a frequency change of 5×10^{-6}. The power output is only one milliwatt.

Note that, as in Circuit 5-5, a bias voltage for the transistor base is provided by resistors R_1 and R_2. The emitter current is then determined by the voltage drop across R_3. This method of d.c. stabilization is adequate for use with germanium transistors over the temperature range given above.

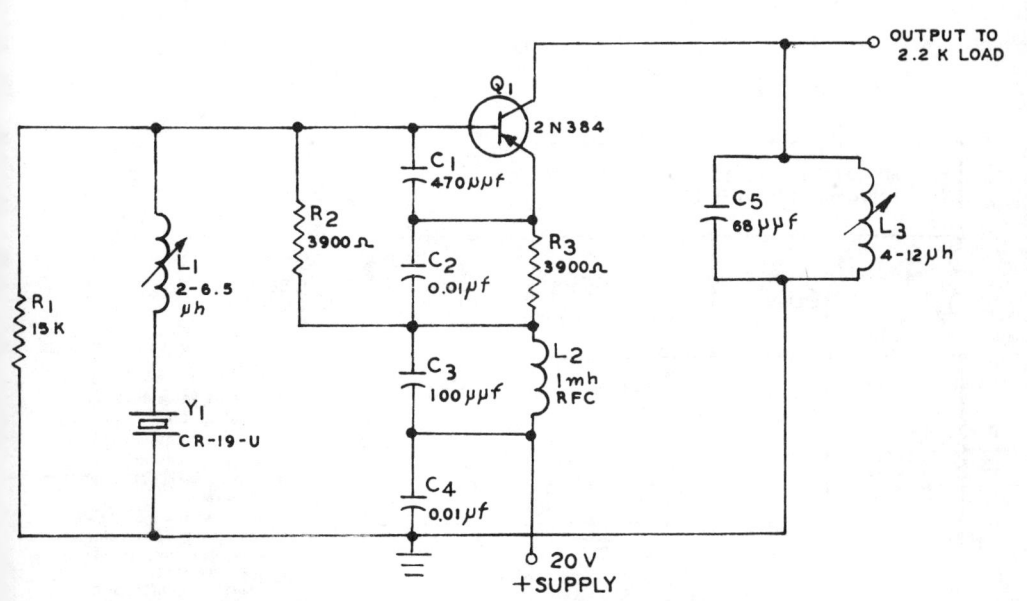

Circuit 5-7 - 9 mc crystal standard oscillator

CIRCUIT 5-8

108 MC CRYSTAL OSCILLATOR

Naval Research Laboratory

Contributed by A. F. Thornhill, Edgar L. Dix, and J. E. Scobey, Jr.

Circuit 5-8 which is a 108 mc crystal oscillator employing a diffused base transistor was designed for use in Vanguard I. It also formed a portion of the telemetry transmitters for Vanguard III and Explorer IV. The basic circuit is similar to Circuit 5-4, although the precautions necessary to obtain proper operation at a much higher frequency are evident. Thus the crystal holder capacitance is tuned out by a shunt inductor, and a small capacitor is connected in series with the crystal to compensate for the phase lag in the transistor. As mentioned above, the use of an overtone crystal requires the use of an additional tuned circuit, which is placed in the collector here.

As shown, the oscillator has a separate emitter supply voltage. This is convenient when batteries are being used as a power source, and with a suitable chosen emitter resistor, leads to practically perfect stabilization of the operating point. If a single power supply is to be used, R_1 could

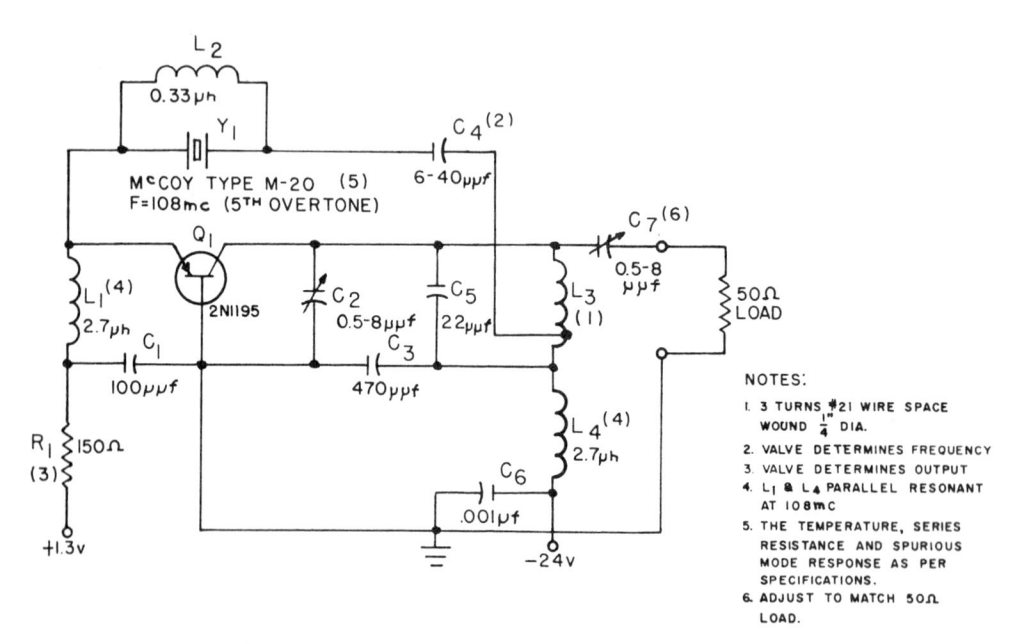

Circuit 5-8 - 108 mc crystal oscillator

Circuit 5-8, 108 Mc Crystal Oscillator (cont'd.)

be connected directly across C_1, and the base could be supplied from a voltage divider connected from ground to the negative supply.

The efficiency of the oscillator is between 30 and 35% when delivering 15 to 30 milliwatts to a 50 ohm load; however, the oscillator is capable of delivering 100 milliwatts at an efficiency of 40 to 45%. With constant loading and temperature the frequency stability is approximately $\pm 1 \times 10^{-8}$, for a 10 minute period.

CIRCUIT 5-9

20 KC AUDIO CRYSTAL OSCILLATOR

James Knights Company   Contributed by John Silver

Circuit 5-9 is a crystal oscillator for use in the audio-frequency range. The circuit constants shown are suitable for use at 20 kilocycles with a 100,000 ohm crystal unit. The relatively high resistance of such a crystal suggests that it should be connected directly at the collector circuit, rather than at the emitter or base. A 100:1 impedance transformation is placed between the crystal unit and the transistor base. The pi-section transformer also provides the necessary phase reversal. Stabilization of the operating point by a constant base potential and an emitter resistor was used because the circuit was intended for use with a wide range of silicon transistors without selection.

Inspection of the circuit will show that its design is not optimum from the standpoint of frequency stability. Thus, the transformed input impedance of the transistor plus the collector load resistance will decrease the working Q of the crystal unit to slightly less than one-half its maximum

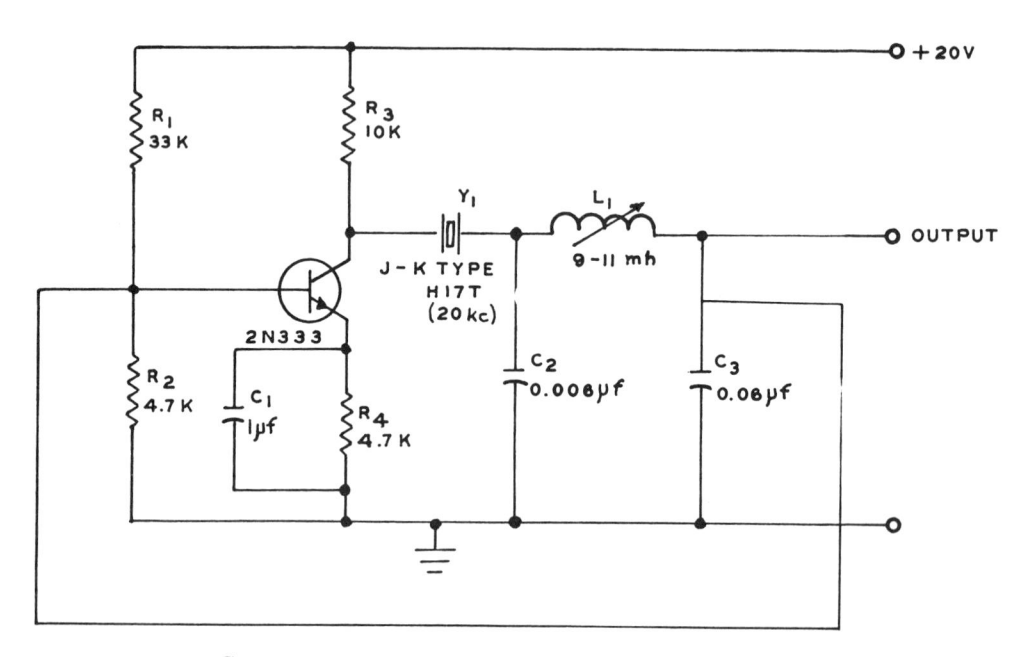

Circuit 5-9. 20 kc audio crystal oscillator

Circuit 5-9, 20 Kc Audio Crystal Oscillator (cont'd.)

value. Also, there has been no attempt to transform the crystal resistance to a low impedance at the collector. It is fortunate that the collector capacitance will have a relatively small effect at audio frequencies. The voltage coefficient of frequency is approximately 3×10^{-6} per volt at 20 volts. The output at the base connection is 0.4 volt at an impedance of 330 ohms.

CIRCUIT 5-10

AUDIO-FREQUENCY OSCILLATOR

Airtronics, Inc.   Contributed by D. Gleason

Circuit 5-10 is a tuned-collector audio-frequency oscillator of a type whose design has been discussed by Oakes (8). The oscillator is useful as a source for bridge measurements at a frequency of one kilocycle. An interesting feature of the design is the insertion of a high impedance in series with the transistor base, which is accomplished here merely by omitting the capacitor from the base-bias voltage divider. It has been stated by Melehy and Reed (9) that such an impedance will help to linearize the operation of the oscillator, improving both the waveform and the frequency stability. Actually, with the circuit constants shown, the efficiency was also improved. As mentioned in the design philosophy section, the use of a high base source impedance will make the limiting occur in the collector circuit. In this particular oscillator, excellent waveform is obtained because the collector circuit is tuned. This should be compared with Circuit 5-11, immediately following.

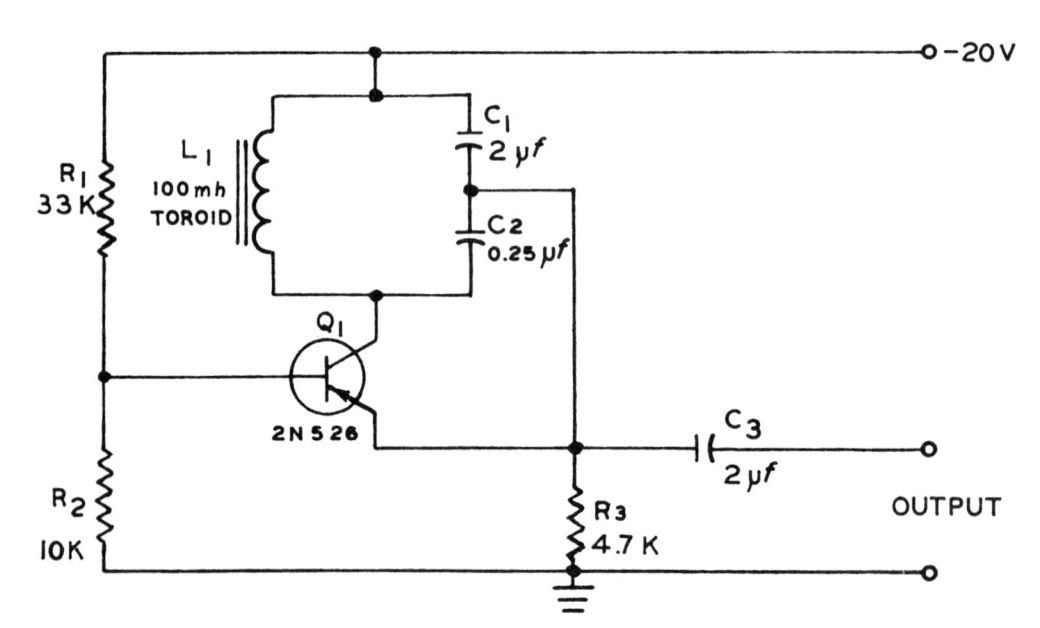

Circuit 5-10 - Audio-frequency oscillator

Circuit 5-10, Audio-Frequency Oscillator (cont'd.)

The frequency of the oscillator changes only three cycles as the supply voltage is varied from -15 to -30 volts. The output is 1.5 volts with a source impedance of less than 300 ohms. The frequency increased 9 cycles when the load impedance was decreased from a very high value to 000 ohms. The distortion is approximately 1%, and is relatively independent of loading.

CIRCUIT 5-11

PHASE-SHIFT OSCILLATOR

Airtronics, Inc. Contributed by D. Gleason

Circuit 5-11 is a 650 cycle audio oscillator of the current-derived phase-shift type (10). It is included because it should be of value for use in light-weight, portable equipment where inductors may be undesirable. The three-section phase-shift network is designed to make the base current lead the collector current by 180 degrees at the operating frequency. This is a natural consequence of the fact that the transistor is basically a current amplifier. A current gain of at least 29 is required. Although it might at first appear that the phase-shift network is driven and loaded in such a way that an operating current gain greatly in excess of 29 would be obtained, inspection will show that the first section of the network, which includes the 10K collector load resistance, is driven by the transistor collector impedance which is of the order of 500,000 ohms. The third section is terminated by the base impedance, which is about 1K ohm, in parallel with a voltage divider. Since the network impedance is approximately 10,000 ohms, the desired conditions of having an infinite-impedance

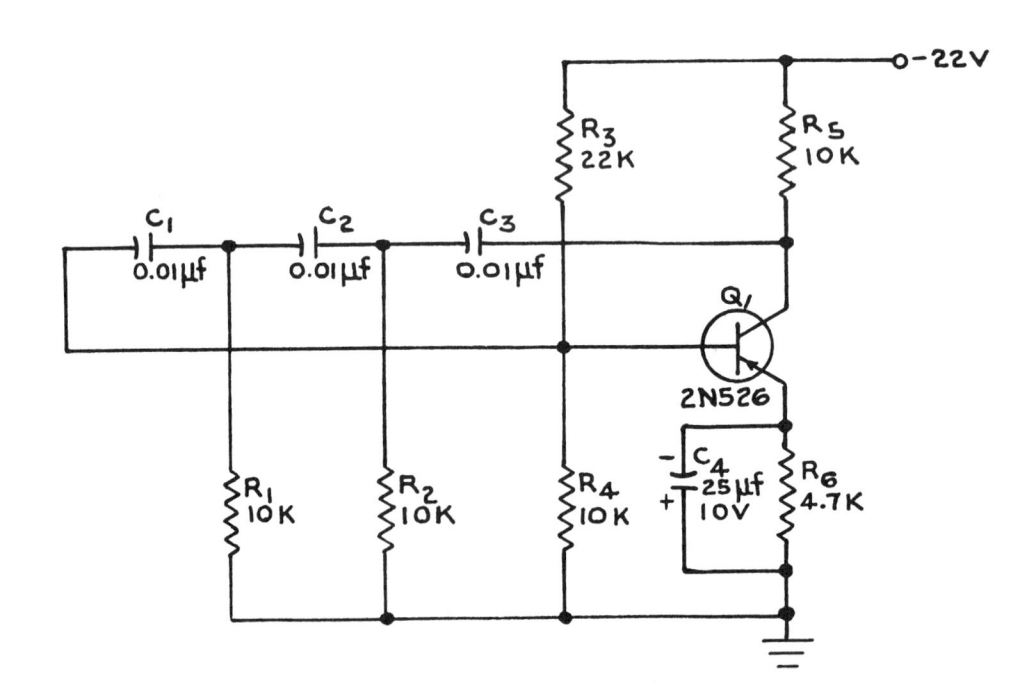

Circuit 5-11 - Phase-shift oscillator

Circuit 5-11, Phase-Shift Oscillator (cont'd.)

generator and a short-circuit load for the network are reasonably well satisfied. If a voltage-derived network had been used, a low collector load resistance would have been required, and an emitter-follower amplifier would have been used between the output of the network and the input of the amplifier. For purposes of comparison, it should be mentioned that voltage-derived networks are used in vacuum-tube phase-shift oscillators. The frequency of oscillation is given by $f = 1/2\sqrt{6}\,\pi RC$, assuming that the reactance of C is much greater than the input impedance of the transistor.

The performance of the circuit is interesting because the frequency passes through a maximum at a collector supply of -22 volts. The frequency stability is surprisingly good at this point; changes of ±5 volts will decrease the frequency 3 cycles. The frequency is rather sensitive to load, and will increase approximately 10% as the load is decreased to 10,000 ohms. The open-circuit output is 2 volts at an impedance of 12,000 ohms. The distortion is 10% at no load.

CIRCUIT 5-12

TWIN TEE OSCILLATOR

Baird-Atomic, Inc.   Contributed by William R. Lamb

The oscillator in Circuit 5-12 was designed to have low distortion, good frequency stability and an output unaffected by power supply variations.

The circuit comprises a two-stage amplifier with positive and negative feedback loops. The frequency sensitive element, a twin tee network, is in series with the negative feedback loop. The gain of the positive loop is dependent on the output amplitude. In operation, the total loop gain adjusts itself to unity at a specified output. Thus oscillations of constant amplitude at the rejection frequency of the twin tee are generated. Changes in amplifier forward gain, due to variation in components or supply voltage, will alter the gain of the positive feedback loop in the correct direction to restore the original output.

Referring to the schematic, Q_1 is a common emitter stage coupled to a common collector stage, Q_2, which furnishes the necessary low output impedance to the twin tee negative feedback path. The open loop voltage gain of this combination is approximately 100. Q_3 serves as a detector which provides a d.c. bias current through the diode, CR_1, proportional to

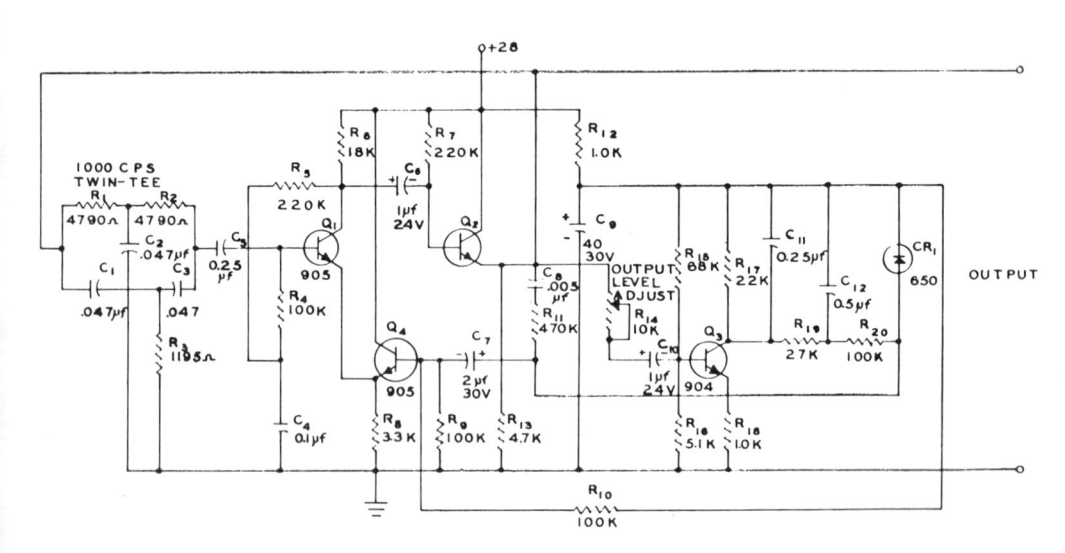

Circuit 5-12 - 1000 cps stabilized oscillator

Circuit 5-12, Twin Tee Oscillator (cont'd.)

the output voltage. This diode also serves as part of a voltage divider for the positive feedback loop. Since the diode is operated in a region where its impedance is inversely proportional to the bias current, the a.c. gain of this network is also inversely proportional to the output voltage. The output of the positive feedback network is fed into Q_4, which is emitter-coupled to Q_1 and hence to the forward gain path of the oscillator. The inputs of Q_1 and Q_4 are out of phase; therefore, positive feedback is developed between the output and Q_4, while the negative feedback is between the output and Q_1. The RC time constants of the amplifier and detector circuits have been designed to give good stability. Instability causes a low frequency motor-boating which modulates the output.

As previously mentioned, the twin tee determines the frequency stability and distortion characteristics of the oscillator. The twin tee network is noted for a theoretically infinite rejection ratio at the null frequency, and a rapid phase shift with frequency as the null is approached. The most usual applications of twin tees has been with vacuum tubes which implies a low impedance drive and infinite load impedance. However, it is possible to design the twin tee to be compatible with transistor impedances. The insertion loss of the network is increased, but the symmetrical characteristics about the null frequency are maintained. The method is described in Figure 5-12.1. R_1 is the generator resistance, and R_2 is the load resistance. F_o is the null frequency (11). $R_2 = 18.75\ R_1$; $R_3 = 7.5\ R_1$; $R_4 = 1.875\ R_1$; and $C = 30,000/R_1\ F_o$. Using this criteria, the parameters for the twin tee are as follows: $F_o = 1000$ cps; $R_1 = 169$ ohms; $R_2 = 28K$; $R_3 = 4790$ ohms; $R_4 = 1195$ ohms; and $C = .047$ mfd.

In practice, the twin tee resistances are deposited carbon low temperature coefficient units. The capacitors are accurate to within 0.5%, and are also temperature stabilized. The final trimming may be done by padding the resistances with high value 5% carbon resistors in parallel. Temperature tests on the twin tee revealed no appreciable change (one cycle out of 1,000) from 20 to 60°C.

Overall characteristics of the oscillator are as follows:

1. <u>Distortion</u>

   0.18% at 1 volt output

   0.16% at 2 volt output

   0.20% at 2.7 volt output

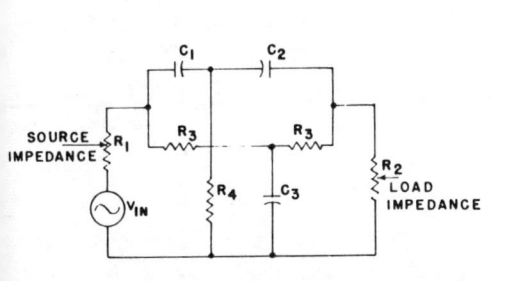

Fig. 5-12.1 - The twin-tee network

Circuit 5-12, Twin Tee Oscillator (cont'd.)

2. Regulation with respect to supply voltage change (supply voltage varied from 30 to 20 volts)

Output voltage at 28 volts	Output change for 10 volts variation
1.0 volt	-0.1 volt
2.0 volt	-0.25 volt
2.7 volt	-0.4 volt

   Maximum output voltage with AGC saturated is 4 volts.

3. Frequency shift

   a. With change in supply voltage (30 to 20 volts): -1 c.p.s.

   b. With change in transistors: 4 cycle spread when input transistor changed. (Beta variation from 55 to 85.) No change when other stages substituted.

4. Temperature Characteristics

Temperature Range	Output Voltage	Frequency Change
25 to 40°C	-0.1	-1 cycle
25 to 60°C	-0.1	-2 cycles

5. Output Loading

   The oscillator should not be loaded with less than 3,000 ohms, which corresponds to 1 milliwatt at 2 volts output, approximately.

CIRCUIT 5-13

VOLTAGE-CONTROLLED WIEN BRIDGE OSCILLATOR

Northeastern University   Contributed by J. A. Lapointe

This circuit is a modification of the standard Wien bridge oscillator. The principle difference is that silicon diodes are used as controllable resistance elements to make the oscillator frequency vary in proportion to a control voltage.

The control diodes have a forward conduction characteristic such that their dynamic resistance is inversely proportional to current over a wide range. It varies from about 1500 ohms at 30 microamperes to 160 ohms at 300 microamperes. By using a large resistance (R_1), between the diodes and the control source, diode resistance is made inversely proportional to control voltage. It also serves to isolate the oscillator from the control source.

For the type of bridge chosen (made up of the components with the dashed lines plus resistors R_6 and R_7), oscillation frequency is $1/2\pi RC$.

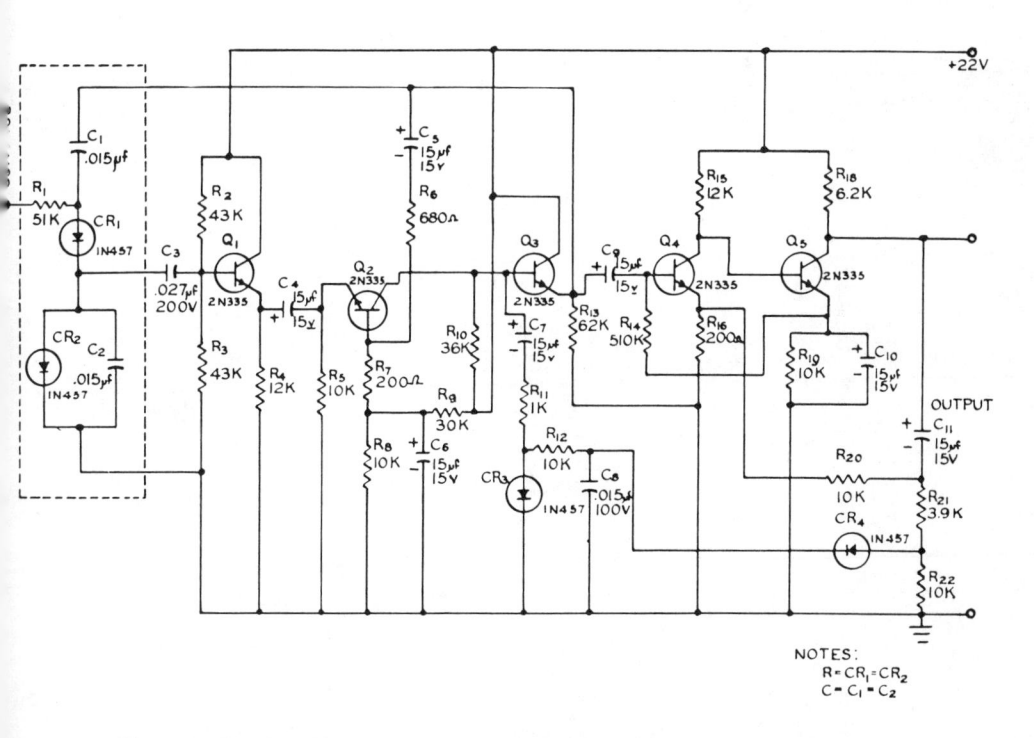

Circuit 5-13 - Voltage controlled Wien bridge oscillator

Circuit 5-13, Voltage-Controlled Wien Bridge Oscillator (cont'd.)

By using resistance elements that vary inversely proportional to a control voltage, the oscillator frequency is made to vary directly proportional to this voltage. A gain of 3 is required for oscillation and is provided by adjustment of feedback resistors R_6 and R_7.

The three stage oscillator amplifier need only provide high open loop gain with no phase shift. The two stages of output amplification provide a 1 volt rms signal at 1000 ohms. Automatic gain control to compensate for bridge dynamic variations is accomplished by controlling the resistance of the shunt diode CR_3.

With this circuit, 5% linearity of frequency control from 10 kc to 70 kc was effected with control voltage of 2-1/2 volts to 27 volts respectively. Frequency range is determined by bridge capacitors C_1 and C_2 and any required frequency in the audio or low RF range may be generated; consisten with amplifier requirements and diode capabilities.

CIRCUIT 5-14

SERIES-TUNED EMITTER, TUNED-BASE OSCILLATOR

General Electric Co. Contributed by W. F. Chow

Series resonant LC networks can be effectively utilized in frequency stable transistor oscillators. In this circuit configuration, the oscillator frequency is dependent almost entirely on the LC tank circuit elements and is only very slightly affected by transistor properties or oscillator supply voltages. The frequency of oscillation is largely determined by a tank circuit composed of the inductances L_1 and L_2 in series with capacitors C_1 and C_2. It is necessary that the reactance of C_2 be larger than the reactance of L_2. Since the LC components can be made very stable, a high degree of frequency stability can be achieved without the use of elaborate nonlinear compensation techniques. The stability with respect to supply voltage variations was found to be of the order of seven parts per million per 12 percent variation of voltage.

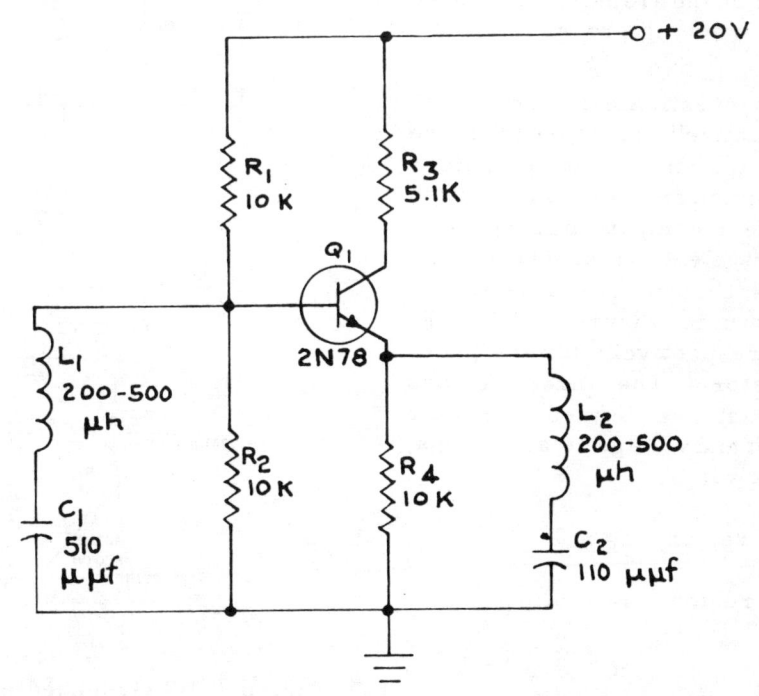

Circuit 5-14 - Series tuned emitter, tuned base oscillator

PART 6

SWITCHING CIRCUITS

Section 1. Design Philosophy

A. Transistor Characteristics

Switching circuits represent a class of circuits that are used extensively in digital computers, control applications, radar systems and communications. These switching circuits essentially require an "on" and "off" condition with some form of control of these conditions. The transistor is similar to a classical switch in that it may be biased in either a high impedance state, cut off, or in a low impedance state, "on," and consequently is ideally suited for applications in switching circuits. The characteristics exhibited by the transistor in the "on" and "off" states are important and will be described in detail, since they determine the efficiency with which switching is accomplished. The characteristics of the active region, that is, the region between cut off and the "on" state are equally important, since they determine the transient response of the switching process.

To investigate the "on," "off" and "active" regions in more detail, it is convenient to consider the common emitter configuration with the accompanying large signal equivalent circuit (1) and the voltage current characteristics, as shown in Figure 6-0.1A, B, and C respectively, for the p-n-p transistor. The three regions that exist are defined from the current and voltage relationships, as follows:

cut off region $\quad i_E < 0$

active region $\quad i_E > 0, \ v_C < v_B,$

$$i_C = \alpha_{FB} \ i_E + I_{CO}$$

on region
(saturation) $\quad i_E > 0, \ v_C \gtreqless v_B.$

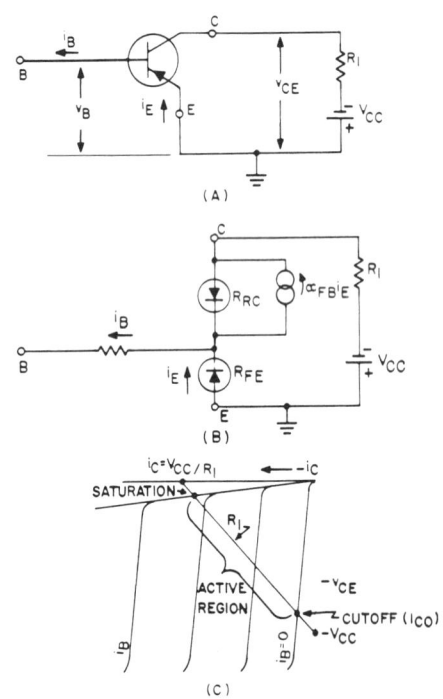

Fig. 6-0.1 - Grounded emitter configuration, large signal equivalent circuit and characteristics

192

At cut off, the base and emitter terminals are reverse-biased and no base or emitter current flows. Under these conditions, the collector current is essentially zero, except for the small leakage current I_{CO}. This leakage current represents one of the primary circuit design limitations and will be discussed in more detail later. The voltage from collector to emitter is almost equal to the supply voltage V_{CC} at cut off.

When the base-emitter potential is such that a base current exists, the transistor operates in the active region and the general current and gain relations may be described as follows.

$$\alpha_{FB} = -\frac{\Delta i_C}{\Delta i_E}\bigg|_{v_C} = \text{constant} \qquad 6\text{-}0.1$$

$$\alpha_{FE} = \frac{\Delta i_C}{\Delta i_B}\bigg|_{v_C} = \text{constant} \qquad 6\text{-}0.2$$

$$i_C = \alpha_{FB} i_E + I_{CO} \qquad 6\text{-}0.3$$

$$i_B = (i_E - i_C) \qquad 6\text{-}0.4$$

$$i_B = i_E (1 - \alpha_{FB}) - I_{CO} \qquad 6\text{-}0.5$$

$$i_C = \left(\frac{\alpha_{FB}}{1 - \alpha_{FB}}\right) i_B + \frac{I_{CO}}{1 - \alpha_{FB}} \qquad 6\text{-}0.6$$

$$i_C = \alpha_{FE} i_B + \frac{I_{CO}}{1 - \alpha_{FB}} \qquad 6\text{-}0.7$$

Operation in the active region can be continuous but generally exists only during the transient period between the "off" and "on" state. When the collector voltage becomes equal to the internal base voltage, the saturated region is reached and the collector to emitter voltage is very low as indicated by the characteristic curves. The collector equivalent diode is assumed forward biased and the current generator is short circuited during saturation.

From the voltage current characteristic it should be clear that only the small I_{CO} current exists in the load resistor R_1 at cut off and a maximum current exists at saturation. The control of the load current is effectively accomplished by the smaller base current and represents a gain factor. The transistor acts as a high impedance during cut off and a low impedance during saturation as evidenced by the voltage drop across the transistor in each state.

It can be seen from the previous discussion that the equivalent diode circuit is an accurate representation of the transistor characteristics and can be used with confidence in the design of switching circuits.

The dependence of the leakage currents on temperature is perhaps the most important limitation of transistors, and must be considered in every design. The leakage current I_{CO} may be expressed as (2)

$$I_{CO} = I_0 \, e^{[\alpha(T+HP_C-T_0)]} \qquad 6\text{-}0.8$$

where I_0 is the leakage current at T_0, T_0 is the reference temperature.

$H = \dfrac{\Delta T_J}{MW}$ the junction temperature rise per milliwatt dissipation, $6\text{-}0.9$

T is the ambient temperature,

α is a constant equal to 0.06 - 0.08 for germanium transistors.

The rate of change of I_{CO} with temperature may be related to the circuit parameters as (3)

$$\frac{dI_{CO}}{dT} = f(I_0, V, K_1, K_2) \qquad 6\text{-}0.10$$

where
V = bias voltages V_{CC}, V_{BB}, V_{EE},

$K_1 = \dfrac{\text{Emitter resistance}}{\text{Base resistance}}$

$K_2 = \dfrac{\text{Collector resistance}}{\text{Emitter resistance}}$.

This relation indicates that circuit designs may be made thermally stable by choosing a proper relation between the circuit constants.

The response times generally of interest in switching circuits are rise time, t_r, storage time, t_s, and fall time, t_f. The storage time exists only for saturated operation and is proportional to the degree of saturation and the length of time the unit is saturated. The three basic connections used in switching circuits are common base, common emitter, and common collector and they exhibit different response times (4). The response times associated with the three connections, as shown in Figure 6-0.2, are common base configuration:

$$t_r = \frac{1}{\omega_{afb}} \ln \frac{i_E}{i_E - \dfrac{0.9\, i_C}{\alpha_{FB}}} \qquad \text{6-0.11}$$

$$t_s = \frac{\omega_{afb} + \omega_{arb}}{\omega_{afb}\, \omega_{arb}\, (1 - \alpha_{FB}\, \alpha_{RB})} \ln \left[\frac{i_{E_1} - i_{E_2}}{\dfrac{i_{C_1}}{\alpha_{FB}} - i_{E_2}} \right] \qquad \text{6-0.12}$$

$$t_f = \frac{1}{\omega_{afb}} \ln \left[\frac{i_{C_1} + \alpha_{FB}\, i_{E_2}}{0.1\, i_{C_1} + \alpha_{FB}\, i_{E_2}} \right] \qquad \text{6-0.13}$$

common emitter configuration:

$$t_r = \frac{1}{(1 - \alpha_{FB})\, \omega_{afb}} \ln \left[\frac{i_B}{i_B - \dfrac{0.9\,(1 - \alpha_{FB})\, i_C}{\alpha_{FB}}} \right] \qquad \text{6-0.14}$$

$$t_s = \frac{\omega_{afb} + \omega_{arb}}{\omega_{afb}\, \omega_{arb}\, (1 - \alpha_{FB}\, \alpha_{RB})} \ln \left[\frac{i_{B_1} - i_{B_2}}{i_{C_1}\, \dfrac{(1 - \alpha_{FB})}{\alpha_{FB}} - i_{B_2}} \right] \qquad \text{6-0.15}$$

$$t_f = \frac{1}{(1 - \alpha_{FB})\, \omega_{afb}} \ln \left[\frac{i_{C_1} - \dfrac{\alpha_{FB}\, i_{B_2}}{(1 - \alpha_{FB})}}{0.1\, i_{C_1} - \dfrac{\alpha_{FB}\, i_{B_2}}{(1 - \alpha_{FB})}} \right] \qquad \text{6-0.16}$$

common collector configuration:

$$t_r = \frac{1}{(1 - \alpha_{FB})\, \omega_{afb}} \ln \left[\frac{\alpha_{FB}\, i_B}{i_B - 0.9\,(1 - \alpha_{FB})\, i_E} \right] \qquad \text{6-0.17}$$

$$t_s = \frac{\omega_{afb} + \omega_{arb}}{\omega_{afb}\, \omega_{arb}\, (1 - \alpha_{FB}\, \alpha_{RB})} \ln \left[\frac{i_{B_1} - i_{B_2}}{i_{B_2} + (1 - \alpha_{FB})\, i_{E_1}} \right] \qquad \text{6-0.18}$$

$$t_f = \frac{1}{\omega_{afb}(1-\alpha_{FB})} \ln\left[\frac{i_{E_1} - \frac{i_{B_2}}{(1-\alpha_{FB})}}{0.1\, i_{E_1} - \frac{i_{B_2}}{(1-\alpha_{FB})}}\right] \qquad 6\text{-}0.19$$

where

α_{FB} and α_{RB} are the normal and reverse short-circuited current transfer ratio respectively,

ω_{afb} and ω_{arb} are the normal and reverse common base cutoff frequencies respectively. The dimensions are in $\dfrac{\text{radians}}{\text{sec}}$

The current subscripts 1 and 2 refer to initial and final values of current at the time the turn off voltage is applied.

The times t_r and t_f are defined as the time required to reach 90% of its final value.

The equations for rise, storage and fall times are useful in understanding in a general way the effect of the circuit configuration and current amplitudes on the response characteristics. For example, in the common emitter configuration of Figure 6-0.2 the transient response times may be calculated in the following manner. First, the parameters ω_{afb} and ω_{arb} must be measured and are taken for this example as 10^7 and 10^6 radians per second respectively. The parameters α_{FB} and α_{RB} are taken as 0.99 and 0.70 respectively. The maximum collector current possible exists just at saturation and is calculated as V_{CC}/R_2. If the maximum collector current is 15 ma and the base current is taken as 2 ma, the rise time is calculated from Equation 6-0.14, as

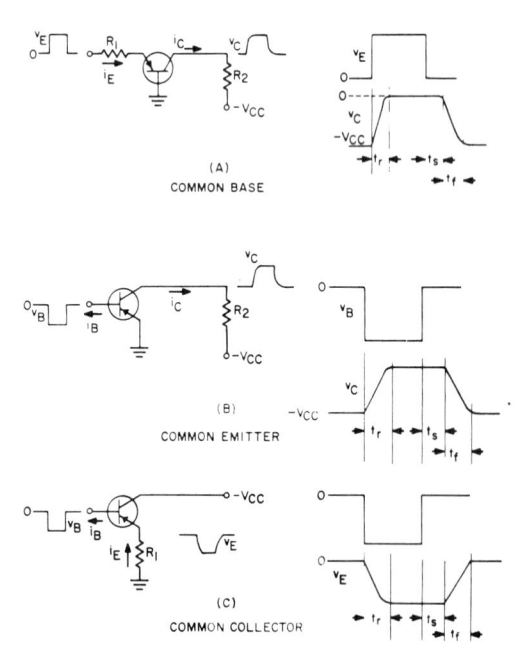

Fig. 6-0.2 - Circuit response times

$$t_r = \frac{1}{(1-0.99)\,10^7} \ln\left[\frac{2}{2 - \frac{0.9(1-0.99)\,15}{0.99}}\right] \qquad \text{6-0.14a}$$

$$t_r = .70\ \mu s. \qquad \text{6-0.14b}$$

If a reverse base to emitter bias voltage is included in the circuit such that a reverse base current i_{B_2} of 0.8 ma exists, the storage time from Equation 6-0.15 is

$$t_s = \frac{10^7 + 10^6}{10^7 \times 10^6 (1 - .693)} \ln\left[\frac{2 - (-0.8)}{\frac{15(1-0.99)}{0.99} - (-0.8)}\right] \qquad \text{6-0.15a}$$

$$t_s = 3.9\ \mu s. \qquad \text{6-0.15b}$$

The fall time from Equation 6-0.16 is

$$t_f = \frac{1}{(1-0.99)10^7} \ln\left[\frac{15 - \frac{0.99(-0.8)}{1-0.99}}{0.1 \times 15 - \frac{0.99(-0.8)}{1-0.99}}\right] \qquad \text{6-0.16a}$$

$$t_f = 1.5\ \mu s. \qquad \text{6-0.16b}$$

If the base current i_{B_1} is increased in the example, there will be a decrease in rise time but an increase in storage and turn off time. By increasing the reverse base current i_{B_2}, the storage and turn off time may be reduced, improving the overall circuit response time. It is instructive to make similar calculations of the response times of the other two basic circuit configurations, common base, and common collector so that the relative response times of each configuration will be understood. In general, it may be stated that the response time of each circuit configuration is proportional to the ratio of input current difference to output current difference.

Calculations of transient times for a particular circuit provide approximate values and should be used only as a guide to circuit design procedures. The collector capacitance has been neglected, and will affect the response times unless the product of load resistance and collector capacitance is much less than the basic circuit response time. In addition, the storage equation assumes that equilibrium conditions exist relative to the carrier density in the base region.

B. Circuit Design Methods

Since transistors have been in the development phase during the last several years, various circuit techniques have been originated to lessen the effect of undesirable characteristics. Some of these techniques are worthwhile and will be illustrated. In addition, transistors have some characteristics not found in vacuum tubes that allow new circuit variations not previously possible. For these two reasons transistor switching circuitry has been in a fluid state.

The design procedure discussed in this section will include designs to overcome transistor variations or deficiencies, even though many present transistors are free of these deficiencies. In addition, circuits that were developed utilizing unique transistor characteristics will also be included, since they are currently in use, although they eventually may disappear in favor of more standard methods.

C. Basic Flip-Flop Design Procedure

Historically, the early transistor switching circuits were simply direct equivalents of the vacuum tube circuits. These basic circuits were subsequently modified to avoid the unique problems associated with the transistors.

1. Saturated Flip-Flop

The unclamped saturated flip-flop was one of the earliest circuits of this sort. Figure 6-0.3 is an example of an Eccles-Jordan flip-flop that may give saturated operation.

The saturated flip-flop has the following advantages over the nonsaturated type to be described in the next section: 1) the output voltage levels are practically independent of the transistors, 2) the circuit is simple, and 3) the design is not complicated.

The one disadvantage of the saturated flip-flop, in comparison to the nonsaturated type, is that the turn off transient is delayed

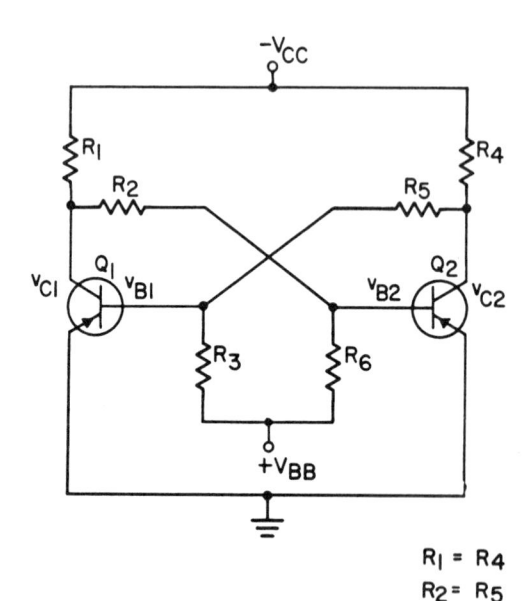

Fig. 6-0.3 - Saturated flip-flop

$R_1 = R_4$
$R_2 = R_5$
$R_3 = R_6$

by the storage time. For a given transistor type, nonsaturated flip-flops may be operated at higher pulse rates than the saturated flip-flop.

For low frequency operation, the use of the saturated flip-flop is indicated. At higher frequencies one frequently finds it desirable to make a critical comparison of the two types of flip-flops. A choice of the type to be used depends upon economic, technical and other considerations for the particular application.

As transistors have been improved with respect to storage time, the operating frequency range, where nonsaturated flip-flops have held unquestioned advantage over the saturated types, has been pushed upwards to such an extent that placing a high frequency limit on the conditions of the decision referred to above would be imprudent.

The following relationships are useful in the d-c design of the saturated flip-flop of Figure 6-0.3. In deriving these expressions, it is assumed that the transistor internal base and emitter resistances (when the transistor is saturated) are very much smaller than the external resistances, that I_{CO} is negligible, the collector resistance in the cut off condition is very much greater than the external resistances, and the "on" collector voltages are very much smaller than the bias voltages. The circuit is assumed to be symmetrical; i.e., $R_1 = R_4$, $R_2 = R_5$, $R_3 = R_6$.

$$\frac{V_{CC}}{R_1 + R_2} - \frac{V_{BB}}{R_3} > \frac{V_{CC}}{\alpha_{FE} R_1} \qquad \text{6-0.20a}$$

Equation 6-0.20a insures that the "on" transistor is saturated.

The output voltage swing is given by

$$\Delta v_o = \frac{V_{CC} R_2}{R_1 + R_2} \qquad \text{6-0.20b}$$

The reverse bias on the collector-base junction of the "off" transistor is given by

$$V_{BE} = \frac{V_{BB} R_2}{R_2 + R_3} \qquad \text{6-0.20c}$$

Increasing V_{BE} reduces the probability of noise triggering the circuit and increases the amount of I_{CO} that can be tolerated, assuming that the resistance in the base is held constant or decreased, but also increases the amount of trigger required to make the flip-flop change state.

The base bias source, V_{BB}, may be placed in the emitter circuit, with the same results as before, provided the proper polarity is observed to reverse bias the "off" transistors, V_{CC} is increased by the magnitude of V_{BB} and R_3 and R_6 are returned to ground. Single bias source operation may be obtained if the source in the emitter circuit is replaced with a breakdown diode. Single source operation may also be obtained with a common emitter resistor as described in Circuit 6-1.

2. Nonsaturating Flip-Flop

The circuit of Figure 6-0.4 is a nonsaturating flip-flop. The emitter resistor alone will not guarantee nonsaturation, but must be used if nonsaturation is to be achieved. The design procedure may be one of trial and error, but there are several methods that guarantee a solution that is valid for ranges of transistor parameters (5).

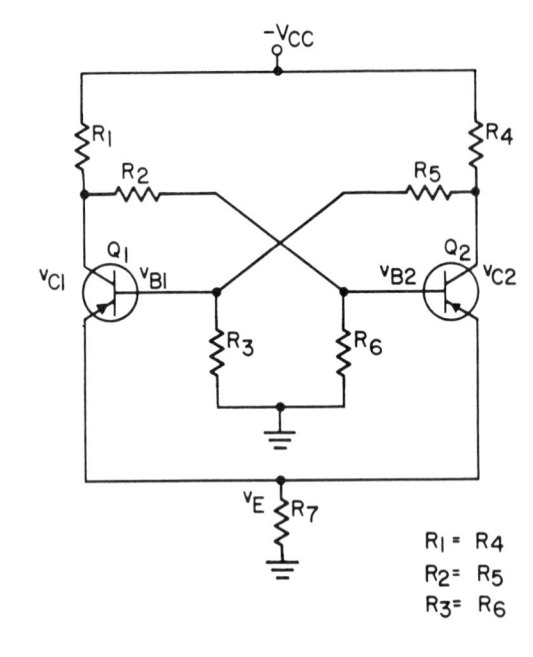

Fig. 6-0.4 - Non-saturated flip-flop

The expression that provides stability for the nonsaturated transistor flip-flop, as well as the saturated flip-flop, is given in Equation 6-0.21.

$$\frac{R_2}{R_1} < \frac{2\alpha_{FB} - 1}{1 - \alpha_{FB}}. \qquad 6\text{-}0.21$$

This may be derived in a straightforward way by assuming one flip-flop transistor is "off" while the other is "on." The condition necessary for this to be true is that

$$\left. \begin{array}{c} v_{B_2} < v_E \\ v_{B_1} > v_E \end{array} \right\} \quad Q_1 \text{ "off"}, \; Q_2 \text{ "on"}. \qquad 6\text{-}0.22$$

By deriving the expressions for v_{B_1} and v_{B_2}, the stability condition (Equation 6-0.21) may be determined.

To avoid saturation, it is necessary to restrict the base voltage for the "on" transistor to values above the collector voltage.

$$v_{B_2} > v_{C_2}. \qquad 6\text{-}0.23$$

This condition must be guaranteed for the range of transistor parameters expected. If the emitter to base voltage of the transistor is neglected, Equation 6-0.23 may be written

$$v_E > v_{C_2}. \qquad 6\text{-}0.24$$

It is convenient to consider that saturation occurs as the emitter current increases above a critical value. This critical value of current is derived by finding the emitter current that exists just at saturation. The normal emitter current is limited to values below this critical or maximum current for any range of transistor parameters, by choosing the circuit resistances properly. By deriving the expression for v_E and v_{C_2} of Equation 6-0.24, the maximum emitter current permissible before saturation occurs is obtained.

$$i_{E\ max} < \frac{\dfrac{V_{CC}(R_6+R_5)}{R_6+R_5+R_4}}{R_7 + \dfrac{\alpha_{FB} R_4 (R_6+R_5)}{R_6+R_5+R_4}}. \qquad 6\text{-}0.25$$

The normal operating emitter current is found as

$$i_{E\ op} = \frac{\dfrac{V_{CC} R_3}{R_3 \quad R_2 \quad R_1}}{R_7 + \dfrac{R_3(1-\alpha_{FB})(R_2+R_1)}{R_3+R_2+R_1}}. \qquad 6\text{-}0.26$$

The normal operating emitter current and the maximum current permissible to avoid saturation vary with alpha as indicated in Figure 6-0.5.

It is necessary to keep $i_{E\ op}$ less than $i_{E\ max}$ to avoid saturation, and hence the design is arranged so that the crossover point (α_{FBO}) is chosen for $\alpha_{FB} = \alpha_{FBO} = 1$.

The output voltage swing is usually of interest, since it must be consistent with the logic design levels. The collector voltage of the "on" transistor may not be assumed close to ground as in the saturated flip-flop case, but is described by Equation 6-0.27.

$$v_{C_2} = \frac{i_C R_4 (R_5 + R_6)}{R_6 + R_5 + R_4} - \frac{V_{CC}(R_5 + R_6)}{R_6 + R_5 + R_4}. \qquad 6\text{-}0.27$$

The collector voltage of the "off" transistor is

$$v_{C_1} = \frac{i_B R_3 R_1}{R_3 + R_2 + R_1} - \frac{V_{CC}(R_2 + R_3)}{R_3 + R_2 + R_1}. \qquad 6\text{-}0.28$$

The output voltage swing then from "off" to "on" is

$$\Delta v_o = \frac{\alpha_{FB}\, i_E\, R_4 (R_5 + R_6)}{R_6 + R_5 + R_4} - \frac{\alpha_{FE}\, R_3 R_1}{R_3 + R_2 + R_1} \qquad 6\text{-}0.29$$

The preceding considerations may be combined to provide a unique design procedure. The circuit values are listed below.

$$R_4 = R_1 = \frac{V_{CC}}{i_E} B \qquad 6\text{-}0.30$$

where B is considered a design factor and equals 0.8 - 0.9.

$$R_6 = R_3 = 4 \left[\frac{V_{CC}}{i_E} - R_1 \right] \qquad 6\text{-}0.31$$

$$R_5 = R_2 < \left(\frac{2\alpha_{FB} - 1}{1 - \alpha_{FB}} \right) R_1 \qquad 6\text{-}0.32$$

$$R_7 = \frac{V_{CC} R_3}{i_E (R_3 + R_2 + R_1)}. \qquad 6\text{-}0.33$$

The factor B is a function of α_{FB}, and the ratio of output voltage to supply voltage. It is described completely in reference 5, but is somewhat complex and is not convenient for quick design methods. B should be chosen less than unity and considered as a design variable. Figure 6-0.6 shows the relation of B and the minimum α_{FB} that the design will accommodate and the ratio of supply voltage to

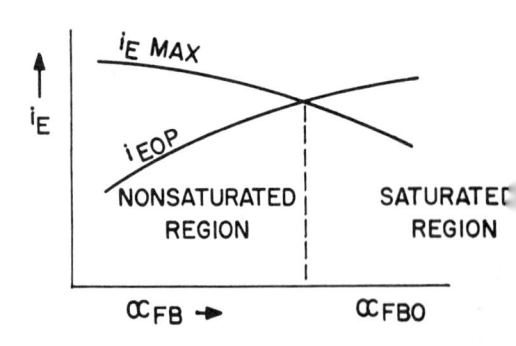

Fig. 6-0.5 - Current variation vs α_{FB}

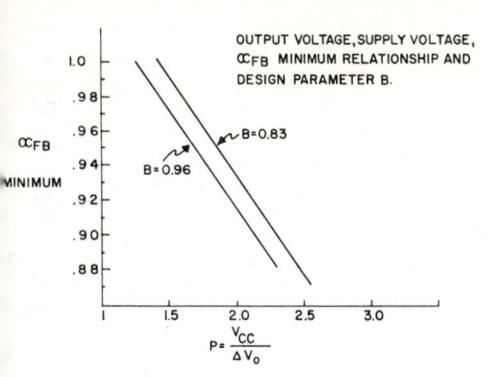

Fig. 6-0.6 - Non-saturated flip-flop

output voltage (ρ). From Figure 6-0.6, the value of B chosen does not affect appreciably $\alpha_{FB\,min}$ or ρ; however, it does directly affect the value of R_3. If B is taken as unity, $R_3 = 0$ and the design is not valid. I_{CO} considerations depend to a large degree on the base resistance value, and B may be used to decrease I_{CO} effects. If B is chosen too low, the supply voltage required to give a particular output voltage swing becomes larger and represents a decrease in efficiency. Figure 6-0.6 also indicates that a design that allows for low α_{FB} values requires a larger supply voltage for a specific output voltage.

The minimum α_{FB} value is chosen based on the spread expected and the end of life considerations.

After a value of B is chosen, the ratio (ρ) of supply voltage and output voltage is read from Figure 6-0.6. The output voltage desired is chosen and the supply voltage is then calculated from

$$\rho = \frac{V_{CC}}{\Delta v_o} \qquad 6\text{-}0.34$$

The collector load resistor (R_1) is calculated from Equation 6-0.30, using the supply voltage required after a suitable emitter current is chosen. The emitter current is chosen from loading considerations. The design automatically establishes the emitter current chosen as the maximum emitter current possible, and the maximum, of course, occurs only when α_{FB} approaches unity. R_3 may then be calculated from Equation 6-0.31. The stability equation, as given in Equation 6-0.32, is not described in terms of an α_{FB} minimum, since it is wise to include a safety factor. In addition, the stability equation affects the triggering requirements directly and should be determined with this in mind. A tentative degree of inequality may be chosen and R_2 calculated from Equation 6-0.32. This can be changed later to allow for more or less trigger sensitivity without upsetting the overall stability. The emitter resistance R_7 is calculated finally from Equation 6-0.33.

This design procedure is useful for many applications, since it allows the design to reflect temperature, triggering or efficiency considerations.

3. Clamping Methods

A further technique to reduce the effect of transistor variation and provide nonsaturation is the clamped circuit. The clamped flip-flop assures uniform levels in spite of transistor variations and, if designed properly, gives better system performance. Clamping increases circuit complexity, however, and requires additional supply voltages. Despite the fact that the levels are specified by the clamping voltages, a unique design procedure for the complete circuit is difficult, and a trial and error approach must be used.

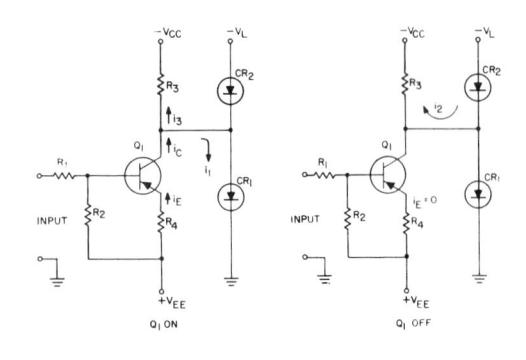

Fig. 6-0.7 - Clamped circuit

Figure 6-0.7 shows a simple diode clamping circuit that fixes both output levels.

The direct clamping method in Figure 6-0.7 fixes the collector levels to ground and $-V_L$ where $-V_L$ is greater (less negative) than the supply voltage $-V_{CC}$. For the case of the transistor in the "on" condition, the collector current must be large enough to supply current through R_3 and an additional current for the diode CR_1. The clamping current is chosen by considering expected loading and the collector current is then calculated from Equation 6-0.36.

$$i_C = i_3 + i_1 \qquad 6\text{-}0.35$$

$$i_C = \frac{V_{CC}}{R_3} + i_1 \qquad 6\text{-}0.36$$

where i_3 is the current through R_3, and i_1 is the current through CR_1.

In the case of the flip-flop, the cross coupling resistor adds another current path that reduces the clamping current and should be included in the design.

The clamping current through CR_2 for the "off" case is simply calculated as

$$i_2 = \frac{V_{CC} - V_L}{R_3} . \qquad 6\text{-}0.37$$

A design procedure for clamped switching circuits, and particularly flip-flops and multivibrators, is conveniently started by deciding on the clamped levels, choosing the proper clamping currents, and calculating the collector current and load resistor R_3 from the above equations.

The clamped circuit does not eliminate completely design problems and, in fact, only assures accurate levels. The response time for the circuit appears improved when clamped, but care must be taken not to load the circuit before the clamping current has reached its final value. Figure 6-0.8 shows the voltage and current relations.

The clamping currents have a delay time which is the delay of a standard unclamped circuit.

There are several variations of clamped circuitry to avoid the additional supply voltages and affect a change in base current to compensate for loading changes. The back clamped circuit of Figure 6-0.9 illustrates this method (3).

The silicon diode (CR_1) provides about a 0.7 volt drop so that point "a" is at -0.7 volt. The germanium diode (CR_2) then will conduct whenever the collector voltage reaches about -0.5 volt, assuming that CR_2 has a voltage knee of about 0.2 volt. The collector to base voltage then can

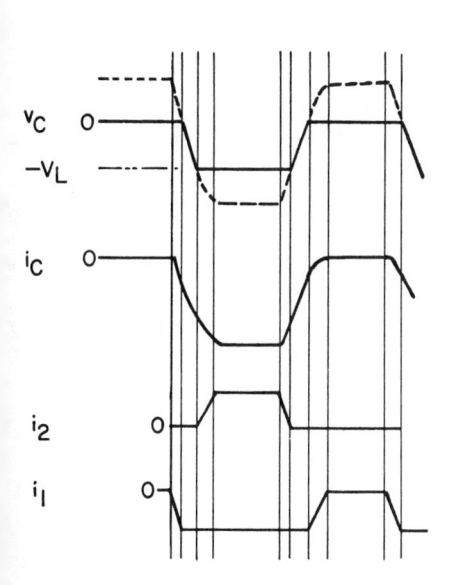

WHERE i_1 IS CURRENT THROUGH CR_1 IN CIRCUIT IN FIG. 6-0.7, i_2 IS THE CURRENT THROUGH CR_2

Fig. 6-0.8 - Clamped current and voltage waveshapes

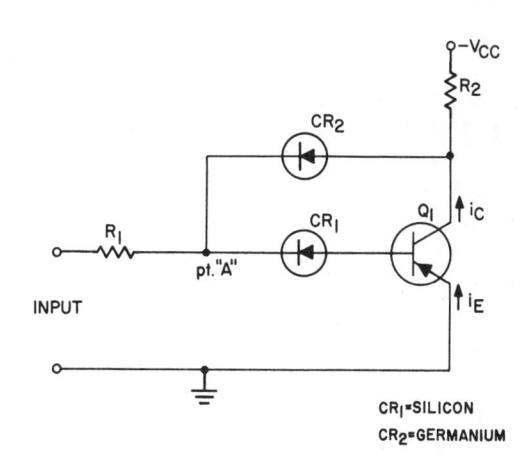

Fig. 6-0.9 - Back clamped circuit

never be less than 0.5 volt and saturation is thus avoided. In addition, whenever the collector loading is small, the base current to the transistor is reduced due to the conduction of the germanium diode. This is a more direct approach to regulate the collector current to the load requirements, and results in improved efficiency. A resistor may be used in place of the silicon diode, and is chosen based on α_{FB} variations and current levels. However, the silicon diode requirements are not stringent and, in fact, the diode is never back biased. The germanium diode must withstand the collector voltage and must have a low storage time, if speed is desired. If saturated circuitry is desired, the silicon diodes, or their equivalent, resistors, may be removed.

D. Symmetrical Circuits

The fact that, unlike vacuum tube circuitry, both p-n-p and n-p-n transistors are available, had led to the development of complementary or symmetrical circuits. Since the manufacturing technology of the p-n-p type has until recently been ahead of the n-p-n technology, this type of circuitry has been employed erratically. Its primary advantage is equal rise and fall times as well as improved efficiency. The emitter follower circuit of Figure 6-0.10 illustrates one of the first circuit uses of complementary designs.

The p-n-p, used alone as an emitter follower, will give characteristically a faster fall time for the type of input shown. The improvement of rise time due to the n-p-n may be analyzed by assuming the emitter rise lags behind the input rise. In this case, the base of the n-p-n becomes more positive than the emitter and the n-p-n must conduct, improving the rise time. The basic circuit configuration described in Part 7 includes the pertinent design considerations for emitter followers and will not be included here. The practical difficulties regarding complementary emitter followers are attributed to a mismatch of frequency response for the p-n-p and n-p-n transistor that allows instability, and the unlimited current flow that exists when either unit fails. In addition, an oscillatory condition occurs for low resistive and especially capacitive loads. The design procedure, when loads of this type are expected, is to parallel emitter follower sections and add sufficient series emitter resistances to prevent oscillations or burnout due to accidental shorting.

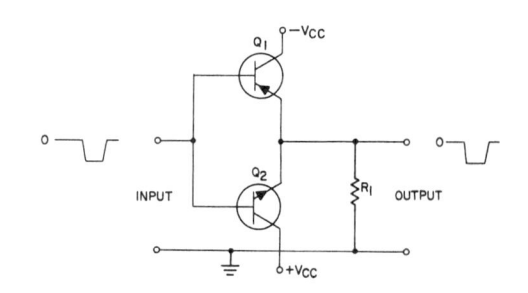

Fig. 6-0.10 - Complementary symmetrical emitter follower

The complementary circuits may be applied to flip-flops and other basic switching circuits. The circuit of Figure 6-0.11 illustrates this design (3). The output collector current is produced by an active source for both directions. The clamping circuit previously described provides collector current during the time the transistor is off by use of an added supply voltage (V_L), and the load resistor is determined by the "off" clamping current required. The clamping current for the "on" case is supplied to the clamping diode after current is supplied to this load resistor, and results in wasted current. In the complementary circuit, no current is wasted in a standby load resistor.

C. Triggering Methods

The trigger circuitry has not been included in the basic designs presented, since it will be discussed as a separate topic. Unfortunately, triggering is usually considered after the basic design is complete, and often becomes the most unreliable section of the circuitry. The situation is aggravated further, since a clock source for a system may be required to trigger a large number of circuits and must have power capabilities. The basic circuits are designed frequently for switching speeds just short of the frequency limitations of the transistors. To obtain maximum speed capabilities, semiconductor devices are necessarily reduced in power capabilities and so the designer is left without a suitable transistor that has the power to be used as the clock at the basic speeds of the system.

It is generally agreed that the trigger circuits should be arranged to affect a turn off rather than a turn on condition. This approach gives improved circuit response and minimum delays. The trigger requirements are best described in terms of base charge that exists and the source charge necessary to oppose it. The base charge is described by Equation 6-0.38.

$$Q_B = \frac{K i_C}{\omega_{afb}} \qquad 6-0.38$$

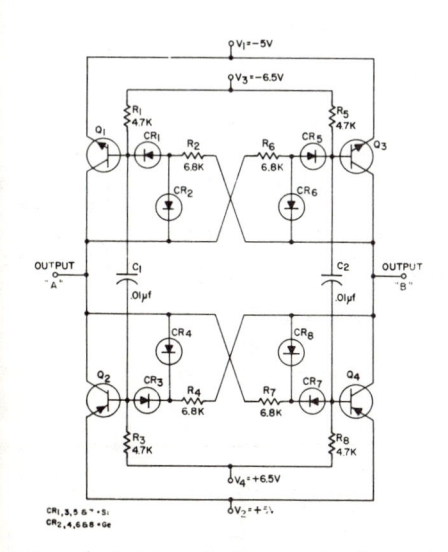

Fig. 6-0.11 - Symmetrical back clamped flip-flop

where i_C is the collector current, ω_{afb} is 2π times the frequency response of the transistor, and K is a proportionality constant. The capacitor used to couple the input trigger should be large enough to reverse the base charge, but small enough to prevent recovery problems at the maximum speeds desired. For the case of a saturated transistor, the stored charge must also be considered and, of course, requires an increase of trigger charge. The turn on time is also important in any design, and is directly proportional to the base current supplied and the speed with which the current is delivered. Internal base resistance and base to emitter capacitance affect the useful charge that can be delivered to the base and these parameters vary considerably for different transistor types. It is desirable to experimentally check the trigger design over the complete range of repetition rates desired.

The trigger requirements for any circuit are related to the trigger pulse width (6) as shown in Figure 6-0.12. A good circuit design will make efficient use of the trigger pulse available. This is accomplished in several ways and Figure 6-0.13 illustrates one method. The base network is chosen to give an overshoot that increases the turn off drive. A considerable improvement in turn off time is achieved using this technique.

When the transistor response time is slow, relative to the trigger rise time, a considerable amount of the trigger pulse is wasted. The use of the network shown in Figure 6-0.14 provides a better utilization of the pulse energy and effectively matches the pulse shape to the transistor response.

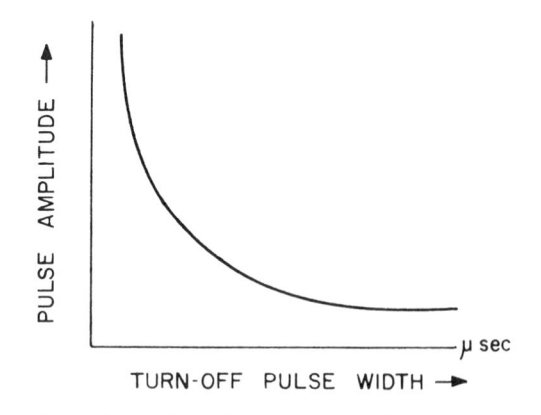

Fig. 6-0.12 - Trigger amplitude and width relation

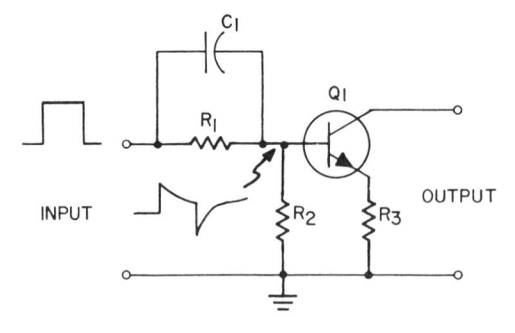

Fig. 6-0.13 - Base network for improved triggering

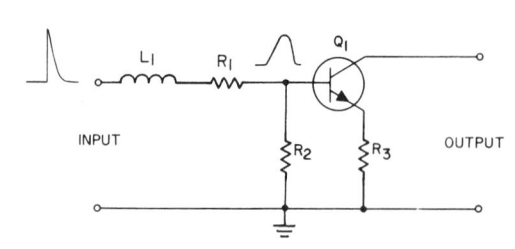

Fig. 6-0.14 - Base network to efficiently utilize trigger energy

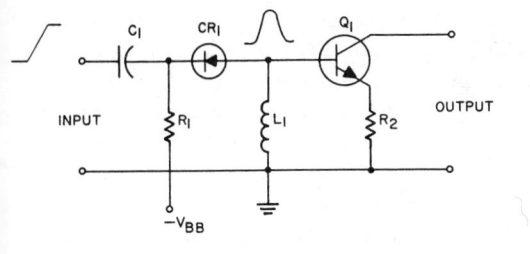

Fig. 6-0.15 - Base network to improve trigger pulse shape

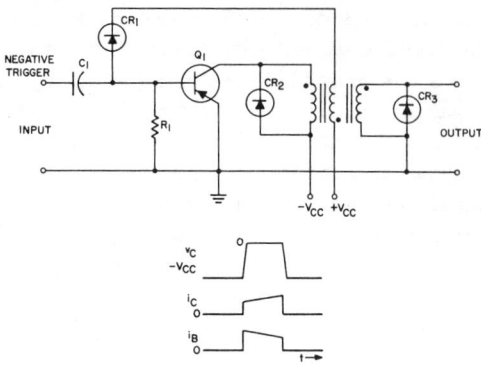

Fig. 6-0.16 - Basic blocking oscillator circuit

Figure 6-0.15 is another method to improve the trigger circuit (7). The diode is normally conducting while the transistor is cut off. The input trigger pulse at the capacitor opens the diode so that the current in the inductance (L_1) is diverted to the transistor. The trigger may be a slow or poorly shaped pulse and is only required to open the diode.

Base triggering is normally employed, because it gives wider margins and greater sensitivity, although emitter and collector triggering are possible and occasionally are used. Base triggering is used generally for noncommutative operation, and frequently a triggering transistor common to each side of the flip-flop is added for more positive triggering. Commutative-type flip-flops frequently employ collector triggering by using diode steering networks and characteristically allow considerable trigger amplitude variation. Commutative flip-flops and the diode steering are described in Part 7.

F. Blocking Oscillator Circuit Design Procedure

The clock source limitations described previously still exist, although the negative resistance devices (p-n-p-n) recently developed may reduce the problem. At present, the blocking oscillator circuit is the circuit used most often as an initial clock generator. There are many excellent papers describing the analytical features of the blocking oscillator circuit (8, 9). The circuit of Figure 6-0.16 is commonly used as the basic blocking oscillator configuration.

The input pulse initiates base current, and hence the collector current begins to build up. The collector voltage rises toward ground, and this rise is coupled to the base by the transformer adding to the base current.

The transistor quickly saturates, keeping the collector voltage fixed, and the base current begins to decrease. The collector voltage begins to drop when the transformer magnetizing inductance accepts enough current to unsaturate the transistor. The voltage fall is coupled by the transformer to the base and aids the turn off process. The back voltage associated with an inductive load, such as the transformer, is usually clamped by a small resistance or diode across the transformer. The back voltage, if not removed, would exceed the voltage breakdown of the transistor, causing an increase in power dissipation and probably burn out.

The collector and base current may be calculated as

$$i_C = \frac{V_{CC}}{r_s}\left[1 - \left(1 - \frac{r_s}{N^2 r_b}\right) e^{\frac{-r_s}{L}t}\right] \qquad 6\text{-}0.39$$

$$i_B = \frac{-V_{CC}}{r_b r_s} r_2 \left[1 - \left(1 + \frac{r_s}{N r_2}\right) e^{\frac{-r_s}{L}t}\right] \qquad 6\text{-}0.40$$

where

r_1 represents the d-c transformer primary resistances and the collector resistance during saturation,

r_2 represents the sum of the saturated emitter resistance and any external emitter resistance,

L is the magnetizing inductance,

r_b is the internal base resistance,

r_s is the sum of r_1 and r_2, and

T_d is the duration of the pulse.

The load resistance is assumed large compared to r_1.

Since the pulse duration continues until saturation no longer exists, (the collector current equals $i_B \alpha_{FE}$ at this point) we may solve for T_d by equating these two currents.

$$T_d = \frac{L}{r_s} \ln \left[\frac{\left(1 - \frac{r_s}{N^2 r_b}\right) + \alpha_{FE} \left(1 + \frac{r_s}{Nr_2}\right) \frac{r_2}{r_b}}{1 + \alpha_{FE} \frac{r_2}{r_b}} \right] \quad 6\text{-}0.41$$

If $\alpha_{FE} \, r_2 \gg r_b$

$$T_d = \frac{L}{r_s} \ln \left(1 + \frac{r_s}{Nr_2}\right). \quad 6\text{-}0.42$$

The pulse duration then, if the assumptions of Equation 6-0.42 are valid, is independent of the transistor gain and is related only to the saturation and circuit resistance, inductance and turns ratio.

The pulse width is conveniently controlled by the turns ratio, and a large value of N, stepdown between collector and base, decreases the pulse width, while a small value of N increases the width. The collector current however, increases as the turns ratio is decreased and should be calculated to avoid excessive power dissipation. The maximum collector current reached is given by Equation 6-0.43.

$$i_C = \frac{V_{CC} \left(1 + \frac{r_s}{Nr_b}\right) \alpha_{FE} \frac{r_2}{r_b}}{Nr_2 \left[1 - \frac{r_s}{N^2 r_b} + \alpha_{FE} \frac{r_2}{r_b}\left(1 + \frac{r_s}{Nr_2}\right)\right]} \quad 6\text{-}0.43$$

If the current gain is large,

$$i_{C \, max} = \frac{V_{CC}}{Nr_2 + r_s} \quad \begin{cases} \alpha_{FE} \, r_2 \gg r_b \\ Nr_b \gg r_2 \end{cases} \quad 6\text{-}0.44$$

The operation of the grounded base blocking oscillator is essentially the same as that described for the grounded emitter configuration. In the common base circuit the output voltage may exceed the supply voltage and go positive. The collector to base voltage in the grounded emitter blocking oscillator goes positive also. The repetition rate is limited partly by how much the transformer has to be damped at the termination of the pulse in order to protect the transistor. An increase in transistor voltage rating allows a higher duty cycle, for given constant output pulse dimensions, up to the point where damping is required to prevent ringing.

Figure 6-0.17 shows a common base blocking oscillator in which saturation is prevented by clamping with source V_1 and CR_1.

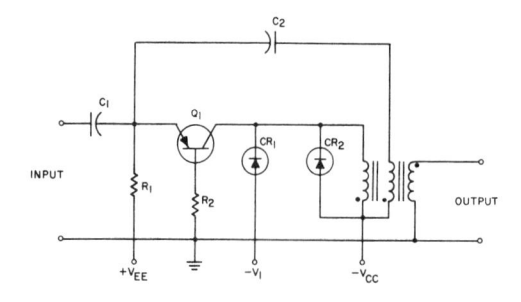

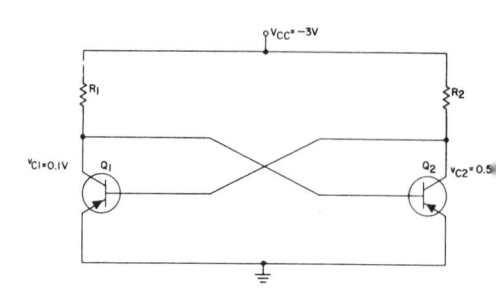

Fig. 6-0.17 - Nonsaturated blocking oscillator with emitter feedback

Fig. 6-0.18 - Direct-coupled transistor flip-flop

Linvill has shown that to obtain a maximum response time for norma transistor parameters, a transformer ratio of about 5 to 1 (stepdown between collector and base) is best, although loading will affect the prope values (3). In addition, the triggering energy is not significantly different for base, emitter, or collector triggering. However, the impedance chang during the initial switching phase at these triggering points varies considerably. The source impedance of the triggering device influences the triggering connection used. For collector triggering the input impedance increases from an initial value of a few hundred ohms to several thousand ohms. The input impedance for emitter triggering decreases from a few hundred ohms to 40-50 ohms and then increases to the original value. Fo base triggering after an initial drop, the input impedance remains constar at a few hundred ohms.

G. Direct-Coupled Transistor Methods

Direct-coupled transistor logic (DCTL) represents a class of circuit that have achieved widespread use. The direct-coupled circuitry characteristically utilizes saturated operation and low levels, and has been accepted because of its great simplicity and complete utilization of transistor characteristics (10).

Figure 6-0.18 illustrates the basic design of a direct coupled transistor flip-flop circuit.

Operation of the flip-flop depends on the saturated collector voltage of the "on" transistor being low enough to prevent conduction of the other transistor when this voltage is applied to its base. Since the flip-flop transistors are never cut off completely, there are strict requirements on the maximum leakage currents that can exist when the base is forward biased slightly. In addition, the saturation voltage must also be low. The transistor characteristics required depend on the system application, to a large degree, and are usually determined on this basis.

The logic networks for DCTL presented in Part 7 will illustrate the design procedures more clearly.

I. Negative Resistance Circuit Methods

Negative resistance devices recently achieved prominence with the availability of the p-n-p-n four-layer diode. The p-n-p-n units are both two and three terminal devices. The negative resistance characteristically results in rise times less than most other present types of semiconductors, and at current levels useful for core drivers and clock generators.

The voltage-current characteristics for the two terminal p-n-p-n are shown in Figure 6-0.19.

The p-n-p-n may be used in a monostable, astable or bistable configuration. The circuit of Figure 6-0.20 indicates the resistance value required for monostable and bistable operation, while the associated load lines are indicated in Figure 6-0.19.

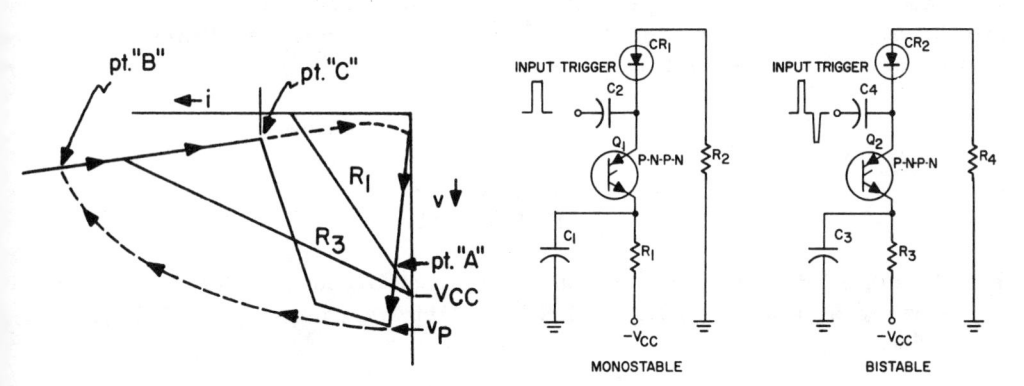

Fig. 6-0.19 - Voltage-current characteristics for the p-n-p-n diode

Fig. 6-0.20 - p-n-p-n circuit configurations

For the monostable case, the static operating point (point "a") is lowered by the input trigger to the breakdown voltage v_p and the capacitor C_1 discharges through the p-n-p-n and CR_1 and follows a locus as shown. The circuit current that flows through R_2 is determined by the supply voltage (the charged voltage of the capacitor) and the circuit resistance

213

excluding R_1. Currents as large as 10-20 amperes, with rise times of 10-30 mμs, are possible.

The current reduces after it reaches its peak value at point "b," since the capacitor cannot maintain the current. Eventually, the operating point reaches point "c" where it returns to the original static point "a."

In the bistable circuit the "off" condition, at point "a," allows a few microamperes of leakage current to flow. The circuit may be triggered on in the same manner as the monostable circuit. The diode CR_2 prevents the load R_4 from reflecting back to the trigger. Only a few microamperes of input current are required to turn on the device to point "b." However, the turn off current of the trigger must reduce the operating current to a value below i_h. The capacitor C_1 conveniently provides the stored charge but also limits the maximum repetition rates possible. R_1 is typically about 1K-5K for monostable operation, and C_1 is about 2000 $\mu\mu$f for currents of 1-3 amperes. The recovery time for C_1 to recharge (the p-n-p-n is off during this time) is about 10 μs. The p-n-p-n devices are also finding considerable use in low level logic networks, since their "on" to "off" resistance ratio is in the order of 10^6.

I. Other Circuit Methods

Multivibrators and other switching circuits are designed in a manner similar to those illustrated. The precise considerations of these circuits will be described for each of the specific selected circuits.

REFERENCES

1. Adler, R. B., "A Large Signal Equivalent Circuit for Transistor Static Characteristcis," M.I.T., RLE Transistor Group Report T-2, August 1951.

2. Hurtig, C. R., "Bias Stabilization of Junction Transistors," Transistor Group Report T-9, RLE, M.I.T., August 1953.

3. Baker, R. H., "Maximum Efficiency Transistor Switching Circuits," Lincoln Lab., M.I.T. Technical Report 110, March 1956.

4. Ebers, J. J. and J. L. Moll, "Large-Signal Behavior of Junction Transistors," Proc. IRE, 42, 1761, December 1954.

5. McMahon, R. E., "Designing Transistor Flip-Flops," Electronic Design, Vol. 3, October 1955.

6. Lebow, I. L. and R. H. Baker, "The Transient Response of Transistor Switching Circuits," Proc. IRE, 42, 938-943, June 1954.

7. Sferrino, V. J., "One Megacycle Silicon Transistor Circuits," T.R. 177, Lincoln Laboratory, M.I.T.

8. Linvill, J. G. and R. H. Mattson, "Junction Transistor Blocking Oscillators," Proc. IRE, 1632-1638, November 1955.

9. Carlson, A. W., "Blocking Oscillators," Transistor Applications, Inc. Internal Report #3, 1958.

10. Beter, R., et al, "Directly-Coupled Transistor Circuits," Electronics, Vol. 28, No. 6, June 1955, pp. 132-136.

*11. Yourke, H. S., "Millimicrosecond Transistor Current Switching Circuits," 1957 IRE-AIEE Conference on Transistors and Solid State Circuits, Philadelphia.

*Circuit 6-4.

BIBLIOGRAPHY

Carlson, A. W., "A Four-Layer Diode Flip-Flop," Transistor Applications, Inc. Scientific Report No. 2, "Investigation of the Applications of Negative Resistance Diodes for Switching Circuits," Contract AF19(604)-3066.

Deuitch, D. E., "Transistor Circuits for Digital Computers," Electronics, 29, 160-162, May 1956.

Ebers, J. J. and J. L. Moll, "Large-Signal Behavior of Junction Transistors," Proc. IRE, 42, 1761-1772, December 1954.

Gott, E., "High Speed Counter Uses Surface Barrier Transistors," Electronics, 29, 174-178, March 1956.

Henle, R. A., "Multistable Transistor Circuits," Elec. Eng., 74, 570-572, July 1955.

Kingston, R. H., "Switching Time in Junction Diodes and Junction Transistors," Proc. IRE, 42, 829-834, May 1954.

Lebow, I. L. and R. H. Baker, "Transient Response of Transistor Switching Circuits," Proc. IRE, 42, 938-943, June 1954.

Linvill, J. G., "Nonsaturating Pulse Circuits Using Two Junction Transistors," Proc. IRE, 43, 826-834, July 1955.

Linvill, J. G., and R. H. Mattson, "Junction Transistor Blocking Oscillator," Proc. IRE, 43, 1632-1639, November 1955.

Moll, J. L., "Large-Signal Transient Response of Junction Transistors," Proc. IRE, 42, 1773-1784, December 1954.

Pederson, D. O., "Regeneration Analysis of Junction Transistor Multivibrators," IRE Trans., CT-2, 171-178, June 1955.

Shea, R., "Transistor Circuits Engineering," John Wiley & Sons, Inc., New York, 1957.

Statz, H., et al, "Design Considerations of Junction Transistors at Higher Frequencies," Proc. IRE, 42, 1620-1628, November 1954.

Suran, J. J. and E. Keonjian, "A Semiconductor Diode Multivibrator," Proc. IRE, 43, July 1955.

Suran, J. J. and F. A. Reibert, "Two-Terminal Analysis and Synthesis of Junction Transistor Multivibrators," IRE Trans., CT-3, 26-38, March 1956.

Sylvan, T. P., "Unijunction Transistor Simplifies Bistable Circuits, Electronics, November 24, 1958.

CIRCUIT 6-1

SATURATED RC COUPLED FLIP-FLOP

Convair-Pomona   Contributed by J. W. Flaherty

The saturated flip-flop of Circuit 6-1 is a simple and direct approach to the design of switching circuits. The emitter resistance, R_7, is often omitted to establish the "on" voltage level at near ground and simplify the design. The "off" voltage level is not well established and may vary from one circuit to another, depending on the loading.

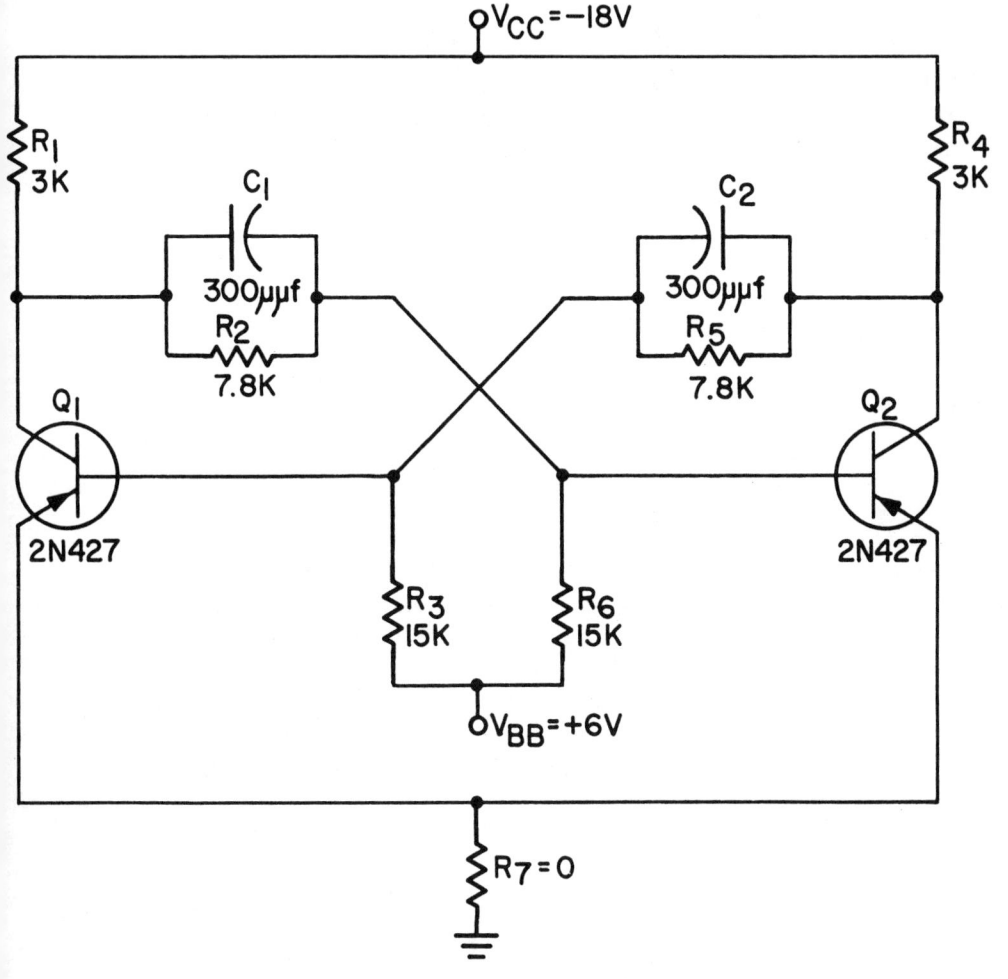

Circuit 6-1 - Saturated RC coupled flip-flop

Circuit 6-1, Saturated RC Coupled Flip-Flop (cont'd.)

The design procedure will include an emitter resistance to give a general design method.

The use of the emitter resistor in the saturated flip-flop provides degeneration and improves the stability. The voltage drop across the emitter resistor determines the level at which saturation occurs and hence the "on" level of the collector. However, variations of transistor characteristics cause this voltage level to change. This fact also complicates the design procedure.

The actual design method should depend on the circuit characteristics that are considered important. The design method described here assumes that saturated operation is desired for a range of transistor parameters, that a specific minimum amount of saturated current is desired, and that the "on" collector level is important and the temperature limits are to be specified. Initially V_{BB} will not be included in the design and the base resistors are returned to ground.

For saturation to exist, the emitter current must meet the following inequality:

$$i_E \geq \frac{V_{CC}}{R_7 + \alpha_{FB} R_1'}, \qquad 6\text{-}1.1$$

where

$$R_1' = \frac{R_1 (R_6 + R_2)}{R_6 + R_2 + R_1}. \qquad 6\text{-}1.2$$

The above equation is useful in checking the final design. A minimum α_{FB} should be used in the equation, since saturation becomes less positive as the value of α_{FB} decreases, and it is assumed saturation is desired at some minimum α_{FB}.

The design method is conveniently carried out by first specifying the saturation voltage level of the "on" collector ($v_{C\;on}$).

$$v_{C\;on} = i_E R_7 + v_{CE}. \qquad 6\text{-}1.3$$

The v_{CE} is normally specified by the transistor manufacturer, but is approximately 0.2 volt for most transistors.

Circuit 6-1, Saturated RC Coupled Flip-Flop (cont'd.)

After selecting the "on" collector voltage, the "off" base voltage is chosen on the basis of triggering and stability. It is the difference (Δv) between the "off" base voltage and emitter voltage that is important and should be specified.

$$\Delta v = v_E - \frac{v_{C\ on}\ R_6}{R_6 + R_2} \qquad 6\text{-}1.4$$

where

v_E is the emitter voltage ($i_E\ R_7$),

$v_{C\ on}$ is the saturated collector voltage,

$v_{C\ on} = v_E + 0.2$ volt.

After selecting Δv, Equation 6-1.4 may be used to obtain a value for the ratio of R_6 and R_2.

The average I_{CO} is determined at the maximum operating temperature. This may be found from the manufacturer's specifications, or measured directly. The I_{CO} flows through the equivalent base resistance in such a direction to reduce Δv. R_3 and R_5 (and hence R_2 and R_6) are chosen so that Δv is not reduced to zero before the maximum temperature is reached.

$$\Delta v = \frac{R_6\ (R_2 + R_1)}{R_6 + R_2 + R_1}\ I_{CO\ max} \ . \qquad 6\text{-}1.5$$

From this equation, a second relation for R_6 and R_2 is obtained, in terms of R_1.

To maintain saturation, the base current must agree with the inequality of Equation 6-1.6.

$$\alpha_{FE}\ i_B > \frac{V_{CC}}{R_1} - \frac{v_{C\ on}}{R_1'} \qquad 6\text{-}1.6$$

where

$$R_1' = \frac{R_1\ (R_6 + R_2)}{R_6 + R_2 + R_1}$$

Circuit 6-1, Saturated RC Coupled Flip-Flop (cont'd.)

or

$$\alpha_{FE} \left[\frac{V_{CC} R_6}{R_6 (R_2 + R_1) + \frac{R_7}{(1 - \alpha_{FB})} (R_6 + R_2 + R_1)} \right] > \left(\frac{V_{CC}}{R_1} - \frac{v_{C\,on}}{R_1'} \right) \cdot \qquad 6\text{-}1.7$$

To continue the design, R_1 is chosen on the basis of loading considerations and using Equations 6-1.4 and 6-1.5, R_2 and R_6 are calculated. The emitter resistance is then calculated from Equation 6-1.7 using the minimum α_{FE} anticipated. This value of R_7 will provide saturation only for α_{FE} values above this minimum value.

The base supply voltage (V_{BB}) is often used to provide a larger "off" base voltage without requiring a large "on" collector voltage. The base bias voltage becomes necessary if the emitter resistance is zero. The design procedure is not changed appreciably for the inclusion of V_{BB} and the design equations become

$$v_{C\,on} = i_E R_7 + v_E \qquad 6\text{-}1.8$$

$$\Delta v = v_E - \frac{v_{C\,on} R_6}{R_6 + R_2} - V_{BB} \left(1 - \frac{R_6}{R_6 + R_2} \right) \qquad 6\text{-}1.9$$

$$\Delta v = I_{CO} \frac{R_6 (R_2 + R_1)}{R_6 + R_2 + R_1} \qquad 6\text{-}1.10$$

$$\alpha_{FE\,min} \left[\frac{V_{CC} R_6 - V_{BB} R_6}{R_6 (R_2 + R_1) + \frac{R_7}{(1 - \alpha_{FB})} (R_6 + R_2 + R_1)} \right]$$

$$> \left(\frac{V_{CC}}{R_1} - \frac{v_{C\,on}}{R_1'} + \frac{V_{BB}}{R_1} - \frac{V_{BB}}{R_1'} \right) \qquad 6\text{-}1.11$$

Circuit 6-1 is capable of operating at repetition rates of about 200 KC and is intended for use as a stable, low speed flip-flop. The transistor characteristics are not an important consideration, except that α_{FB} must be consistent with the limits imposed by the design procedure.

CIRCUIT 6-2

NONSATURATED RC COUPLED FLIP-FLOP

Lincoln Laboratory M.I.T. Contributed by R. E. McMahon

Circuit 6-2 is a basic flip-flop that is designed for nonsaturating operation. The circuit allows a range of alpha values from 0.90 to a theoretical unity, without saturation or instability. The output voltage swing has been designed for a maximum of 10 volts while the power is well within the transistor rating.

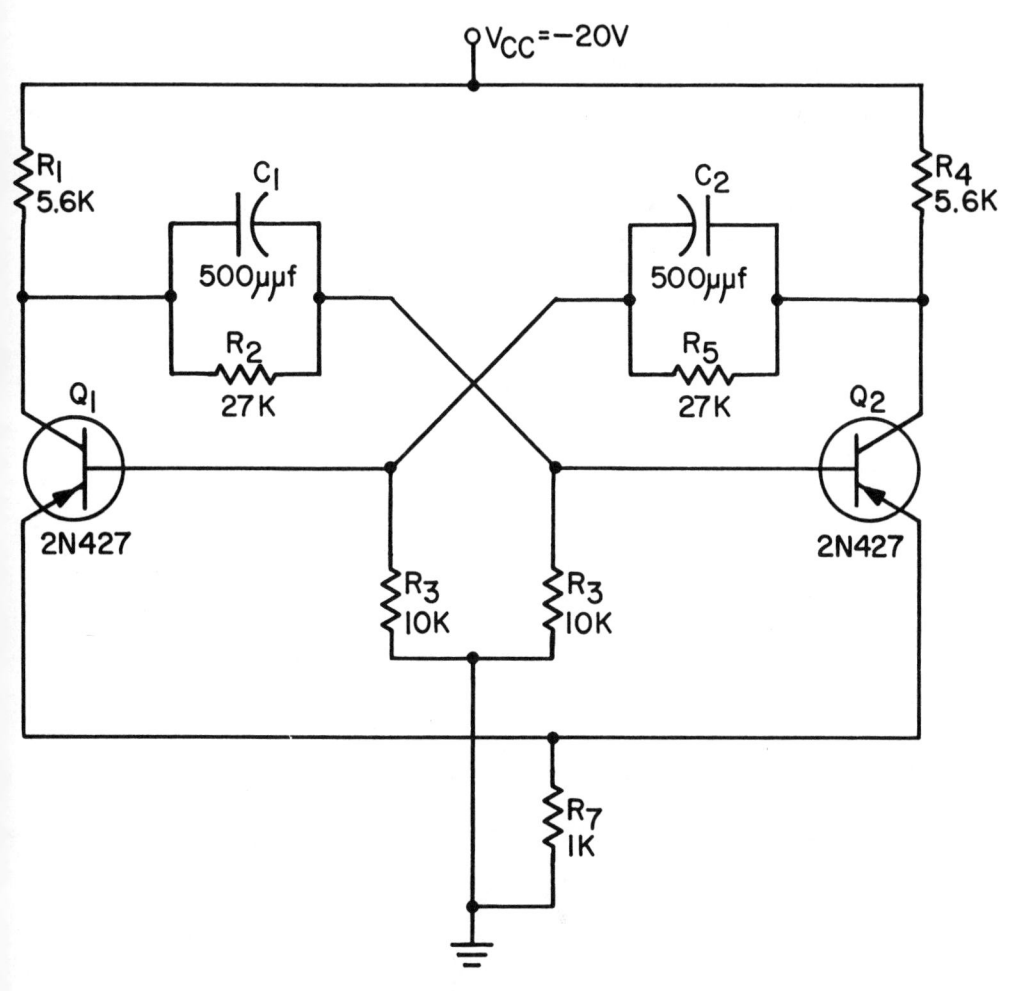

Circuit 6-2 - Nonsaturated RC coupled flip-flop

Circuit 6-2, Nonsaturated RC Coupled Flip-Flop (cont'd.)

The circuit is not intended for high speed operation and the cross coupling capacitors are chosen for over driving characteristics at the expense of speed.

A bypass capacitor may be added from the common-emitter point to ground to improve the triggering and gain characteristics. It is possible to obtain an optimum capacitor value that improves the triggering without a degeneration of the circuit response.

The circuit is designed to operate from -20° C to +70° C with adequate margins and stability. The output rise time is less than 0.1 μs while the fall time is about 0.2 μs. The output loading is capable of supplying moderate loads.

The circuit is unsophisticated and employs no novel techniques; however, it has been life-tested over long periods and displays excellent long term reliability.

The variation of output voltage level restricts the circuit to special applications or simple systems.

CIRCUIT 6-3

DIRECT-COUPLED TRANSISTOR FLIP-FLOP

Philco Corporation — Contributed by J. B. Angell

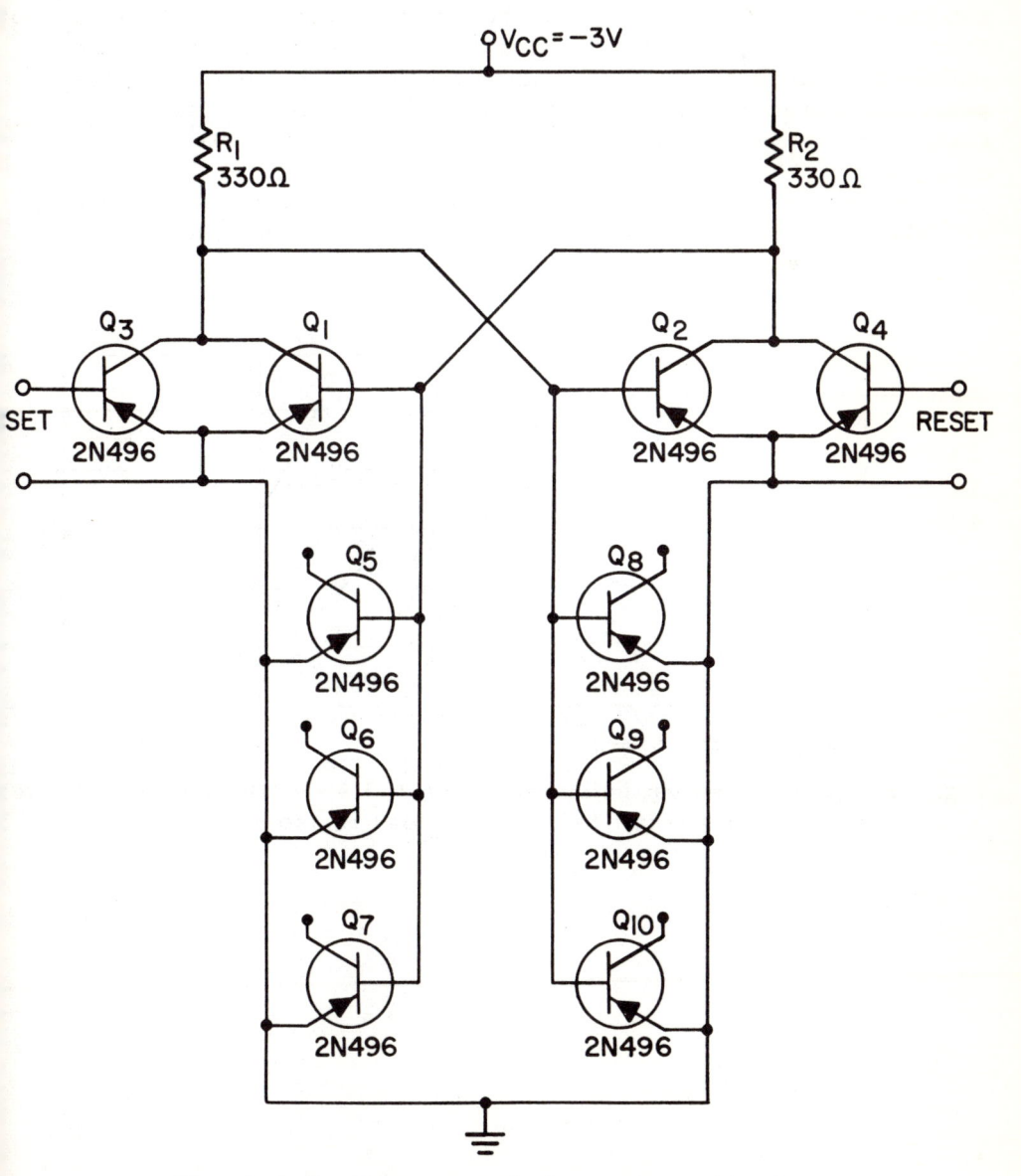

Circuit 6-3 - Direct-coupled transistor flip-flop

Circuit 6-3, Direct-Coupled Transistor Flip-Flop (cont'd.)

Circuit 6-3 is a direct-coupled transistor flip-flop. The circuit includes the load and trigger transistors, since they are an integral part of the basic flip-flop. The simplicity of the circuit is obvious and the design procedure must be based primarily on the transistor parameters.

The transistor requirements for Direct-Coupled Transistor Logic (DCTL) circuitry are briefly described in the design philosophy section of Part 6 - "Switching Circuits," and more fully described in the design philosophy section of Part 7 - "Logic Circuits." The general design problems, however, are related to I_{co} and the use of the 2N496 silicon transistors extends the temperature range of this circuit to 140° C.

The Figures 6-3.1 and 6-3.2 indicate the relation of turn on and turn off delay time to the load resistors R_1 and R_2 and temperature.

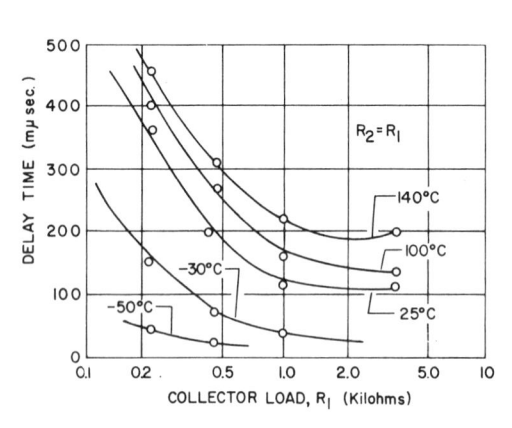

Fig. 6-3.1 - Delay time vs. collector load resistor for Circuit 6-3

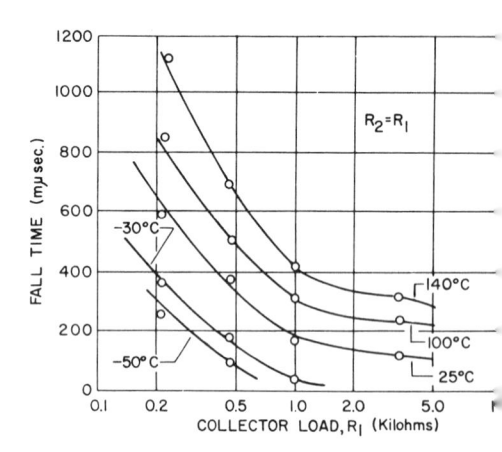

Fig. 6-3.2 - Fall time vs. collector load resistor for Circuit 6-3

CIRCUIT 6-4

CURRENT SWITCHING FLIP-FLOP

International Business Machines Corp.   Contributed by H. S. Yourke

Circuit 6-4 is a flip-flop that illustrates an important class of transistor circuits that utilize current steering techniques to achieve maximum response and stability (11). The circuits are designed to provide low voltage swings to minimize the effect of capacitance. In addition, nonsaturated operation is used to avoid the delay time associated with saturation. In effect, well-specified currents are switched reliably with small voltage swings. The circuits impose no upper limit of α_{FB} and have a d-c stability factor of unity.

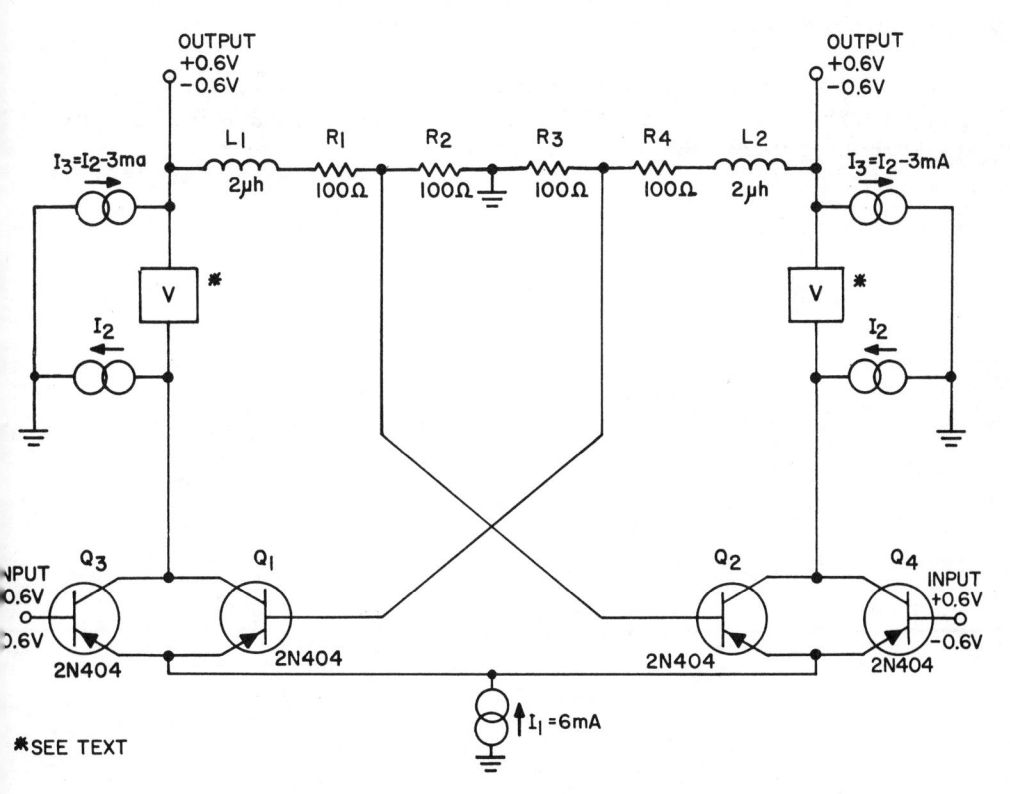

Circuit 6-4 - Current switching flip-flop

Circuit 6-4, Current Switching Flip-Flop (cont'd.)

The basic logic is described in Part 7 - Logic Circuits, and complements the discussion here on the flip-flop. In Circuit 6-4, assume Q_2 is "on" while Q_1 is "off." The voltage translate blocks (V) are batteries or any device that simulates a battery, such as a Zener diode operating entirely in the Zener region and provides a voltage drop equal to the desired nominal collector voltage. With Q_1 "off" a current equal to the current source I_2 flows through the voltage translate box. The current source I_3 provides a current I_2 - 3 mAdc and 3 mAdc flows out of the load resistor, establishing an output voltage of -0.6 volt. The opposite side of the flip-flop has Q_2 "on" so there exists a collector current equal to αI_1 or approximately 6 mAdc. In this case, I_2 - 6 mAdc flows through the voltage translate box and since I_2 - 3 mAdc is provided on the load side of the box, 3 mAdc flows into the load resistor establishing an output voltage of +0.6 volt. The current I_2 is set greater than I_1 for normal operation. The cross coupling network for the condition chosen holds Q_1 "off" and Q_2 "on." The input transistors Q_3 and Q_4 receive inputs of ±0.6 volt and reverse the state of the flip-flop. The regenerative loop has little effect on the switching speed of the circuit. Switching is primarily accomplished through the switching of emitter currents.

The circuit shown operates with rise and fall times in the 10 mμs region with transistors of the drift type. The temperature limits and transistor interchangeability are excellent.

CIRCUIT 6-5

BASE GATED DIRECT-COUPLED FLIP-FLOP

Philco Corporation　　　　Contributed by A. K. Rapp and M. M. Fortini

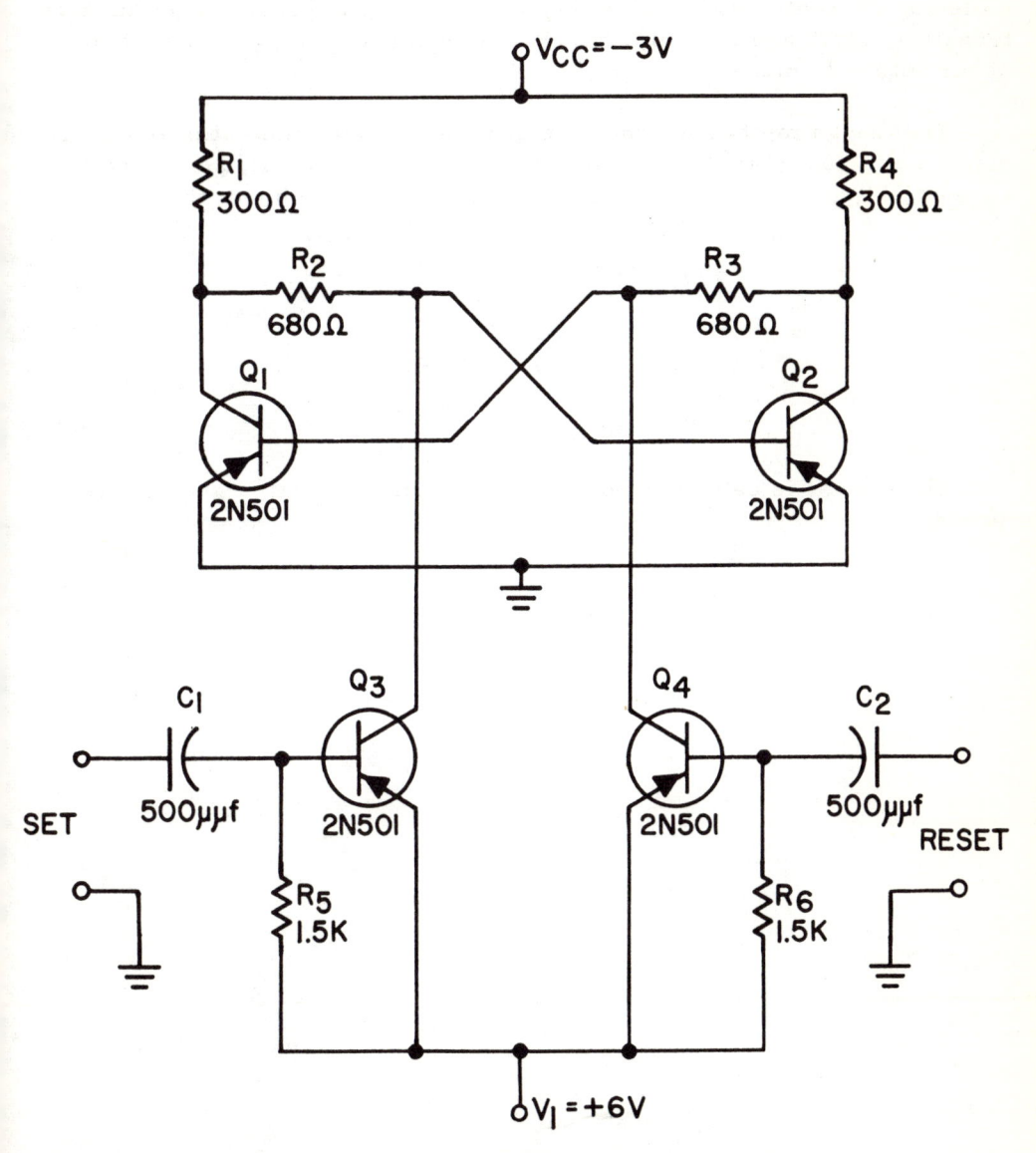

Circuit 6-5 - Base gated direct-coupled flip-flop

Circuit 6-5, Base Gated Direct-Coupled Flip-Flop (cont'd.)

The Circuit 6-5 improves the switching speed of the basic direct-coupled flip-flop by reverse biasing the base to emitter diode during turn off. If the voltage V_1 is chosen properly, the gate not only turns off the conducting transistor, but supplies an amount of current to the load resistor of the non-conducting transistor equal to the amount the transistor will conduct after the switching transition is complete. This pseudo-collector current reduces the delay which normally precedes the turn-on transient. Both are thus minimized so that the switching transient will occur within 20 mμs.

The design method for the circuit first assumes the saturated transistor collector voltage is zero. The supply voltage, V_{CC}, the collector current, I_C, and the saturated current gain are determined, then

$$R_1 = R_4 = \frac{V_{CC}}{i_C} \qquad 6\text{-}5.1$$

$$R_2 = R_3 = (\alpha_{FE} - 1) R_1 - R_{in} \qquad 6\text{-}5.2$$

$$V_1 = R_2 \, i_C . \qquad 6\text{-}5.3$$

The circuit is capable of operation at repetition rates greater than 40 mc.

CIRCUIT 6-6

SATURATED CLAMPED FLIP-FLOP

Radio Corporation of America

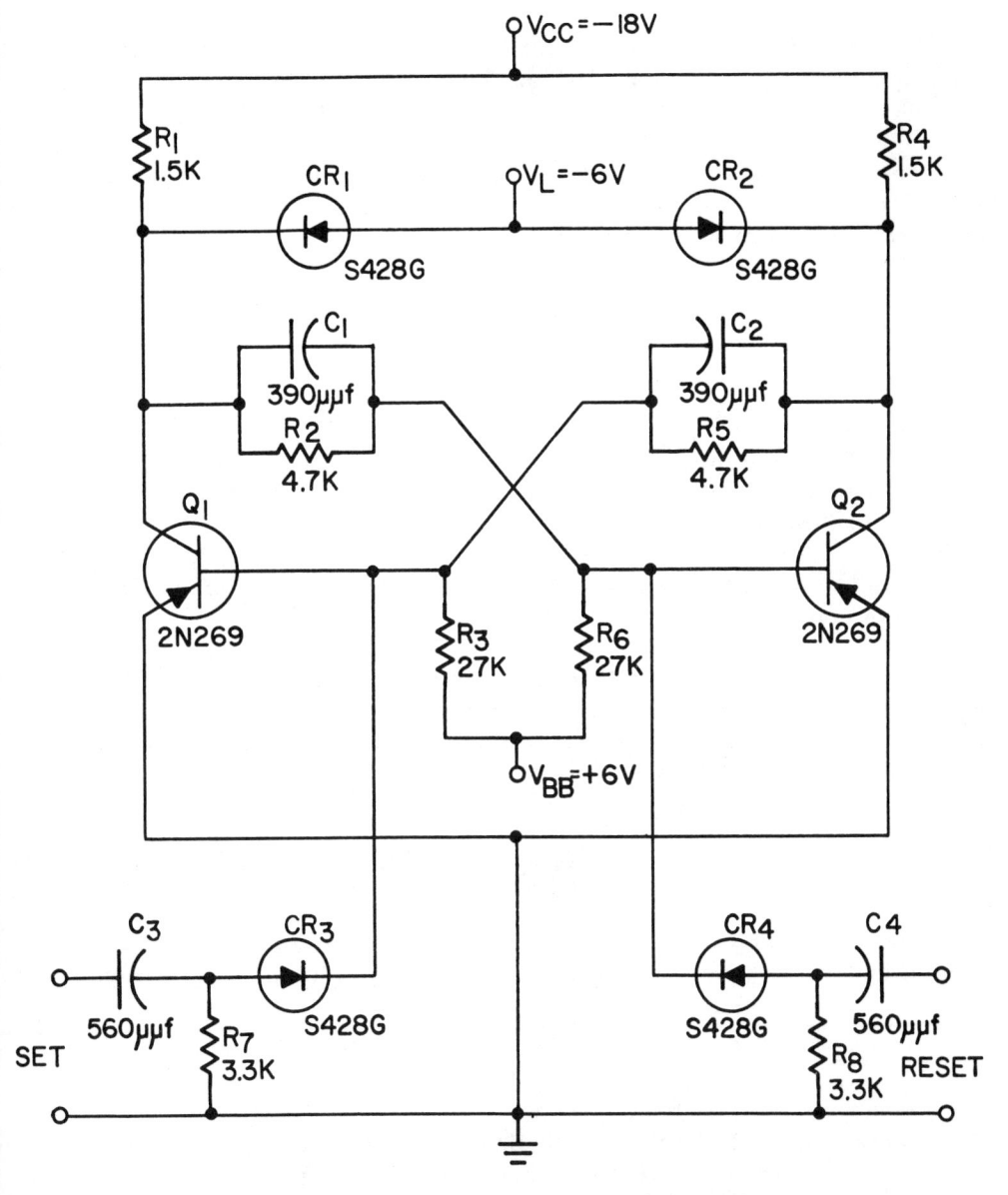

Circuit 6-6 - Saturated clamped flip-flop

Circuit 6-6, Saturated Clamped Flip-Flop (cont'd.)

Circuit 6-6 provides clamping only on the low or "off" condition and relies on saturation to establish the "on" level. The technique reduces the need for additional clamping circuitry and additional voltage supplies, but the delay time of saturation exists and must be considered in the overall system design.

The circuit speed is moderate (150 kc) and emphasis is placed on simplicity and loading capabilities. The temperature range is -30°C to +60°C.

The clamping current and voltage level are chosen first based on the expected loading. The value of R_1 to provide the clamping current may be approximated from Equation 6-6.1.

$$i_{CL} = \frac{V_{CC} - V_{CL}}{R_1} \qquad 6-6.1$$

where

V_{CL} is the clamping voltage

i_{CL} is the clamping current.

This is not quite the true clamping current because of the effect of the cross coupling network.

The base resistances, R_3 and R_6, and the cross coupling resistances, R_2 and R_5, are chosen as described in Circuit 6-1. In this circuit, however, the emitter is returned to ground so that the voltage difference Δv between the "off" base and the emitter is simply

$$\Delta v = V_{BB} \left(1 - \frac{R_6}{R_6 + R_2} \right). \qquad 6-6.2$$

The collector to emitter voltage drop is considered zero for the design equations here. The relation between R_6, R_2, and I_{CO} simplifies to,

$$\Delta v = \frac{I_{CO} R_6 R_2}{R_6 R_2}. \qquad 6-6.3$$

After determining R_6 and R_2 from Equation 6-6.3 and Equation 6-6.2, there may not be sufficient collector current to provide saturation. The condition for saturation is derived as

Circuit 6-6, Saturated Clamped Flip-Flop (cont'd.)

$$\alpha_{FE} \left[\frac{V_L}{R_2} - \frac{V_{BB}}{R_6} \right] > \frac{V_{CC}(R_6 + R_2 + R_1)}{R_1(R_2 + R_6)} \qquad 6\text{-}6.4$$

If this inequality is not met, the design must be reconsidered. The collector resistance, R_1, may be increased to provide saturation at the existing collector current level, but this decreases the clamping current. Another choice is in a re-design of the base circuit to provide a greater base current at the expense of stability or I_{CO} limits. This method will also reduce the clamping current available for a fixed clamping voltage and a particular R_1 value.

CIRCUIT 6-7

SATURATED CLAMPED HIGH SPEED FLIP-FLOP

Lincoln Laboratory, M.I.T.   Contributed by Vincent J. Sferrino

The Circuit 6-7 is intended for high speed operation in the 20 mc range. The circuit impedances are generally lower than previous design methods would indicate to reduce the effect of stray capacitance. The saturation delay time is less than 10 mμs for the transistor types used, but must be considered in the specifications for turn off time. The total turn off time is about 30 mμs, which includes the saturation delay time and the normal fall time. The rise time is less than 15 mμs. It is common to specify the minimum time between set and reset pulses as the

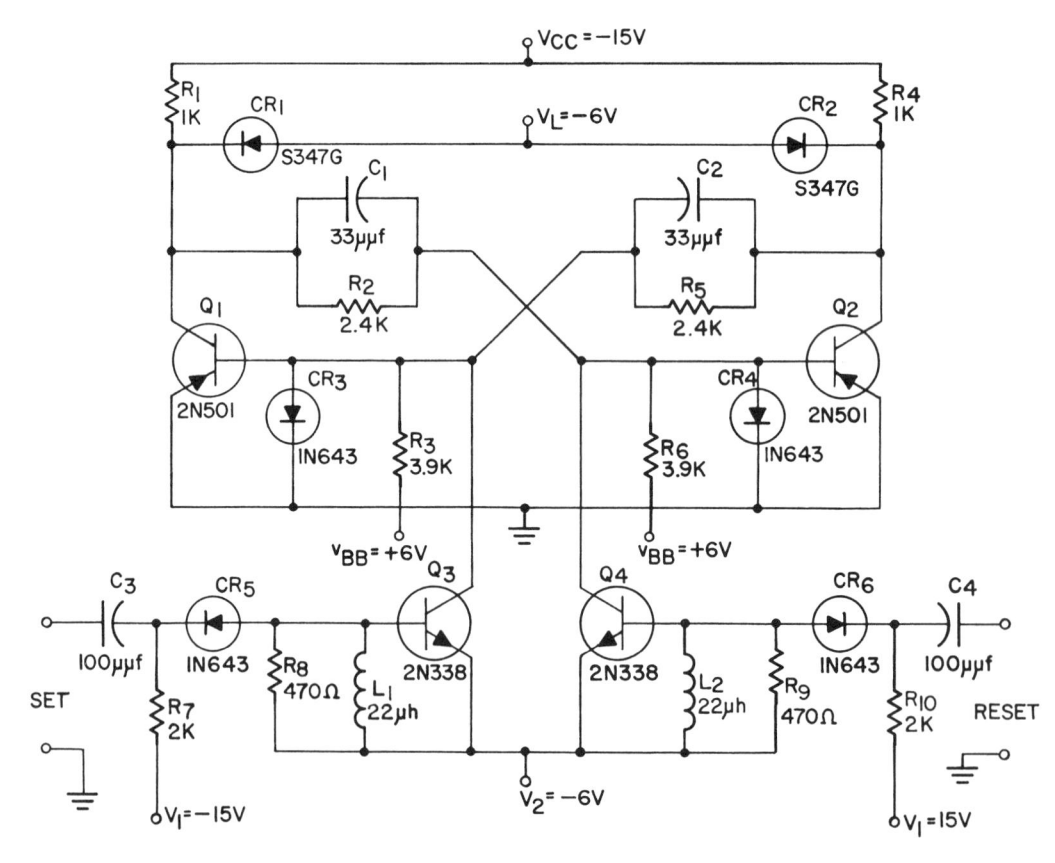

Circuit 6-7 - Saturated clamped high speed flip-flop

Circuit 6-7, Saturated Clamped High Speed Flip-Flop (cont'd.)

resolution time. For this circuit, the resolution time is less than 50 mμs. The base silicon diode is used to prevent excessive reverse base to emitter voltage during the transient periods. The diode establishes an exact off base voltage (0.7 volt for the silicon diode) and simplifies I_{CO} and stability considerations. The inductor triggering scheme renders the circuit less sensitive to variations in the trigger shape and amplitude.

The design procedure for this circuit is described in Circuit 6-6, the Saturated Clamped Flip-Flop.

CIRCUIT 6-8

HIGH SPEED GRADED BASE TRANSISTOR FLIP-FLOP

Lincoln Laboratory, M.I.T.   Contributed by
R. H. Baker, E. Chatterton, A. Parker

The circuit is designed to reduce the usual dependence of the output current and stability on the transistor characteristics.

In analyzing, the circuit assumes Q_3 and Q_4 are "on" and Q_1 and Q_2 are "off" as shown in the simplified circuit of Figure 6-8.1.

The circuit may be triggered to the opposite condition, that is, Q_3 and Q_4 "off" by applying a pulse to the base of Q_2. The emitter current of Q_2 increases while the emitter current of Q_4 decreases. The collector current of Q_2 reduces the base voltage of Q_1, and the simultaneous decrease

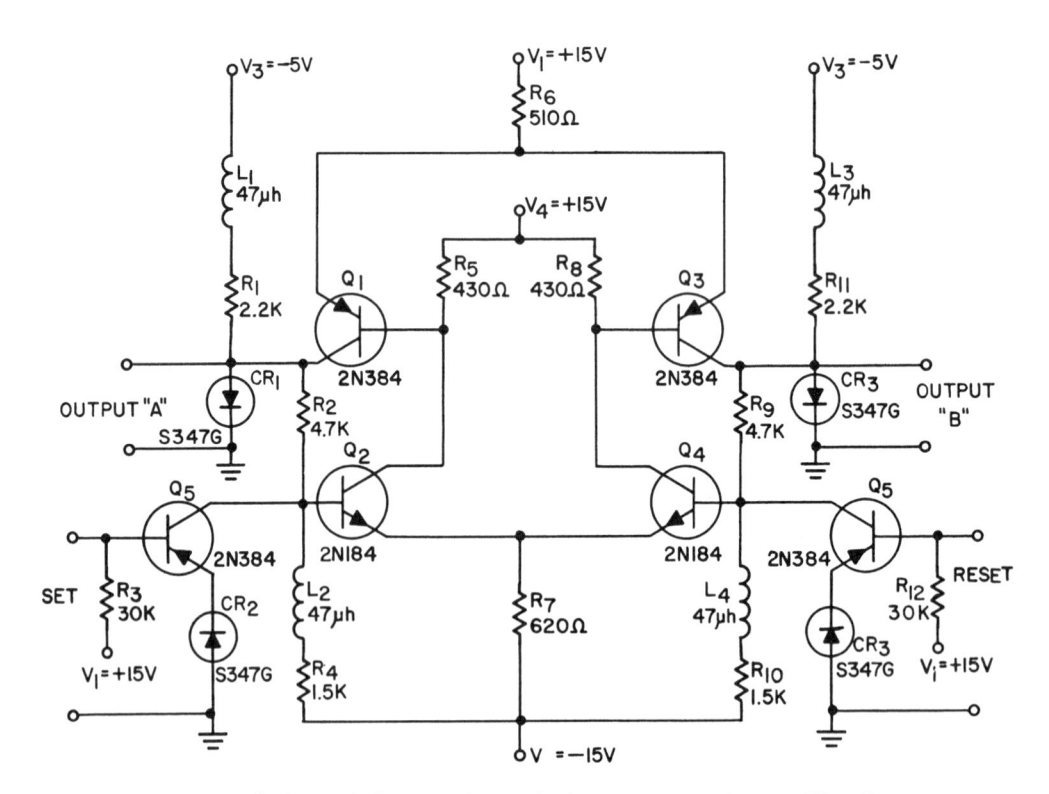

Circuit 6-8 - High speed graded base transistor flip-flop

Circuit 6-8, High Speed Graded Base Transistor Flip-Flop (cont'd.)

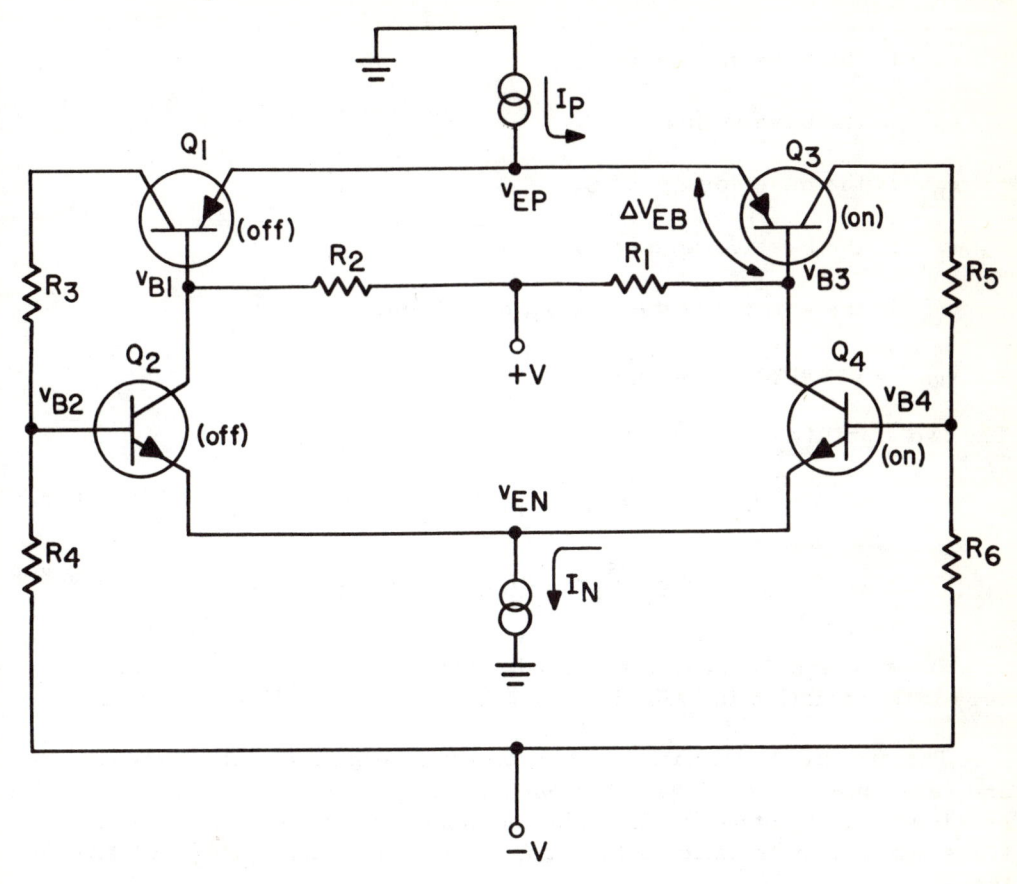

Fig. 6-8.1 - Simplified equivalent circuit for Circuit 6-8

of collector current in Q_4 causes the base voltage of Q_3 to rise. When the base of Q_1 drops below the base voltage of Q_3, the current flowing in the emitter of Q_3 (assumed to be from a constant current source) is transferred to the emitter of Q_1 and hence turns "on" Q_1.

The stability condition assuming Q_3 and Q_4 are "on" while Q_1 and Q_2 are "off" is derived by requiring the voltage conditions following.

$$v_{B_3} < v_{EP}, \quad v_{B_1} > v_{EP} \qquad 6\text{-}8.1$$

$$v_{B_4} > v_{EN}, \quad v_{B_2} < v_{EN} \qquad 6\text{-}8.2$$

Circuit 6-8, High Speed Graded Base Transistor Flip-Flop (cont'd.)

where

v_{B_1} is the base voltage of Q_1,

v_{B_2} is the base voltage of Q_2,

v_{B_3} is the base voltage of Q_3,

v_{B_4} is the base voltage of Q_4,

v_{EN} is the emitter voltage of Q_4, or Q_2, and

v_{EP} is the emitter voltage of Q_3, or Q_1.

The two stability equations that result are,

$$\{\alpha_{FB} I_N - [2I_{CO} + I_P (1 - \alpha_{FB})]\} R_1 > \Delta v_{EB} \qquad 6\text{-}8.3$$

$$-V \frac{R_5}{R_5 + R_6} + 2I_{CO} R_6 < -\Delta v_{EB}. \qquad 6\text{-}8.4$$

Since I_P and I_N may be assumed reasonably good current sources, very little variation in alpha is expected, and the stability is constant.

The circuit shown has a resolution time (minimum delay between set and reset pulses) of 0.2 μs. The output voltage levels are 0 to 5 volts with excellent loading capabilities. The independence of the circuit on the transistor characteristics allows complete interchangeability of transistor types.

CIRCUIT 6-9

MEDIUM SPEED SILICON FLIP-FLOP

Lincoln Laboratory, M.I.T.

Contributed by
V. J. Sferrino and W. G. Schmidt

The silicon flip-flop of Circuit 6-9 utilizes the design features of Circuit 6-8, The High Speed Graded Base Transistor Flip-Flop, combined with the temperature stability features afforded by the use of silicon transistors. In addition, the design has been simplified and the operating speed intended is about 1.5 mc when commutative. The output rise and fall times are less than 0.1 μs with voltage levels of ground and 3 volts. The temperature range over which the circuit meets these specifications extends from -55°C to 100°C.

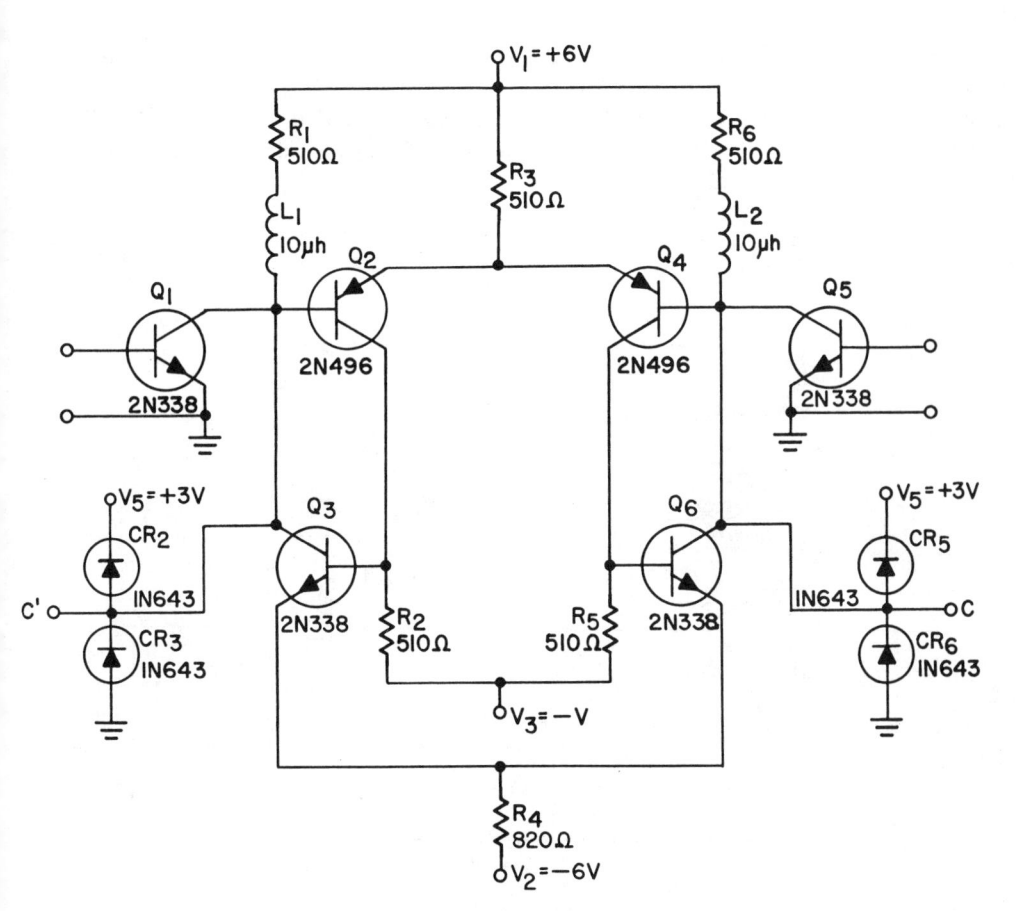

Circuit 6-9 - Medium speed silicon flip-flop

CIRCUIT 6-10

EMITTER FOLLOWER CROSS COUPLED SATURATED FLIP-FLOP

Sanders Associates   Contributed by P. Billings

This circuit is included because it is an example of improving the stability and speed by utilizing a direct circuit approach. The emitter followers in the cross coupling network allow a reduction of the resistance and capacitance normally needed, and improve the recovery time of this network. The collectors of the flip-flop transistors are loaded less using the emitter followers and in addition, the actual outputs may be taken from the emitter followers. The design procedure is similar to the basic non-saturating flip-flop except that the cross coupling circuit provides gain and allows a reduction in the base circuit resistance to improve the temperature stability.

The cross coupling circuits also allow additional overdrive to improve the speed. Repetition rates of 5 mc may be obtained for this circuit. The output voltage difference is somewhat less than 3 volts with good loading capabilities. The circuit rise time is about 20 mμs and the temperature range is -35°C to 55°C.

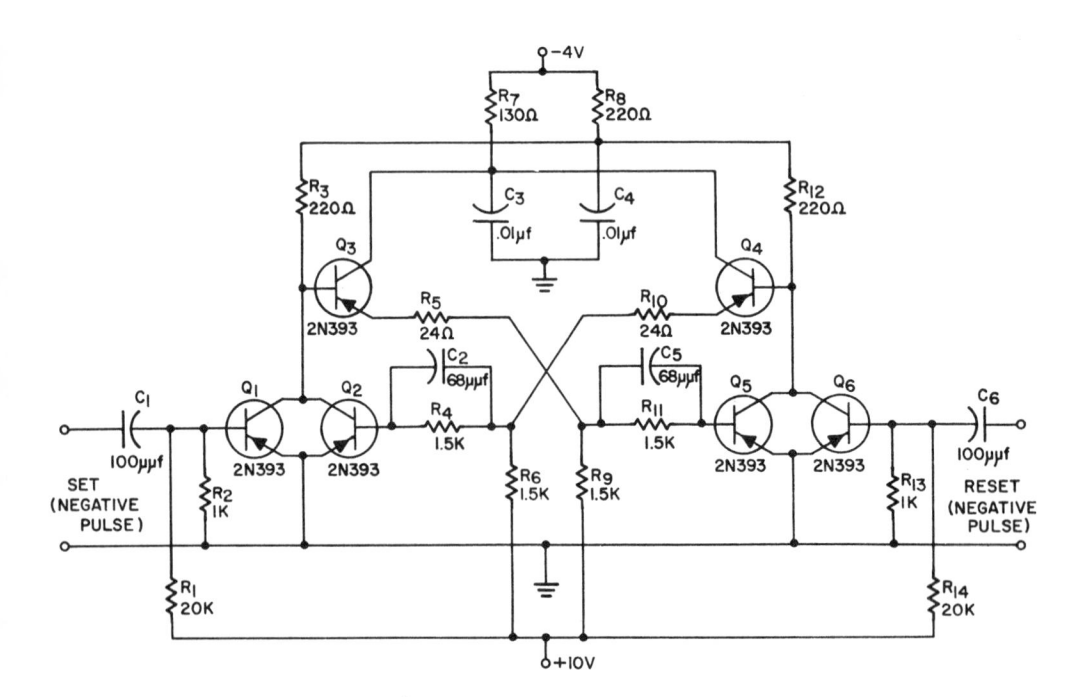

Circuit 6-10 - Emitter follower cross coupled saturated flip-flop

CIRCUIT 6-11

NEGATIVE RESISTANCE DIODE FLIP-FLOP

Transistor Applications, Inc.   Contributed by A. W. Carlson

Circuit 6-11 is a flip-flop that uses four-layer p-n-p-n diodes. The circuit is capable of providing a high output voltage swing and operates over a wide range of voltage settings.

If NCR_2 (negative resistance p-n-p-n diode) is conducting initially, it receives a current from V_2 through CR_4 (20 ma) and an additional current from V_1. The other p-n-p-n diode (NCR_1) is cut off since V_1 and R_1 establish

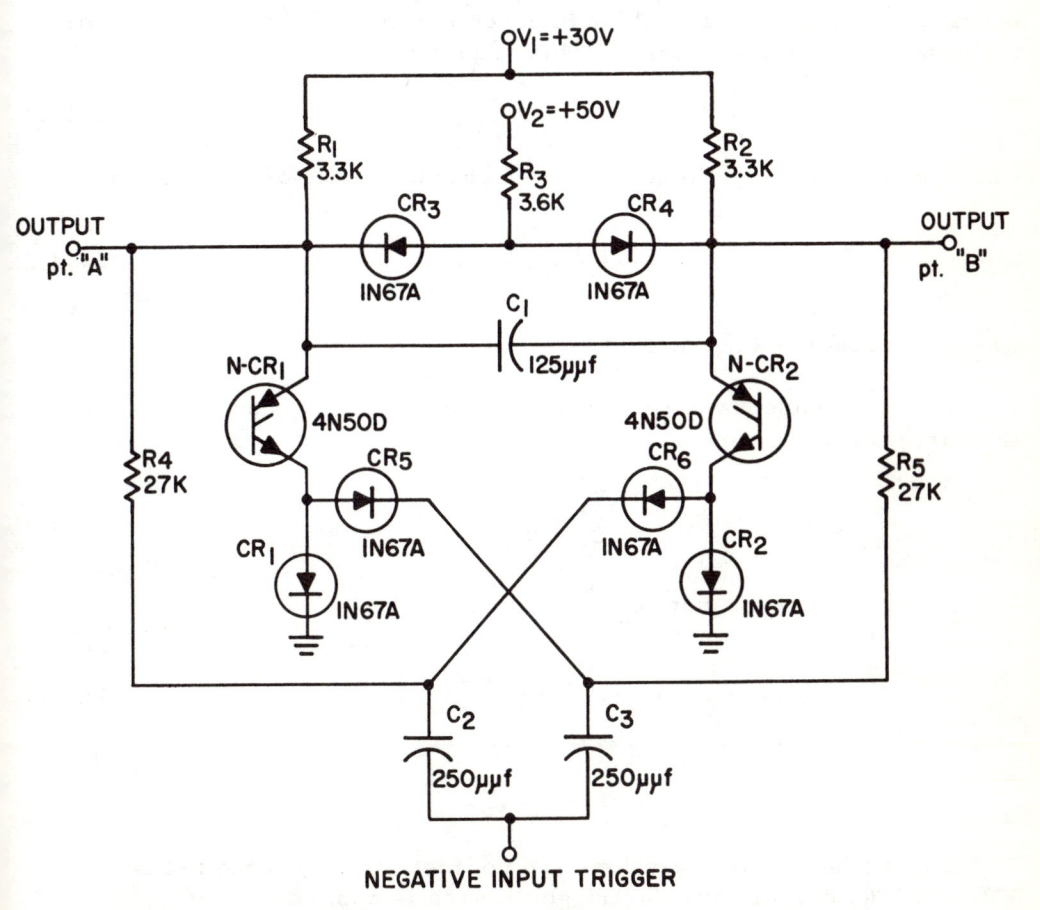

Circuit 6-11 - Negative resistance diode flip-flop

Circuit 6-11, Negative Resistance Diode Flip-Flop (cont'd.)

the operating point in the cut off region (refer to Section 1, Figure 6-0.20). The voltage at point "A" is at V_1 and diode CR_3 is cut off. The diodes CR_5 and CR_6 are steering diodes and are appropriately biased under the present circuit condition to allow an input pulse to turn NCR_1 "on." The diode CR_1 prevents the trigger from being shorted to ground and allows the trigger to develop a total voltage across NCR_1 sufficient to exceed the peak voltage. NCR_1 turns "on" very rapidly (< 0.1 μs) and diverts the current from NCR_2. In addition, when NCR_1 turns "on" capacitor C_1 couples the voltage fall to NCR_2 and aids in reversing the current through NCR_2. When NCR_2 turns "off" diode CR_4 is reverse biased, and point "B" is at a voltage level V_1. The steering diodes are biased such that the next input pulse will turn "on" NCR_2.

For optimum circuit operation V_1, R_1 and R_2 should establish the operating point at cut off with no intersection in the saturated region or the negative resistance region. This requires

$$V_1 < v_P \qquad \qquad 6\text{-}11.1$$

where v_P is the peak voltage of the v-i characteristic of the p-n-p-n.

$$\frac{V_1}{R_1} = \frac{V_1}{R_2} < i_h \qquad \qquad 6\text{-}11.2$$

where i_h is the holding current.

To provide adequate turn on, the currents from V_1 and V_2 must exceed the current i_h.

$$\frac{V_1}{R_1} + \frac{V_2}{R_3} > i_h \qquad \qquad 6\text{-}11.3$$

It is desirable but not necessary that

$$R_1, R_2, > 2\left(\frac{v_P - V_1}{V_2 - v_P}\right) R_3 \qquad \qquad 6\text{-}11.4$$

and

$$V_2 > v_P \qquad \qquad 6\text{-}11.5$$

If these latter two conditions are not met, it is possible to have both NCR_1 and NCR_2 cut off until the trigger source is applied.

Circuit 6-11, Negative Resistance Diode Flip-Flop (cont'd.)

The circuit shown will operate with input pulses greater than 6 volts. The turn off time is about 7 μs while the turn on is less than 0.1 μs. The capacitor C_3 may be made smaller to reduce the turn off time but a higher input pulse will be required. V_2 may be varied from 30 to 150 volts with V_1 at 30 volts and with V_2 at 100 volts V_1 may be varied from 17 to 40 volts.

CIRCUIT 6-12

ASTABLE CIRCUIT FOR DIRECT-COUPLED TRANSISTORS

Bell Telephone Laboratory   Contributed by J. S. Mayo

Circuit 6-12 provides a simple clock source or generator for DCTL type circuitry and operates at speeds up to 5 mc.

The feedback loop provided by the delay line circulates any initial circuit disturbance with increasing amplitude until saturation is reached. Once saturation occurs, the storage time of the transistor will stretch the circulating pulse each time it circulates around the loop. As a result of the addition of storage time to the pulse each cycle, it is soon stretched to a length equal to the time length of the delay line. No further stretching can occur, since the base voltage will have negative feedback.

The output voltage is quite square with an amplitude of about 3 volts. The period is twice the delay line value.

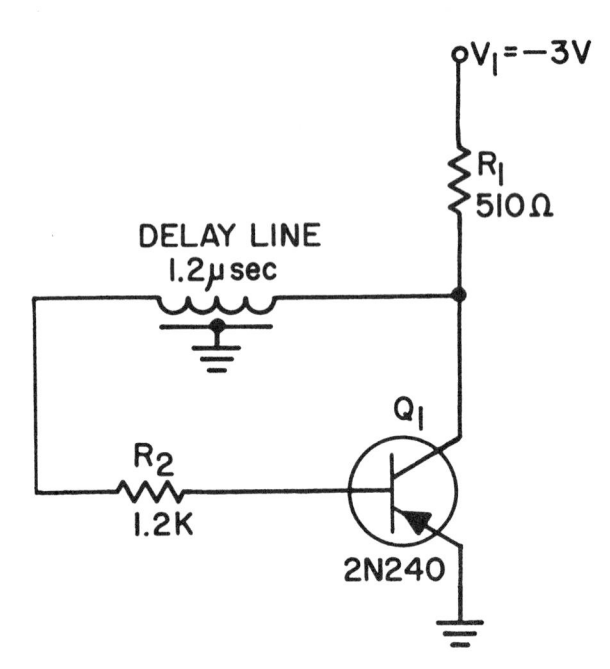

Circuit 6-12 - Astable circuit for direct-coupled transistors

CIRCUIT 6-13

BASIC SATURATED ASTABLE MULTIVIBRATOR

Transistor Applications, Inc.   Contributed by A. W. Carlson

Circuit 6-13 is a conventional astable multivibrator that is designed for reliability and simplicity.

The operation of the circuit is described briefly as follows. Assume that Q_1 has just switched "on," driving the base of Q_2 to a positive voltage

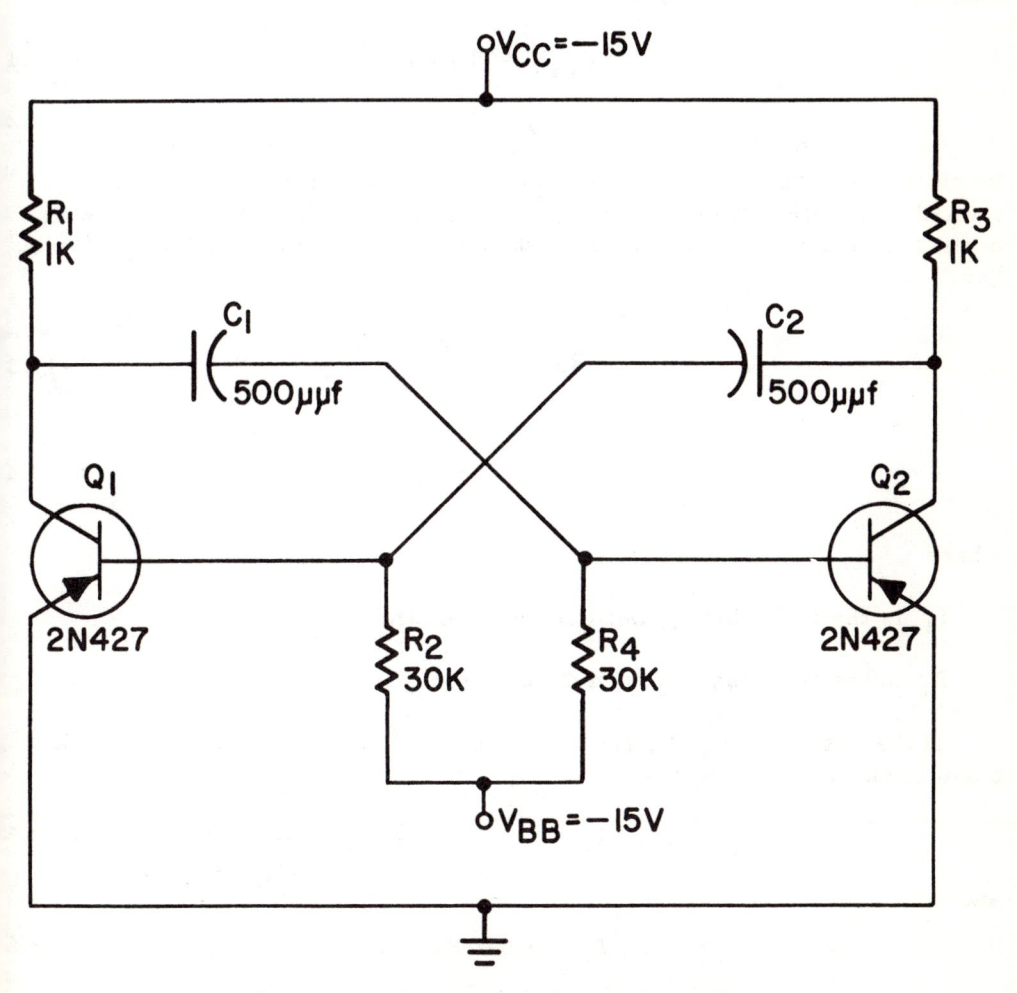

Circuit 6-13 - Basic saturated astable multivibrator

Circuit 6-13, Basic Saturated Astable Multivibrator (cont'd.)

and turning Q_2 "off." The base voltage of Q_1 (v_{B_1}) is a few tenths of a volt negative and C_2 quickly charges to nearly $-V_{CC}$. The collector of Q_1, the "on" transistor, is at nearly ground potential and C_1 is discharging toward $-V_{BB}$ with a time constant R_4C_1. When the voltage at the base of Q_2 starts to go negative, Q_2 begins to conduct with the positive going waveform at the collector transferred to the base of Q_1 through C_2 turning Q_1 "off" and driving Q_2 "on" harder. Q_1 remains "off" until C_2 has discharged with a time constant R_2C_1 from V_{CC} to a slightly negative value at which time Q_1 turns "on" and the cycle repeats.

The time intervals of conduction are easily determined, and in this analysis, the assumptions are that

$$R_1 C_1 \ll R_4 C_2 \qquad \text{6-13.1}$$

$$R_3 C_2 \ll R_2 C_1 \qquad \text{6-13.2}$$

When Q_1 turns "on," the base of Q_2 is driven toward V_{CC}. C_1 discharges toward $-V_{BB}$ with a time constant of R_4C_1. Q_1 turns "off" when the voltage across C_1 becomes zero. Similar statements apply to the situation for Q_2 "on."

$$T_1 = R_4 C_1 \ln \frac{V_{CC} + V_{BB}}{V_{BB}} \qquad \text{6-13.3}$$

$$T_2 = R_2 C_2 \ln \frac{V_{CC} + V_{BB}}{V_{BB}} \qquad \text{6-13.4}$$

where

T_1 is the time that Q_1 is in conduction, and

T_2 is the time that Q_2 is in conduction.

If the resistors R_2, R_4 are returned to $-V_{CC}$, the expressions for the conduction times become

$$T_1 = 0.692 \, R_4 C_1 \qquad \text{6-13.5}$$

and

$$T_2 = 0.692 \, R_2 C_2 . \qquad \text{6-13.6}$$

The repetition frequency is the reciprocal of the sum of T_1 and T_2.

Circuit 6-13, Basic Saturated Astable Multivibrators (cont'd.)

It is not necessary that $T_1 = T_2$. The conduction times may be adjusted by variation of time constants and the base voltages. If inequalities (6-13.1) and (6-13.2) are not met, the coupling condensers do not acquire a maximum charge equal to the supply voltage and the expressions for T_1 and T_2 as given above do not apply.

(Note in this case $V_{BB} = V_{CC}$ for single battery operation.)

The output voltage amplitude is approximately 15 volts and has rise and fall times of less than 1 μs. If capacitors C_1 and C_2 are 500 $\mu\mu$f, the multivibrator period is 20 μs. The circuit will utilize any recommended transistor type with a maximum collector breakdown voltage rating of over 15 volts.

CIRCUIT 6-14

MONOSTABLE MULTIVIBRATOR

Stanford Research InstituteContributed by D. P. Masher

The Circuit 6-14 is an example of a basic saturated monostable multivibrator. The usual design method is to position the charging circuit that determines the delay interval in the "off" transistor base circuit. The time constant is then independent of the transistor parameters, except for the reverse base to emitter resistance, and depends on the circuit resistance and capacitance. The delay time is,

$$T_D = R_5 C_3 \ln \frac{V_{CC} + V_{BB_2}}{V_{BB_2}}. \qquad 6\text{-}14.1$$

In this circuit, $V_{BB_2} = V_{CC} = -6$ volts. The static design procedure is similar to the flip-flop design, except that it may be more direct. The "on" transistor base circuit is independent of the "off" transistor and can be

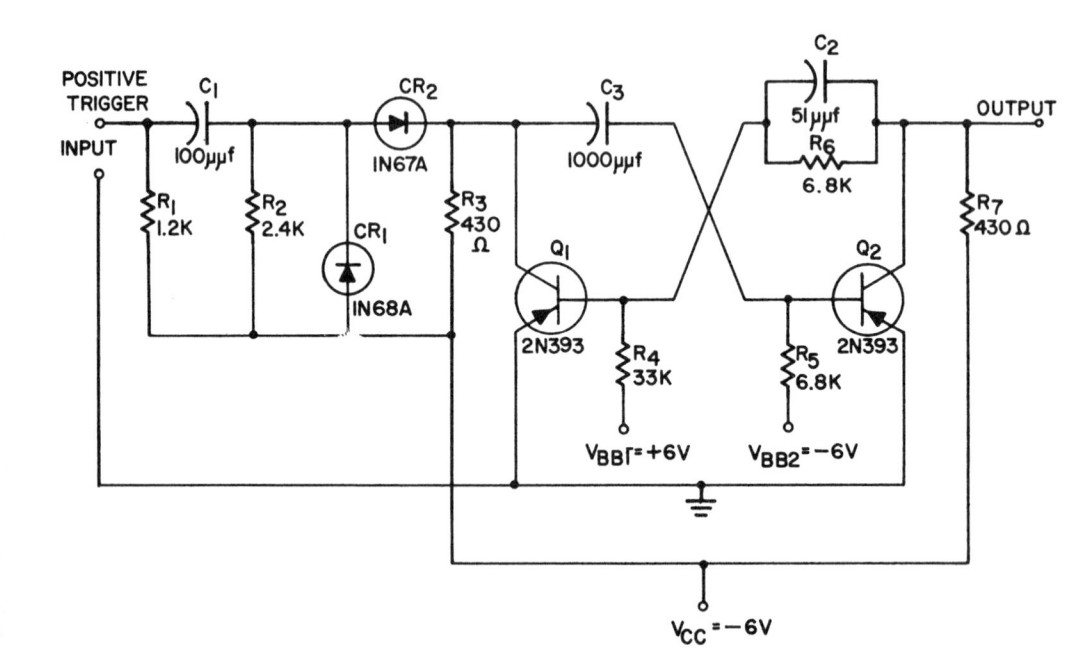

Circuit 6-14 - Monostable multivibrator

Circuit 6-14, Monostable Multivibrator (cont'd.)

designed to give a specific base current. The "off" base circuit is designed with I_{CO} stability or triggering the primary factors.

Circuit 6-14 has a transition time of less than 0.1 μs, and a minimum delay time of 1 μs. The maximum delay time is determined by the amount of jitter tolerable. Delay times of several hundred microseconds are possible with small jitter and a reasonable capacitance value. The input trigger required is greater than 4 volts and 0.1 μs wide. The temperature range is from -55°C to +55°C.

CIRCUIT 6-15

ASTABLE UNIJUNCTION TRANSISTOR MULTIVIBRATOR

General Electric Company   Contributed by E. Keonjian and J. J. Suran

The unijunction transistor multivibrator of Circuit 6-15, provides a considerable reduction of circuit components over the conventional circuit. Circuit 6-15 will be described briefly to illustrate the unique characteristics of the unijunction transistor. During astable operation, the capacitor C_1 is charged from the supply voltage V_{CC} through R_2 and the diode CR_1. During the charging cycle, the diode conducts but the unijunction transistor

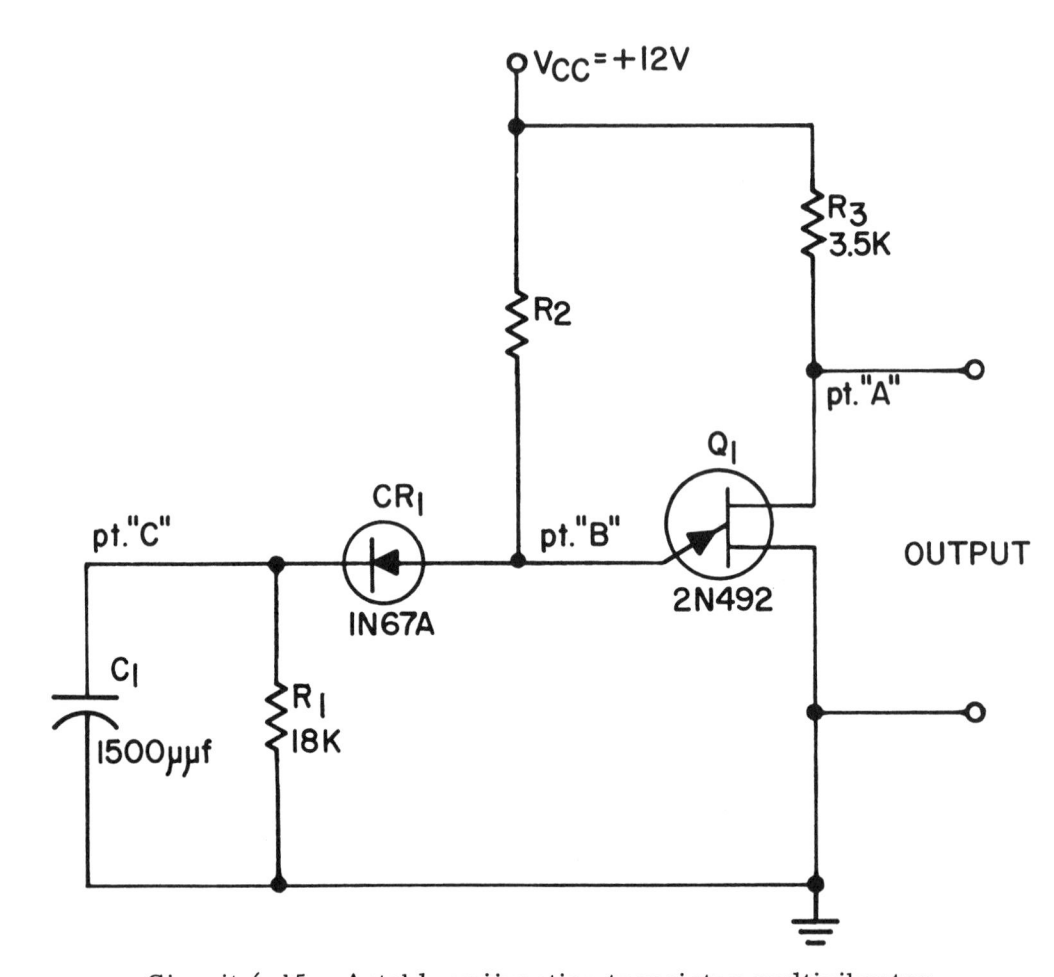

Circuit 6-15 - Astable unijunction transistor multivibrator

Circuit 6-15, Astable Unijunction Transistor Multivibrator (cont'd.)

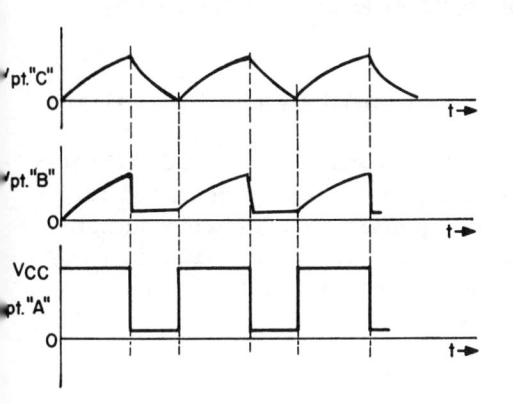

Fig. 6-15.1 - Waveshapes for Circuit 6-15

is in the cut-off state. When the voltage across the capacitor becomes equal to the peak point potential of the unijunction transistor, the latter becomes unstable and switches to the conducting state. The junction voltage at point "B" is then clamped almost to ground, causing the diode to become cut off. The capacitor must then discharge through R_1 until the potential at point "C" is approximately equal to the junction potential of the unijunction transistor. At this instant, the diode conducts again. The current through the unijunction transistor then decreases and the transistor is driven to cut off. The capacitor then begins to recharge and the cycle repeats. During the time the unijunction is cut off, the current through R_3 is low. However, when the unijunction transistor switches to the "on" state, its bar resistance decreases by an order of magnitude and the current through R_3 increases. Figure 6-15.1 shows the circuit waveshapes.

The unijunction transistor is a negative resistance device and its operation and the design considerations in an astable, bistable, and monostable circuit are similar to the negative resistance devices described in the design procedure. At present, the circuit speed using unijunction transistors is below 500 kc.

CIRCUIT 6-16

DIRECT-COUPLED SILICON MONOSTABLE MULTIVIBRATOR

Philco Corporation   Contributed by R. W. Carney

Circuit 6-16 is a direct coupled type silicon monostable multivibrator. It is useful for moderate or high speeds and this circuit is designed for low values of pulse width. The values given provide a 1 μs wide output and the circuit can be triggered at a 12 μs rate. The pulse width is controlled by C_1 (470 μμf) and may be changed to provide various pulse widths.

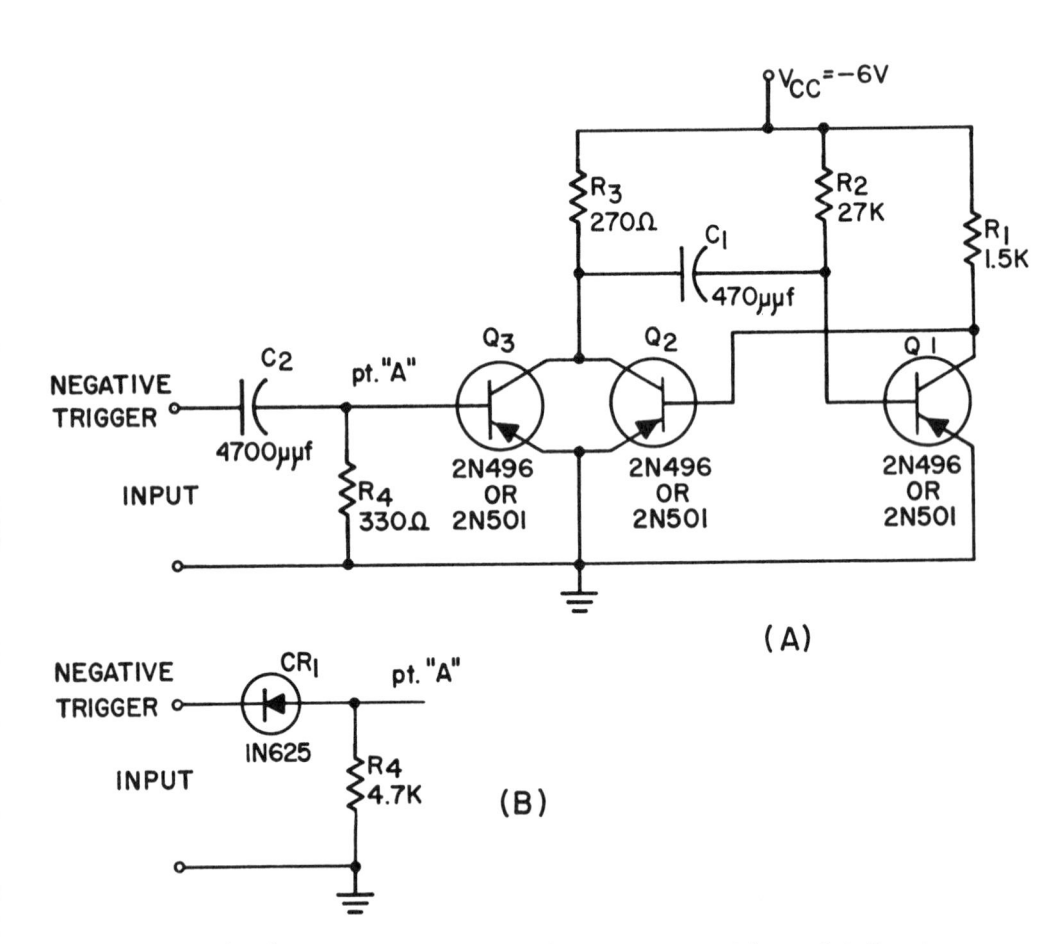

Circuit 6-16 - Direct-coupled silicon monostable multivibrator

Circuit 6-16, Direct-Coupled Silicon Monostable Multivibrator (cont'd.)

The design equations and delay times are similar to those described in Circuit 6-14, the Monostable Multivibrator.

By reducing R_2 to 12K, utilizing the diode resistor input circuit and replacing the transistors with the germanium 2N501 type, operation may be increased to 10 mc. The output rise time is less than 18 μs and pulse widths of 100 μs are obtainable by selecting C_1 = 7 $\mu\mu$f. For the high speed circuit described, the output is insensitive to large variations of V_{CC} and will operate up to 65° C.

CIRCUIT 6-17

FAST RECOVERY MONOSTABLE MULTIVIBRATOR

National Security Agency   Contributed by Norris Hekimian

Circuit 6-17 is a nonsaturating monostable multivibrator that is designed to permit very rapid recovery. The circuit is useful in applications where successive trigger pulses may occur very shortly after the delay pulse has terminated. When Q_2 turns off at the end of the pulse delay time, it results in an overshoot at the base of Q_1. The time for the base of Q_1 to recover to its normal level depends on the large time constant involving the capacitor C_2 and the base resistance R_4. Diode CR_1 effectively clamps the base and prevents the unwanted overshoot. Pulse

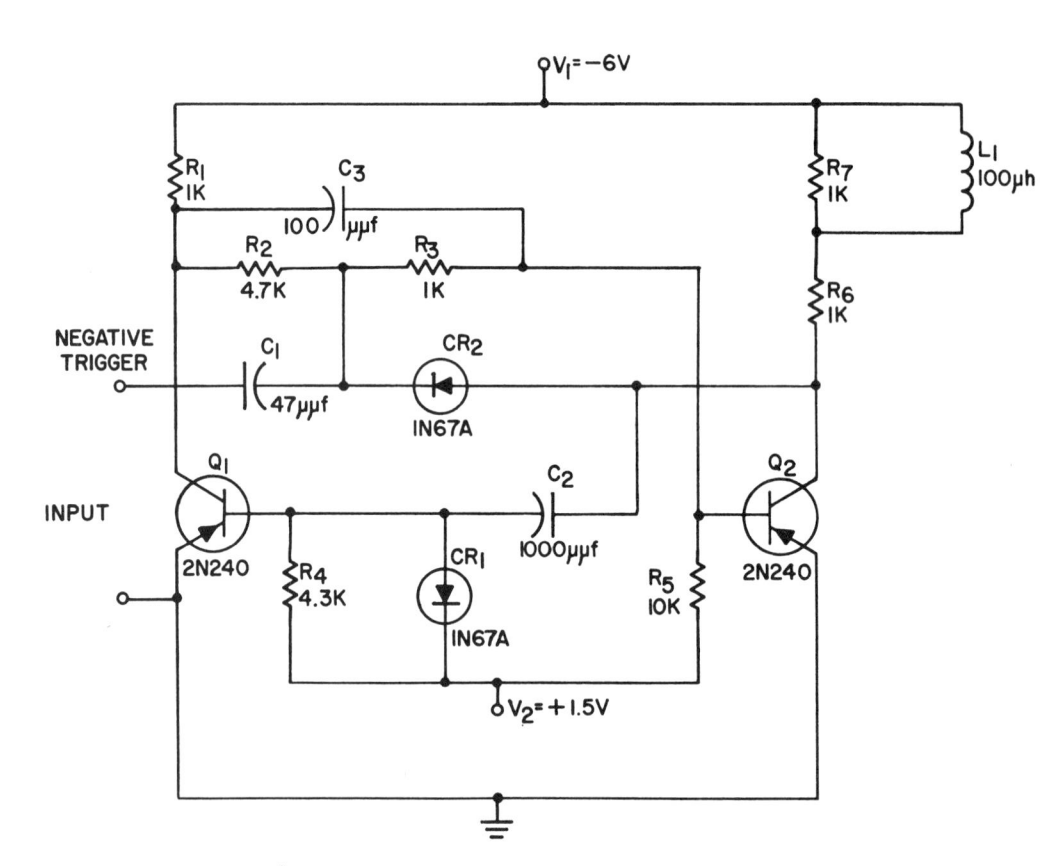

Circuit 6-17 - Fast recovery monostable multivibrator

Circuit 6-17, Fast Recovery Monostable Multivibrator (cont'd.)

widths less than 2 μs are obtainable by selecting the proper value of capacitor C_2.

The output voltage difference is about 6 volts with rise and fall times of less than 1 μs. The circuit allows complete interchangeability of equivalent type transistors (Ex: 2N240, 2N113).

CIRCUIT 6-18

SCHMITT TRIGGER CIRCUIT

Transistor Applications, Inc. Contributed by A. W. Carlson

Circuit 6-18 is a transistor equivalent of the Schmitt trigger circuit. It is used occasionally in switching circuits as a zero crossing detector and may also be used to give fixed output levels for predetermined input levels. The input impedance tends to be high and the sensitivity and stability are excellent.

For stability to exist, the relation of R_3 and R_4 must be related as given by Equation 6-18.1.

$$\frac{R_4}{R_3} < \frac{2\alpha_{FB} - 1}{1 - \alpha_{FB}}. \qquad 6-18.1$$

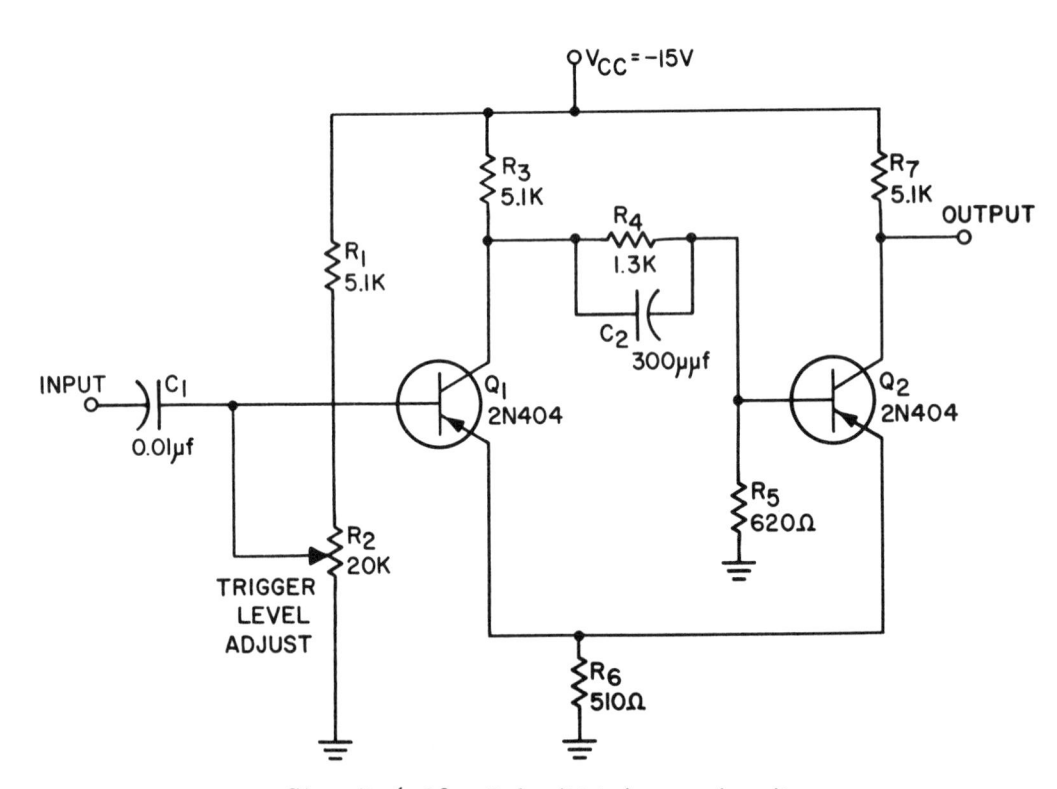

Circuit 6-18 - Schmitt trigger circuit

Circuit 6-18, Schmitt Trigger Circuit (cont'd.)

This is derived in a manner similar to the stability Equation 6-0.21 for the flip-flop. To provide positive regeneration, the relation of Equation 6-18.2 must be valid.

$$\gamma = D \left(A + \frac{C}{C+1} \right) > 1 \qquad 6\text{-}18.2$$

where

$$D = \frac{R_6}{R_5}$$

$$A = \frac{R_4}{R_3}$$

$$C = \frac{R_4}{R_5}.$$

To prevent Q_1 from saturating for large input voltages,

$$D(A+1) > \frac{R_7}{R_3}. \qquad 6\text{-}18.3$$

In addition, Q_2 is restricted from saturating by relating the output voltage Δv_o and other circuit values as given in Equation 6-18.4.

$$\Delta v_o \simeq \frac{R_7}{R_3} \left(\frac{C}{C+1} \right) \gamma V_{CC}. \qquad 6\text{-}18.4$$

The input voltage (Δv_{in}) required to change the state of the circuit is derived as

$$\Delta v_{in} \simeq \frac{ADC V_{CC}}{(C+1)\,\gamma\,(C+\gamma C+\gamma)}. \qquad 6\text{-}18.5$$

A final factor, the valley point voltage of the circuit input voltage-current characteristic establishes the low level switching point.

$$v_v = \frac{-V_{CC} A}{(C+1)\left(A + \frac{C}{C+1}\right)}. \qquad 6\text{-}18.6$$

Circuit 6-18, Schmitt Trigger Circuit (cont'd.)

The upper switching level is simply

$$v_v + \Delta v_{in}. \qquad 6\text{-}18.7$$

The actual design method is determined by the circuit characteristics desired. However, a convenient design procedure is started by selecting R_7 consistent with expected loading.

The emitter resistor R_6 is chosen to provide a high input impedance if that is desirable or to provide a particular output voltage swing. In either case, after R_7 is selected, the resistors R_4 and R_5 are calculated from I_{CO} considerations as described in Circuit 6-1, and the stability of Equation 6-18.1.

The trigger level adjust control R_2 selects the point at the input where the trigger circuit is to operate.

The circuit shown provides an output voltage of 10 volts with a rise time of 0.4 μs. The input may be a sine wave with 0.5 volt peak to peak voltage required for triggering. The maximum operating rate is about 300 kc.

CIRCUIT 6-19

BLOCKING OSCILLATOR

Air Force Cambridge Research Center

Contributed by
Rudolph A. Bradbury

Circuit 6-19 is a standard blocking oscillator circuit which has been designed for good stability over a reasonable range of supply voltages and transistor parameter variations. The circuit is designed to trigger on a 0.5 to 1.0 volt pulse from a source impedance of 1000 ohms.

The R_2, C_2 combination in the emitter return circuit is used to reduce the effect of charge storage variations on the pulse width. R_2 further reduces pulse variations due to changes in transistor input impedance and limits the peak emitter current when C_2 is omitted. Diode CR_1 offers a high impedance to the negative trigger pulse but conducts once triggering action is initiated, producing a low impedance path in the base circuit. This action improves the rise time and triggering sensitivity.

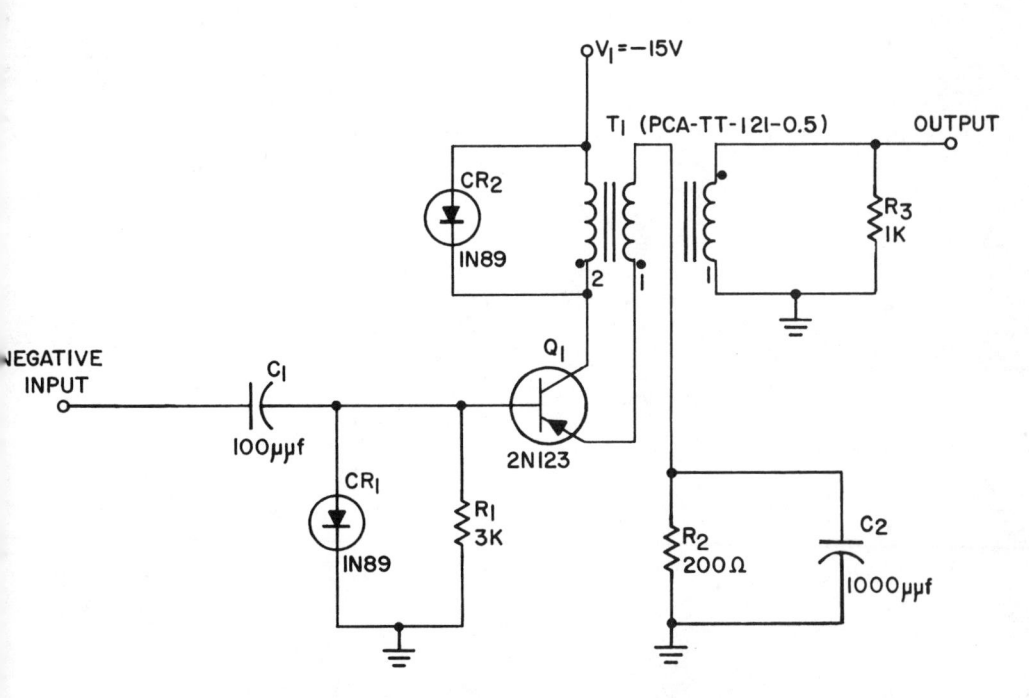

Circuit 6-19 - Blocking oscillator

Circuit 6-19, Blocking Oscillator (cont'd.)

The output pulse width is 0.6 μs with a 0.16 μs rise. The output voltage is 15 volts. Wider pulse widths can be obtained by using larger values of C_2. For example, 10,000 $\mu\mu$f gives a 1.5 μs wide pulse. Smaller values of R_2 will also give wider pulses but care must be taken not to reduce the impedance in the emitter circuit to such a value that excessive peak emitter current flows. With R_2 as low as 10 ohms peak currents of 200 ma have been measured. Whereas this value is not necessarily excessive, smaller values may run into conditions of thermal runaway.

CIRCUIT 6-20

BLOCKING OSCILLATOR WITH SQUARE LOOP TRANSFORMER

Fairchild Semiconductor Corp.   Contributed by Ray Kikoshima

Circuit 6-20 employs standard circuitry except that a square loop transformer is used providing excellent output pulse and uniformity.

The square loop core is normally biased in one direction by the bias current developed by V_1 and R_5. When an input is applied to the base of Q_2 from Q_1, the increase in collector current flows through the 30 turn winding in a direction that begins to switch the core to the opposite state. During the switching transient positive feedback from the core to the emitter provides regeneration turning Q_2 "on." The voltage developed by the core drops to zero rapidly because of the square loop characteristic

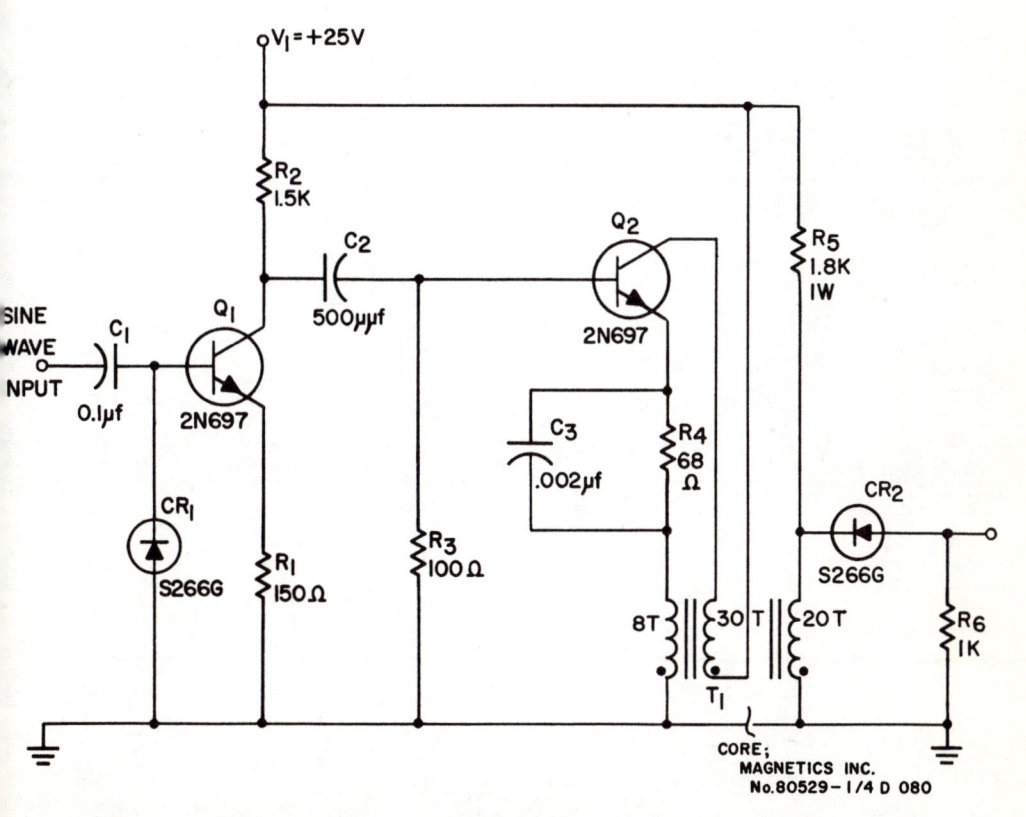

Circuit 6-20 - Blocking oscillator with square loop transformer

Circuit 6-20, Blocking Oscillator with Square Loop Transformer (cont'd.)

and the transistor turns "off." The bias current resets the core to its initial state after Q_2 turns "off." The complete "on" interval is designed to be about 0.1 μs and is determined primarily by the core characteristics. Excellent uniformity of output width exists for different transistors and remains virtually unchanged over wide temperature ranges (-50°C - +85°C).

The output rise time is less than 50 mμs and is typically 35 mμs. The output voltage developed is 16 volts into a 1K load. The circuit is designed to trigger with a sine wave input of 7 volts peak to peak at repetition rates up to 108.5 kc.

CIRCUIT 6-21

BLOCKING OSCILLATOR WITH DELAY LINE WIDTH CONTROL

Librascope   Contributed by D. Christensen, V. Olsen, R. Compton

Circuit 6-21 is a blocking oscillator that provides exact pulse widths over a wide temperature range and is reasonably independent of transistor characteristics. The input pulse initiates the regenerative action by causing emitter current to flow. The collector voltage rises and at the same time provides feedback to the emitter that increases the emitter current. The clamping circuit allows the collector to rise to a 6 volt level. The pulse output would normally remain on until the collector current could no longer supply both the increasing current demands of the transformer magnetizing inductance and the clamping circuit. However, the delay line in the emitter circuit inverts the original input pulse and returns it to the emitter starting a turn off action before the normal turn off time.

The transformer characteristics are chosen based on the design equations presented in the design philosophy to give a larger pulse width than is desired and the delay line then accurately determines the pulse width.

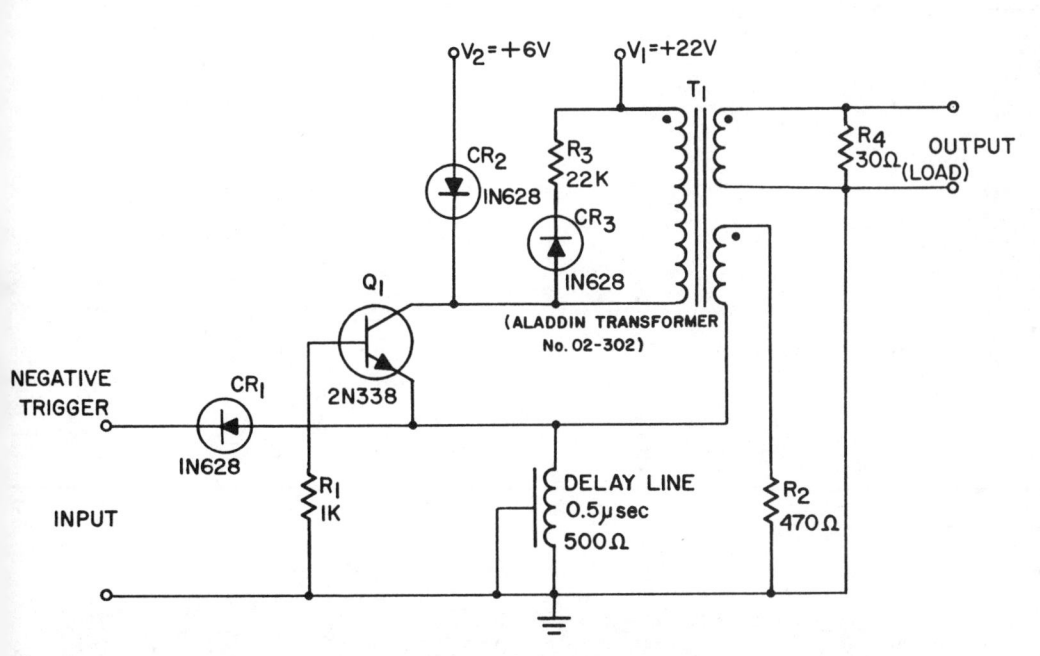

Circuit 6-21 - Blocking oscillator with delay line width control

Circuit 6-21, Blocking Oscillator with Delay Line Width Control (cont'd.)

The pulse width is twice the length of the delay line.

The circuit shown will trigger on a 3 volt input and provides a 1.5 volt output 1 μs wide. The output voltage may be changed by selecting a different output winding ratio. A repetition rate of 250 kc is possible with minor pulse width variation over a temperature range from -20°C to 100°C.

CIRCUIT 6-22

DUAL INVERTER CIRCUIT

I.B.M., Military Products Division   Contributed by W. N. Carroll

Circuit 6-22 is a dual input inverter that is designed for high speed and heavy driving capabilities. Transistors Q_1 and Q_1 are basic inverter stages that provide the dual input feature. The input is nominally ground or -3.5v and the resistor divider R_2 and R_1 are arranged to allow complete cutoff when the input is at ground. The diode CR_1 limits the reverse bias that can be applied to Q_1 to prevent damage to the emitter to base diode of Q_1. CR_1 also provides a fast recovery path for the positive pulses applied to the base of Q_1 during turn off. This action reduces the delay when turning on and allows for a faster repetition rate.

To obtain fast transitions from ground to -3.5 volts, emitter follower Q_2 is used. This transistor removes the capacitive load from the collector of Q_1 and allows the circuit to have a very fast fall time. The collector of Q_1 is clamped at -3.5 volts by the base collector diode of Q_2. During transitions from -3.5 volts to ground CR_2 is employed. This component

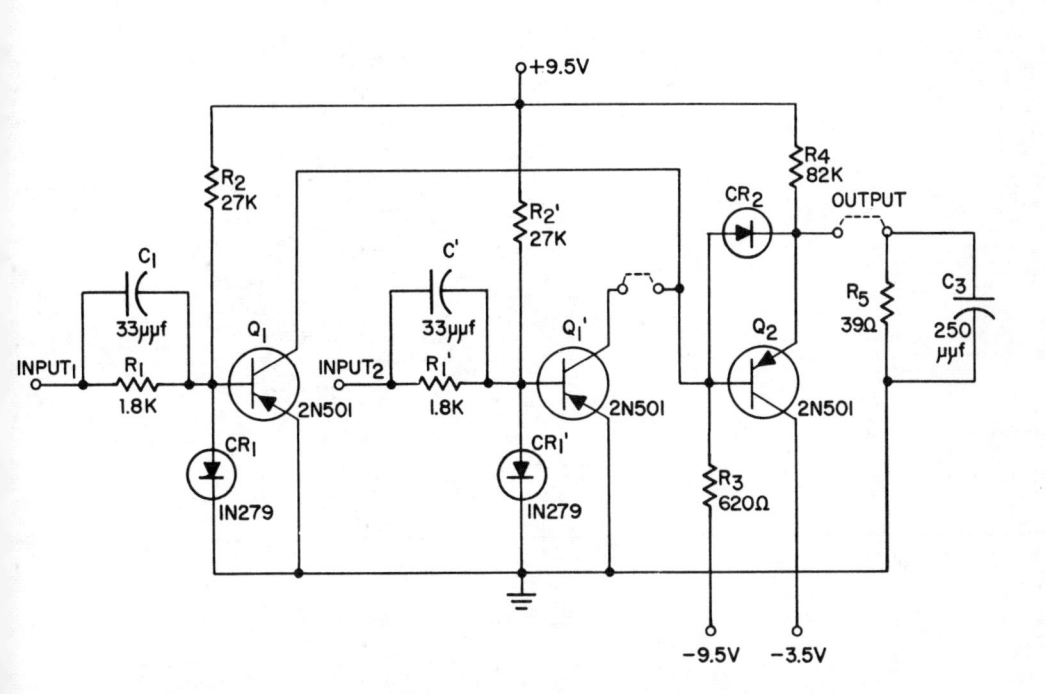

Circuit 6-22 - Dual inverter circuit

Circuit 6-22, Dual Inverter Circuit (cont'd.)

allows the fast rise inherently present at the collector of Q_1 to be coupled to the output. R_4 provides a positive bias to the emitter of Q_2. The positive bias is employed to reverse bias CR_2 during the period the output is at ground potential. Thus, with CR_2 cut off prior to the next transition of Q_1 from ground to -3.5 volts, the circuit is not delayed due to saturation in the diode.

The circuit is capable of driving a load of 39 ohms to ground and a capacitance of 250 $\mu\mu f$. With this load the circuit exhibits transitions from ground to -3.5 volts of 40 mμs and from -3.5 volts to ground at 25 mμs. The repetition rate is 10 mc.

Additional inputs such as Q_1' can be used to form logical circuits, and by cross coupling two inverters a flip-flop can be obtained.

CIRCUIT 6-23

SYMMETRICAL BUFFER INVERTER

Lincoln Laboratory, M.I.T. Contributed by R. H. Baker (3)

Circuit 6-23 uses the complementary circuitry described in the design philosophy of Section 1. The input voltage level is normally +5 volt or -5 volt but the circuit is designed to accept inputs above +2 or below -2 volt. The output voltage levels are ±5 volt with current capabilities of 10 ma for either level. The power consumed during a period when there is no loading is extremely low and represents one of the advantages of this type of circuitry. The output rise and fall time (tends to be equal assuming equivalent transistors) is less than 0.2 μs.

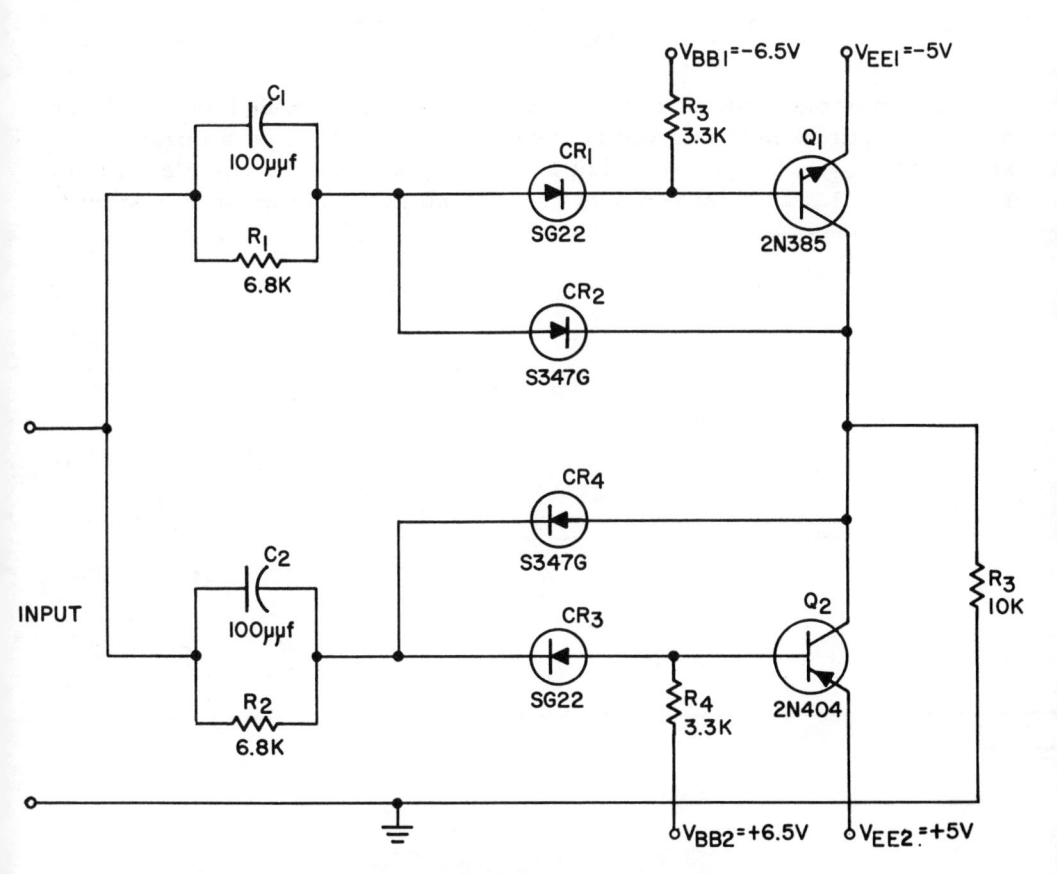

Circuit 6-23 - Symmetrical buffer inverter

Circuit 6-23, Symmetrical Buffer Inverter (cont'd.)

The design procedure is primarily concerned with the current capability and the input voltage level required for conduction. The input level necessary to turn on Q_1 or Q_2 is given by,

$$v_{on} = \frac{V_{BB} R_1 - V_{EE}(R_1 + R_3)}{R_3}. \qquad 6\text{-}23.1$$

The proper voltage polarities should be observed in using the equation. If the input pulse level is ±5 volts, as is the case for this circuit, the level at which turn on occurs is chosen and the ratio of R_1 and R_3 or R_2 and R_4 is obtained from Equation 6-23.1.

The base current when the input level is at its maximum is,

$$i_B = \frac{v_{in}}{R_1} - \frac{V_{BB}}{R_3} - \frac{V_{EE}}{R_1} - \frac{V_{EE}}{R_3}. \qquad 6\text{-}23.2$$

The maximum output current is chosen as determined by the loading and the required base current is then calculated based on a minimum current gain (α_{FE}). The resistances R_1 and R_3 may then be calculated from Equation 6-23.2 and the previous relation obtained for these resistances.

CIRCUIT 6-24

DIFFERENTIATOR AND PULSE SHAPER

Radio Corporation of America

There is frequently a need for a circuit that will shape a degenerated pulse into one that has a proper amplitude and rise time. Circuit 6-24 will accept a pulse or will differentiate a step input and provides a well defined output.

The circuit will accept a 6 volt step having a fall time of 1 μs and produces an output clamped at ground (by transistor saturation) and -6 volt. The normal output rise time and fall time are 0.2 μs. The circuit will operate over a temperature range of -30°C to +60°C.

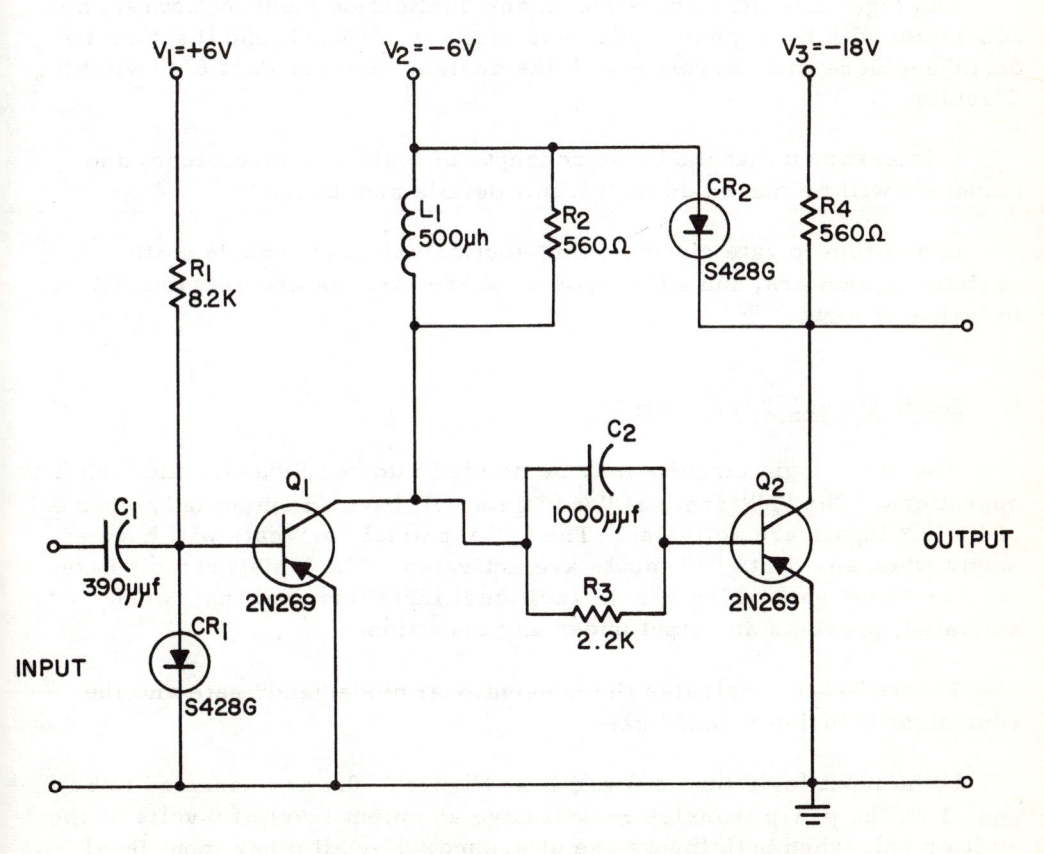

Circuit 6-24 - Differentiator and pulse shaper

PART 7

LOGIC CIRCUITS

Section 1. Design Philosophy

A. Logic Circuits

To perform all the operations demanded of switching networks, it is necessary to have logic circuitry, in addition to the basic switching circuits. Diodes were used primarily as the "and," "or" gates and other logic circuits, before the advent of transistors. The diode characteristics of transistors, however, have allowed transistors to be used directly in the logic circuits. The gain characteristics of transistors have reduced the necessity of additional amplifiers to recover the energy loss associated with passive type gates.

The logic circuitry employed in any application must, of course, be consistent with the type of basic switching circuit used, and the circuits described here will correlate with the designs given in Part 6 - Switching Circuits.

It is assumed that the basic concepts of logic are understood, and emphasis will be placed on the circuit details and design.

In addition to gate circuits, this section will also include shift registers, counters, and adders, since these circuits are best described in terms of logic.

B. Basic Gates. "And"-"Or"

The basic logic circuits may be divided into "or," "and," and "inhibit" operations. The "n" terminal "and" gate will have an output only when all of its "n" inputs are activated. The "n" terminal "or" gate will have an output when any of its "n" inputs are activated. The inhibit circuit is on "or" or "and" gate which has an additional input terminal that, when activated, prevents an output under any condition.

Figure 7-0.1 illustrates the conventional diode "and" gate and the equivalent transistor "and" gate.

If the input logic levels A and B of Figure 7-0.1 are assumed to be 0 and -10v, the p-n-p transistors will have an output level of 0 volts at the emitter only when both inputs are at ground. For all other input level combinations at A and B, the emitter output voltage level is -10v. The current transformation from emitter to base (described by Equation 6-0.5 in Part 6 - Switching Circuits) allows heavier loading at the emitter without reflecting to the circuits which generate the inputs A and B.

Figure 7-0.2 illustrates the conventional diode "or" gate and the transistor equivalent.

Again, if the input levels are ground and -10v, the output will be at ground, if either input A or B is at ground. When both input levels are at ground, then the output is, of course, also at ground level.

The gates commonly used with direct-coupled type circuitry are shown in Figures 7-0.3 and 7-0.4.

In Figure 7-0.3, an input to any of the parallel transistors will divert the current I_1 from the output

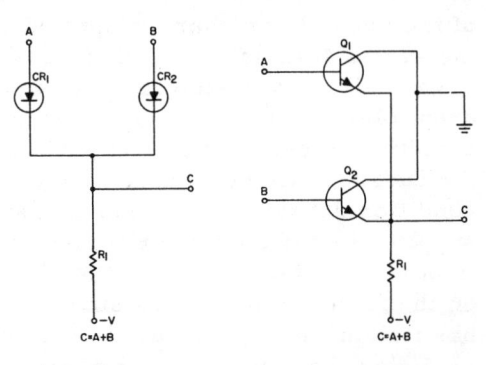

Fig. 7-0.2 - Conventional diode "or" gate with transistor equivalent

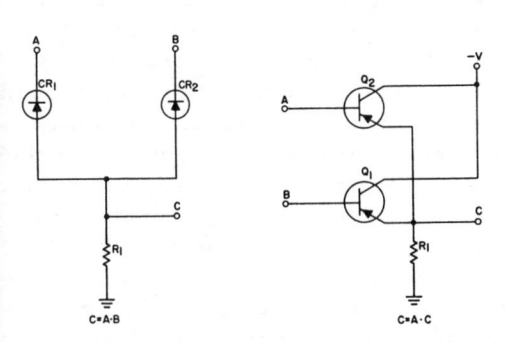

Fig. 7-0.1 - Conventional diode "and" gate with transistor equivalent

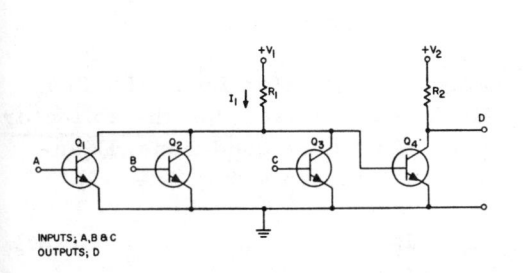

Fig. 7-0.3 - Direct-coupled transistor parallel gate

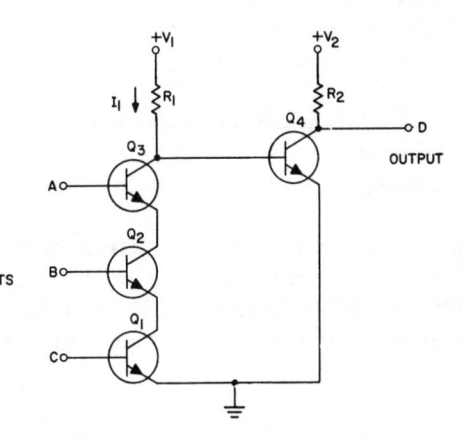

Fig. 7-0.4 - Direct-coupled transistor series gate

transistor Q_4 to the parallel transistor. The transistor Q_4 conducts, and its output level is low or near ground only when all the inputs to the parallel transistors are low. Similarly, when any input is above ground, the output level of Q_4 is above ground, since Q_4 is "off." The circuit operation is easily compared with the previous "or" circuit. The limitation of the number of parallel inputs is based on the leakage current (I_{CO}) of the parallel transistors. If the leakage current is large, part of the supply current normally diverted to Q_4 will be lost.

In Figure 7-0.4 whenever all inputs are activated or above ground, the supply current is diverted from Q_4 to the series path, and the output level at the Q_4 collector is high (above ground). If any one of the series transistors is "off," most of the supply current is fed to the base of Q_4, resulting in an output level near ground. The circuit operation is comparable in its function to the "and" circuit described previously. The limitation of the possible number of inputs depends on the saturation resistance of the series transistors. The total voltage drop for the "on" condition across all the transistors in series must be of the order of the voltage drop tolerated across one transistor in the parallel circuit to achieve the objective of reducing the output current to a few microamperes. Since the current gain decreases as the collector to emitter voltage in the "on" condition decreases, the above restrictions require more base current be supplied to series circuit transistors than to parallel circuit transistors. Another problem in the series circuit is the severe "on" condition imposed on the first and last transistor in the chain. In the last transistor, which has its emitter grounded, the collector current will be the supply current (I_1) plus the sum of all the base currents of other transistors in the chain. The base current, then, for the last transistor must be large to obtain the low collector to emitter voltage required. The first transistor in the chain, with its collector connected to the supply resistor must have a high base voltage to give adequate turn "on," since its emitter potential is above ground by the sum of the collector to emitter voltages of all the other transistors.

The transistor parameters required for stability and proper circuit operation are based on the previous discussion. The design procedure must, of necessity, include these transistor characteristics.

For the "off" condition in Direct-Coupled Transistor Logic (DCTL), the transistor emitter junction is slightly forward-biased, but the collector is reverse-biased by a few tenths of a volt. For these conditions, Equations 7-0.1 and 7-0.2 approximate the terminal currents that exist (1).

$$i_C = -\alpha_{FB} i_E + I_{CO} \qquad 7\text{-}0.1$$

$$i_E = - \frac{I_{EO}}{1 - \alpha_{FB}\alpha_{RB}} e^{\frac{q\phi_E}{kT}} + \frac{(1 - \alpha_{FB}) I_{EO}}{1 - \alpha_{FB}\alpha_{RB}} \qquad 7\text{-}0.2$$

where

I_{EO} is the emitter current that flows when the emitter junction is reverse-biased and the collector is open circuited.

I_{CO} is the collector current that flows when the collector junction is reverse-biased and the emitter is open circuited.

α_{FB}, α_{RB}, are the normal and reverse alpha values.

ϕ_E is the junction voltage.

k is Boltzmann's constant, and

T is the temperature in ° Kelvin.

For the "on" condition, the DCTL circuits operate in the saturated region. The collector to emitter voltage must be low, since it is applied directly to other off stages as a forward bias. The collector to emitter voltage in saturation is given in Equation 7-0.3.

$$v_{CE} = \pm \frac{kT}{q} \ln \left[\frac{1 + \frac{i_C}{i_B}(1 - \alpha_{RB})}{\alpha_{RB}\left(1 - \frac{i_C}{i_B} \cdot \frac{(1 - \alpha_{FB})}{\alpha_{FB}}\right)} \right] + i_C r_C' + (i_C + i_B) r_E' \qquad 7\text{-}0.3$$

where

r_C' is the ohmic body resistance of the collector, and

r_E' is the ohmic body resistance of the emitter.

The primary requirement for stability in DCTL circuits is that an adequate voltage margin (Δv) exist between the maximum "on" collector to emitter voltage v_{CE} and the minimum base to emitter voltage v_{BE} required for an "off" condition.

Since the "on" and "off" conditions are not as positive as in other circuit designs, it is necessary to establish a maximum value of collector current in the "off" state relative to a maximum allowed value of v_{BE}. The "on" state is defined by establishing a maximum value for v_{CE} and the current gain α_{FE} which corresponds to this value of v_{CE}.

To obtain a large value for Δv, the I_{CO} must be low. Low values of I_{CO} permit the base voltage to be large, giving rise to a larger Δv.

The internal collector and emitter body resistance contribute to a reduction of Δv by increasing v_{CE}. This effect is generally negligible in germanium transistors, and may be reduced further by operating at low current levels.

From the previous relation for v_{CE}, and based on other considerations, high values of α_{FB} and α_{RB}, and low values of i_C/i_B; lead to the desired low value for v_{CE}. It is necessary to find a compromise between high α_{FE} values for fan out capabilities and low values of α_{FE} for minimum v_{CE}.

The practical values of v_{CE}, I_{CO} and current gain i_C/i_B considered adequate for the DCTL circuits are,

$$v_{CE} = 75 \text{ mVdc with } \frac{i_C}{i_B} > 10$$

$I_{CO} < 100$ μAdc with a forward bias of 75 mVdc.

These characteristics are intended as examples of acceptable parameters and complete specifications, of course, depend on the system requirements and compromises that may be desirable.

In general, the DCTL circuits require more uniform parameters than other circuit types. Large internal base resistance and high values of α_{RB} allow some relaxation of the control of the other general parameters. Symmetrical properties (as well as high values of α_{FB}, low I_{CO} values, and low internal resistances (r_C', r_E') constitute optimum transistor characteristics for this type of circuitry. A reduced current gain for current levels corresponding to the "off" condition is a further advantage.

An extension of external resistances for improved stability has lead to another class of circuits commonly referred to as Resistor-Transistor Logic (RTL). In addition, a simplification of the general gate circuitry is possible with RTL, as shown in Figure 7-0.5 (2).

If any of the inputs A to D are activated (raised), the transistor

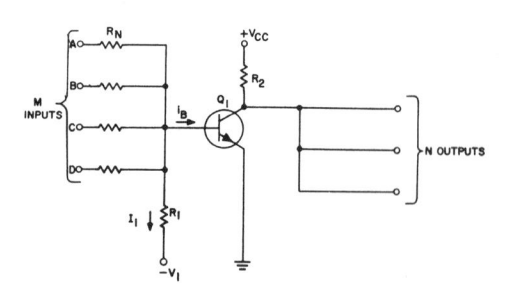

Fig. 7-0.5 - Resistor-transistor gate

conducts. The resistors perform the "or" gating while the transistor amplifies and inverts.

In the design procedure for this type of logic, the worst case is chosen to provide adequate stability. A severe condition is the case where one output is required to drive a certain number (N) of output resistors. From Reference 2, the number of outputs N that may be driven is

$$N = \frac{i_C}{Si_B + I_{CO} + \frac{V_{BE}(M-1)}{R_N}} - \frac{i_C R_N}{V_{CC}} \qquad 7\text{-}0.4$$

where

S is a safety factor that provides for a reduction in gain and other circuit variations.

I_{CO} is the base to collector leakage current with the emitter open circuited,

v_{BE} is the emitter to base "on" voltage,

M is the number of input circuits, and

R_N is the circuit input resistor.

In using Equation 7-0.4 in the design of the gate circuit, several transistor parameters must be established. First, the base current i_B required to provide a collector current i_C in the expected range of operation is determined. Second, the I_{CO} term must be measured at the maximum operating temperature and voltage level desired. The base to emitter voltage v_{BE} may be measured, but an approximate value can be taken as 0.25 volts. The design factors M the number of inputs, and S the safety factor are chosen based on the system requirements. The supply voltage V_{CC} is chosen as large as possible, consistent with the transistor rating, since this reduces the effect of the negative term in the equation.

Once the transistor parameters, design factors, and V_{CC} are determined, Equation 7-0.4 is a function of the input resistor R_N only. It is desirable to make a graph of N versus collector current values for different values of R_N. The final resistance value for R_N is selected from the graph to give the desired value of N at a reasonable collector current level. After R_N is determined, the corresponding collector current is noted and the collector resistor R_2 may be calculated from

$$R_2 = \frac{V_{CC}}{i_C}. \qquad 7\text{-}0.4a$$

The base bias network is designed to provide a reverse current (I_1) at least equal to the maximum I_{CO} value. The base supply voltage is chosen and R_1 calculated as

$$R_1 = \frac{V_1}{I_1}. \qquad 7\text{-}0.4b$$

The power dissipation and cross talk problems should be checked before any final design is accepted. In some cases, it is possible to design the resistor-transistor gate circuit to give desired operational characteristics and, in addition, impose minimum power dissipation on the transistor. To obtain the proper design equations for minimum power, Equation 7-0.4 may be rewritten in terms of R_2 as (3)

$$R_2 = \frac{R_N}{\dfrac{1}{\dfrac{S}{\alpha_{FE}} + \dfrac{I_{CO}}{i_C} + \dfrac{v_{BE}(M-1)}{R_N\, i_C}} - N}. \qquad 7\text{-}0.5$$

From Reference 3, a minimum value of R_2 and consequently a possible minimum power dissipation exists when R_N is

$$R_N = \frac{v_{EB}(M-1)}{i_C \left\{ \pm \sqrt{\dfrac{1}{N}\left(\dfrac{I_{CO}}{i_C} + \dfrac{S}{\alpha_{FE}}\right)} - \left(\dfrac{I_{CO}}{i_C} + \dfrac{S}{\alpha_{FE}}\right) \right\}}. \qquad 7\text{-}0.6$$

The collector current for this minimum is given as

$$i_C = \frac{I_{CO}\left(-1 + \dfrac{4_N S}{\alpha_{FE}} + \sqrt{1 + \dfrac{8_N S}{\alpha_{FE}}}\right)}{2\left[\dfrac{S}{\alpha_{FE}} - N\left(\dfrac{S}{\alpha_{FE}}\right)^2\right]}. \qquad 7\text{-}0.7$$

Substituting Equation 7-0.7 in Equation 7-0.6 gives

$$R_N = \frac{v_{EB}(M-1)}{I_{CO}}. \qquad 7\text{-}0.8$$

Starting with Equation 7-0.8, the input resistor R_N is determined and the correct collector current may be calculated from Equation 7-0.7, after deciding on the other factors and parameters. This design procedure does not guarantee minimum power in all cases and does not give as clearly the design considerations as the previous method. In addition, it is frequently desirable to provide enough base current, either by lower resistance values or speed up capacitors, so that minimum power dissipation is not a primary design consideration.

Another type of logic network is illustrated in Figure 7-0.6. The design utilizes diodes primarily, but provides gain in the form of transistors at the gate output.

The gate has a reasonably high input impedance, but unlike the resistor type network, has excellent speed capabilities.

The design method for this type of logic depends, of course, on the levels desired and maximum loading capabilities of the output. There is reasonable design freedom in choosing the levels, and the loading capabilities are related to the gain characteristics of the transistor.

The voltage drop across the diodes and the transistors tends to compensate, assuming both are germanium or silicon. In effect, there is very little deterioration of the logic level in a long chain of logic networks. However, the circuit is subject to oscillations for long series networks.

An additional gate circuit that has found acceptance recently employs a transformer and diode arrangement, as shown in Figure 7-0.7. It is particularly useful for graded-base transistors, and other transistors that have low emitter to base breakdown voltage. In addition, it requires low input clock power and is generally unaffected by diode leakage current and switching time.

In Figure 7-0.7 the input clock applied at the transformer primary reverses diode CR_3, which initially holds the base level at near ground.

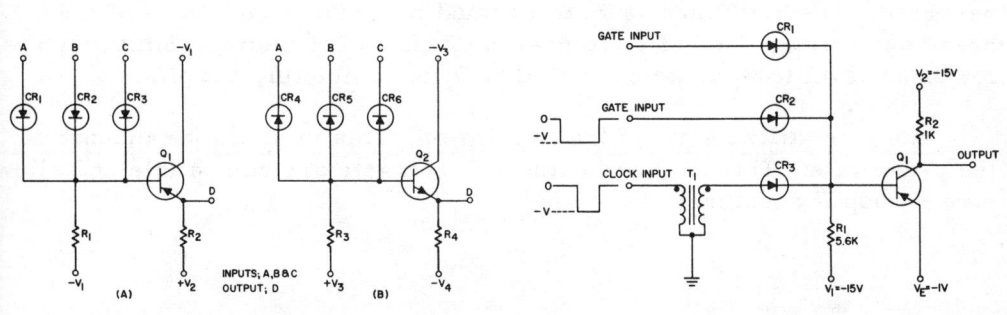

Fig. 7-0.6 - Diode-transistor gates   Fig. 7-0.7 - Transformer-diode gate

If the gate input is at its lower level -v, the transistor Q_1 is allowed to turn "on." If the gate input is at ground level, the transistor base is held at ground and Q_1 remains in an "off" condition.

When the clock is applied coincidentally with a low gate level, the base voltage drops toward V_1 with a time constant related to the stray base capacitance and R_1. This results in a delay time that is inherently small if the voltage V_1 is much more negative than V_E.

A class of logic circuits employing current steering has been developed that achieves optimum response and stability (4). The basic circuit characteristics have been described in Part 6 - Switching Circuits, Circuit 6-4. Fundamentally, the

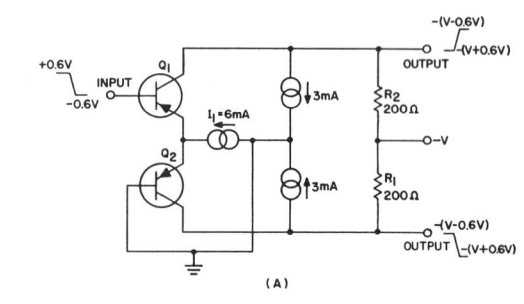

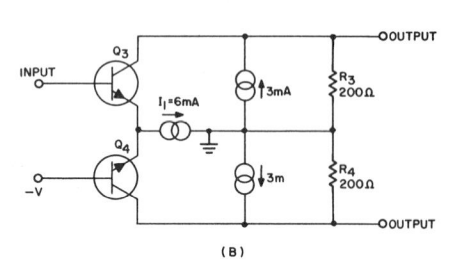

Fig. 7-0.8 - Current switching transistor gates

design technique establishes nonsaturating operation to avoid delays and low voltage operation to reduce the effect of capacitance. The circuit operation employs current sources that establish specific current levels. The gates of Figures 7-0.8A and 7-0.8B are examples of the basic current switching circuits. If the input to Q_1 is at the upper level (+0.6v), Q_1 is "off" and Q_2 accepts the current I_1. The collector current of Q_2 provides I_1 - 3 ma of current into the load resistor R_1, raising the output voltage to 0.6v above -V, if I_1 is taken as 6 ma. Since Q_1 is "off," the current I_1 flows through R_2 in a direction that lowers the output to -(V + 0.6v). When the input to Q_1 is low, Q_1 accepts the current I_1 and the output levels are reversed. The high input level to Q_1 must be greater than the emitter to base "on" voltage drop of Q_2 to prevent Q_1 from conducting. Similarly the low input level to Q_1 must be such that Q_2 is completely cut off.

The gate utilizing n-p-n transistors operates in a similar manner to the p-n-p gate. The input and output of each gate are compatible and allow direct-coupling features.

C. Counters and Shift Registers

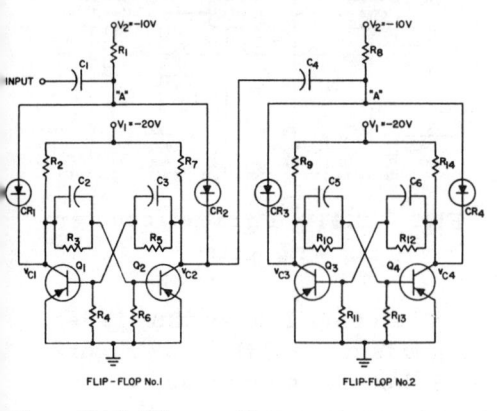

Fig. 7-0.9A - Two-stage serial counter with diode steering

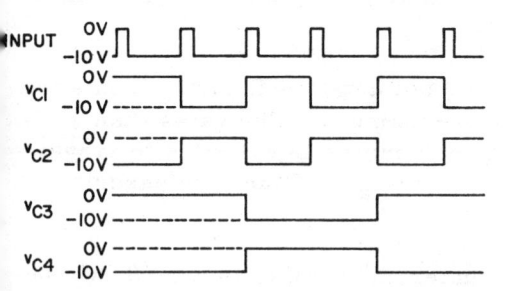

Fig. 7-0.9B - Waveforms for two stage serial counter (Fig. 7-0.9A)

To illustrate the design procedure for combining logic networks and basic circuits, various types of counters and shift registers will be described. Figure 7-0.9A is a two-stage serial counter that utilizes the resistor capacitor coupled saturated flip-flop and diode steering and Figure 7-0.9B is a set of waveforms for the counter.

The input pulse at point "A" is blocked from the diode CR_2 path because of the bias level across the diode if Q_2 is assumed "on" and Q_1 "off." However, the diode CR_1 is biased such that it allows the input pulse to reach the collector of Q_1 and initiates a change of state of the flip-flop. The collector of Q_2 is connected to the input of the second flip-flop, but the negative voltage drop v_{C_2}, as Q_2 turns "off" will not affect the second flip-flop. At the next input pulse CR_1 is cut off while CR_2 conducts again, allowing a change of state of flip-flop one. The rising voltage at the collector of C_2 is coupled to the input of the second flip-flop and passes through CR_3, if Q_3 is "off" and Q_4 is "on," causing a change of state of this flip-flop. At the next input pulse, CR_1 conducts and the first flip-flop reverses its state. Since the collector of Q_2 drops during this time, there is no change in the second flip-flop. At the next input, v_{C_2} rises and is coupled through CR_4 to the collector of Q_4, producing a change in state of the second flip-flop. The diode steering network achieves a commutative operation that is simple and reliable. The waveforms of Figure 7-0.9B illustrate the counter operation completely.

The serial counter may also be designed using series "and" gates as shown in Figure 7-0.10 (5).

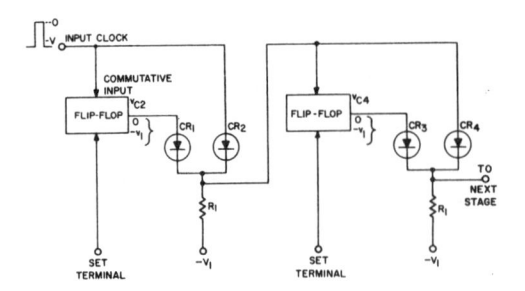

Fig. 7-0.10 - Serial counter with series gate

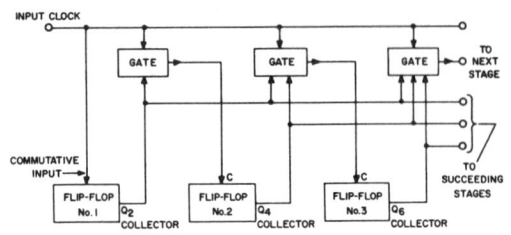

Fig. 7-0.11 - Parallel counter with gates

The overall delay time is less using this method than for Figure 7-0.9, but the input pulse is attenuated by each gate and may require restoration after several stages. The set terminal at each counter stage allows the counter to be preset to a known condition.

Long serial counters with many stages have an accumulative delay from one stage to the next that can cause timing problems. Figure 7-0.11 is a parallel type counter that avoids this delay problem but requires more diode circuitry.

The gate outputs are connected to the commutative flip-flop inputs and each gate is controlled by the flip-flops preceding it. The gates then become progressively more complex as the counter is extended to larger numbers. The gates are basic diode "and" circuits of the type described earlier.

Another type of counter, the ring counter, is similar to a shift register and, as implied from the name, the last stage is connected to the first stage. One of the stages is set to the "1" state with the rest of the stages in the "0" state. The input pulse then moves the "1" state to succeeding positions with each input pulse. An example of a ring counter is given in Circuit 7-9.

It is often necessary to convert a serial code to a parallel one. The shift register represents a class of circuits that perform this function. A series of flip-flops, coupled such that a shift pulse applied simultaneously to all stages moves the information stored in each stage to the succeeding stage, has the essential characteristics of a shift register.

Figure 7-0.12 is a block diagram of a shift register. The reset input of each flip-flop, in addition to being an integral part of the shift operation, is also used to establish a reference state. The gates G_1 and G_2 are controlled by the collectors of the first stage and when a shift pulse is applied,

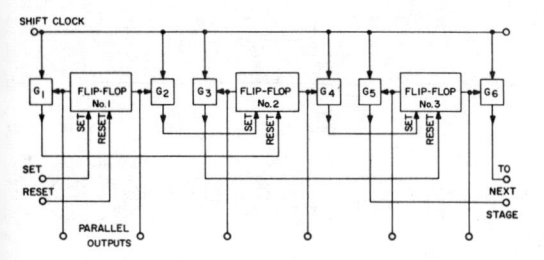

Fig. 7-0.12 - Shift register block diagram

the second flip-flop receives an input, depending on which gate is enabled. The gates are designed to route the shift pulse to the proper input of the succeeding stage in such a manner that the state of each flip-flop is transferred to the following stage.

D. High Speed Gating

The steering and gating operations at high speeds present unique problems not generally encountered at slower speeds and will be described as a separate topic. The logic circuits discussed so far operate satisfactorily at moderate speeds, but "race" problems can exist at high speeds. A race may occur when the circuit to be triggered controls a gate. If the output of the gate, in turn, controls the circuit, successful operation depends on the relative speed of each. For example, in the circuit of Figure 7-0.13, the gates must not be reversed by the flip-flop until the end of the input pulse. This implies a delay, which, in the case of relatively slow circuitry is the built-in delay of the circuit or, in high speed circuits, must be provided externally.

Indiscriminate use of external delays to slow down the circuit must be avoided, as they ultimately defeat the high speed circuit capabilities. The direct approach requires control of the input pulse width to values less than the built-in circuit delay. This generally is expensive, or results in critical circuit operation. The use of trailing edge logic, which initiates circuit triggering at the end of the input pulse avoids any race problems. However, the inherent triggering delay of this method is not permissible for maximum speeds.

Figure 7-0.14 is an example of delay positioning at the gate that avoids a reduction in the response of the basic flip-flop circuit (6). The capacitor C_1 and the resistors R_4 and R_5 form an integrating network that delays the output levels of the gate transistors Q_1 and Q_2 with negligible effect on the collector current rise and fall time. In addition,

Fig. 7-0.13 - Flip-flop and gate with required delay for high speed operation

Fig. 7-0.14 - High speed gate circuit with delay to avoid "race" problem

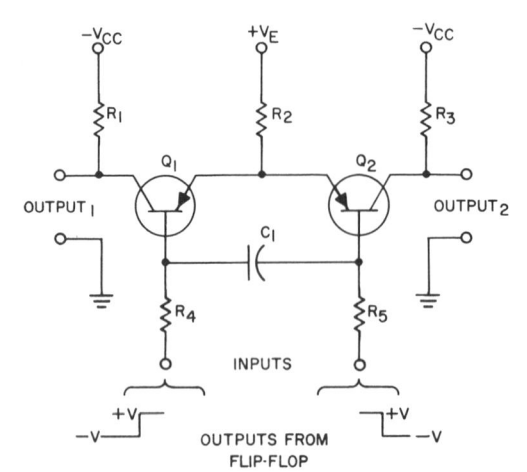

Fig. 7-0.15 - Conditional steering gate and flip-flop

the flip-flop is not slowed down and maximum response capabilities are preserved.

It is possible to avoid race problems in another way, that is referred to as conditional steering (7). In Figure 7-0.15, the differentiated leading edge of the input pulse by the parallel gate circuits Q_1, Q_2, and Q_5, Q_6 provide a trigger for the flip-flop, but the same input pulse inhibits the change in d-c state of the gates. If, for example, Q_4 is "on" and Q_3 is "off," then Q_2 will be "off" and Q_5 will be conducting. When the input pulse is applied, Q_1 will turn "on" and its collector voltage rise is differentiated by capacitor C_1 and applied to the base of Q_4, turning it "off." The flip-flop reverse its state and Q_2 turns "on" but does not affect the gate output, since its collector is already at a saturated level determined by Q_1. Similarly, Q_5 turns "off" but does not affect the gate, since Q_6 is conducting due to the input pulse. When the input pulse terminates, the gates are allowed to resume a normal condition with Q_2 "on" and Q_5 "off." The voltage change at the gate outputs at this time is of a polarity that has no effect on the flip-flop.

The conditional steering described exhibits excellent reliability and wide design tolerances, but decreases the maximum operating frequency because of the recharging time required by the capacitors. Variations of conditional steering to improve the switching speed are found in Circuits 7-11 and 7-13.

REFERENCES

1. Easley, J. W., Transistor Characteristics for Direct-Coupled Transistor Circuits, IRE Transactions EC-7, 6 (1958).

2. Rowe, W. D., The Transistor NOR Circuit, IRE Wescon Convention Record, Part 4, 1957, pp. 231-245.

3. Cox, E. L., The Design of a Transistor NOR Circuit for Minimum Power Dissipation, Symposium on Microminiaturization of Electronic Assemblies, Sept. 30, 1958.

4. Yourke, H. S., Millimicrosecond Transistor Current Switching Circuits, 1957 IRE-AIEE Conference on Transistors and Solid State Circuits.

5. Carlson, A. W., Junction Transistor Counters, Transistor Applications, Inc. Report.

6. Baker, R. H., High-Speed Graded-Base Transistor Switching Circuits, TR 130, Lincoln Laboratory, M.I.T.

7. Clark, E. G., Direct-Coupled Transistor Logic Complementing Flip-Flop Circuits (Two Parts), Electronic Design, June, August, 1957.

BIBLIOGRAPHY

Carlson, A. W., "A New Ring Counter for Junction Transistor and Vacuum Tubes," AFCRC-TN-54-100, June, 1954. AFCRC, Bedford, Mass.

Carroll, W. N. and Cooper, R. A., "Ten Megapulse Transistorized Pulse Circuits for Computer Application," Semiconductor Products, July/August, 1958.

Cox, E. L., "The Design of a Transistor NOR Circuit for Minimum Power Dissipation," Symposium on Microminiaturization of Electronic Assemblies, September 30, 1958.

Emile, P., "Design of a Two Transistor Binary Counter," Symposium on Microminiaturization of Electronic Assemblies, September 30, 1958.

Giguere, W. J., Jamison, J. H. and Noll, J. C., "Transistor Pulse Circuits for 160 MC Clock Rates," 1959 Transistor and Solid State Conference.

Harris, J. R., "Direct-Coupled Transistor Logic Circuitry," IRE Transactions, EC-7, 2, 1958.

Henle, R. A. and Walsh, J. L., "The Application of Transistors to Computers," Proceedings of IRE, 46, pp. 1240-1254, June 1958.

Horn, I., "An Emitter-Follower-Coupled High Speed Binary Counter," IRE Wescon, 1958.

Simkins, Q. W., "Transistor Resistor Logic Circuit Analysis," 1958 Transistor and Solid State Circuits Conference.

Szerly, A., "Transistorized Decade Counter," IRE Wescon, 1958.

Yakelson, B. J., and Ulrick, W., "Engineering Multistage Diode Logic Circuits," Transistor AIEE Commun. & Electronics, 466, (1955).

CIRCUIT 7-1

CURRENT STEERING GATE

Lincoln Laboratory, M.I.T.　　　　　　　　　Contributed by R. H. Baker

Circuit 7-1 is a simple gate that is useful for gating in serial counters, pulse steering networks, and other applications. The circuit utilizes current steering and displays excellent stability and maximum response (4). The emitter resistance R_1 is chosen to give adequate output drive and minimum loading at the base input control.

$$R_1 < \frac{\alpha_{FB}(v_t)}{i_C} \qquad 7\text{-}1.1$$

$$R_1 > \frac{(1 - \alpha_{FB})(v_t)}{i_B} \qquad 7\text{-}1.2$$

where v_t is the trigger amplitude.

The output resistors R_2 and R_3 are chosen to provide a proper impedance and voltage level. When the right side of the gate is "off," the collector level of Q_2 is at $V_1 = -5v$. The supply voltage, $V_1 = -5v$, may be chosen as a desirable logic level. When Q_2 is "on," the collector level is

$$v_{C_2} = V_1 + \alpha_{FB} i_E R_3 . \qquad 7\text{-}1.3$$

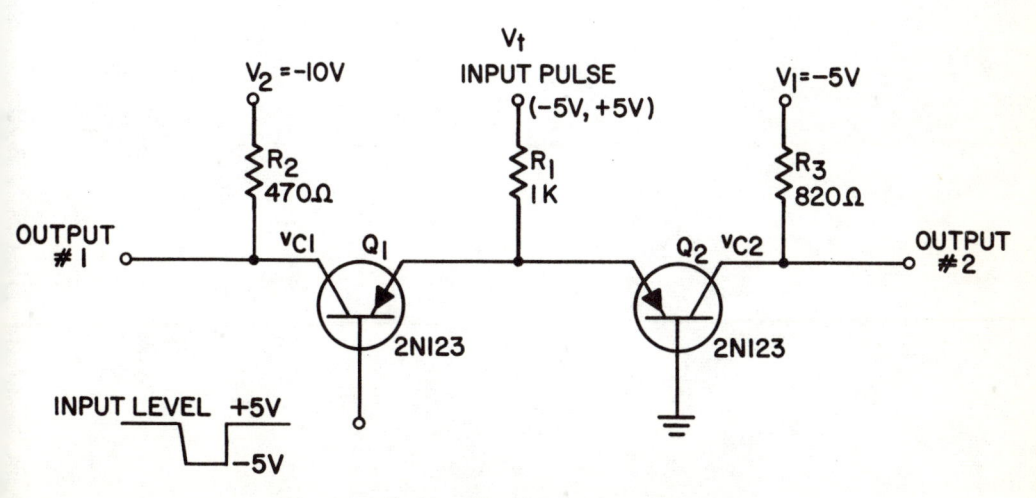

Circuit 7-1 - Current steering gate

Circuit 7-1, Current Steering Gate (cont'd.)

Since the emitter current is set when R_1 is chosen, it is necessary to choose R_3 to give a desirable "on" logic level. To avoid saturation, R_3 is selected to keep v_{C_2} less than ground. The collector voltage "on" level is reasonably insensitive to transistor characteristics, since alpha variations are small.

The design limitations in effect depend on the current gain of the transistors and there is adequate design freedom for present transistor types. The dual gate output may be used as the set-reset inputs to a flip-flop or the gate may be modified to give a single output by connecting the collector of Q_1 directly to the voltage V_1.

The circuit is capable of operation over wide temperature ranges.

CIRCUIT 7-2

DIODE-GATED AMPLIFIER

Radio Corporation of America

Circuit 7-2 is a well-designed diode transistor type gate. The efficient off-on characteristics of the diodes combine with the transistor gain characteristics to provide excellent margins and stability. The circuit gate input levels are ground and -6 volts. The output is clamped at -6 volts by the diode CR_7 and ground by saturation of the transistor. The base resistance and bias voltage are chosen to provide 1 volt or reverse bias and about 1 ma of base current. The base "off" bias level is easily controlled by selecting R_1, R_2, and V_2 properly.

$$v_{B\,off} = \frac{V_2 R_1}{R_1 + R_2}. \qquad 7\text{-}2.1$$

The base current is derived as

$$i_B = \frac{v_L (R_1 + R_2)}{R_1 R_2} + \frac{V_2}{R_2} - \frac{v_L}{R_2} \qquad 7\text{-}2.2$$

where v_L is the lower input level which is -6 volts for this circuit.

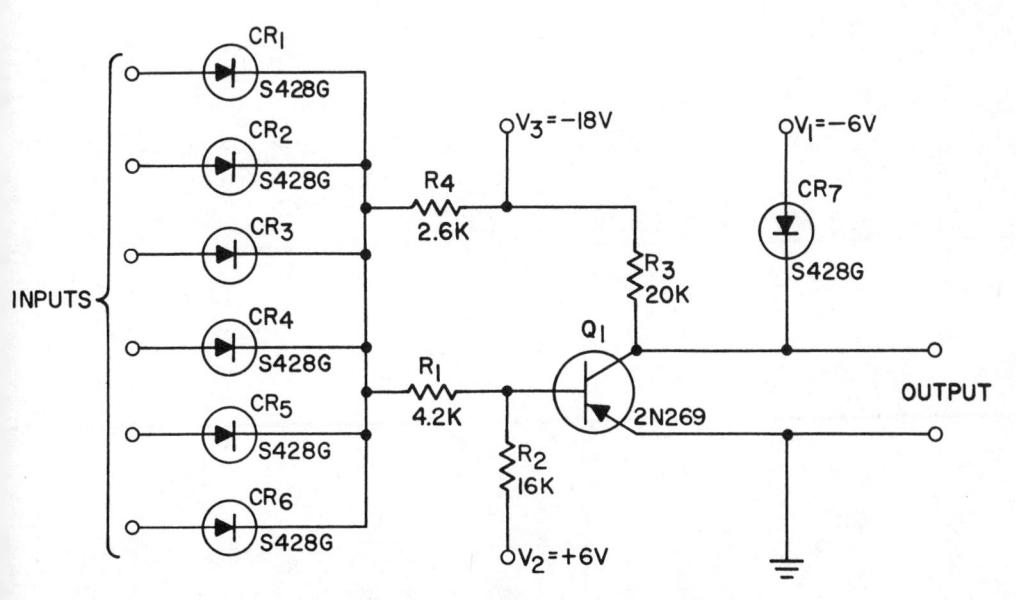

Circuit 7-2 - Diode-gated amplifier

Circuit 7-2, Diode-Gated Amplifier (cont'd.)

The collector loading capabilities depend on the base current level chosen and the current gain of the transistor. For the circuit values given here, the output is capable of driving at least five input circuits.

The diode "or" network shown uses S428G diodes or any equivalent type, and the transistor amplifier displays excellent interchangeability. The output rise time is typically 0.1 μs while the fall time is 0.25 μs. The delay time due to saturation is about 0.1 μs.

CIRCUIT 7-3

DIODE-TRANSISTOR GATED AMPLIFIER, EXCLUSIVE "OR"

Naval Research Laboratory

Contributed by J. M. Hovey
H. H. Levy

Circuit 7-3 is able to compensate for slightly delayed input timing by use of a supplementary gate input. Logically, the exclusive "or" function can be written as $A \cdot B' + A' \cdot B$. Circuit 7-3 includes an additional input, C, to correct timing shifts or delays inherent in A and B and the circuit output then is $(A \cdot B' + A' \cdot B)C$. The inputs A and B are nominally 1 μs wide in phase but can be slightly out of phase due to circuit variations or system delays. The input C is derived from a prime timing source and maintains its reference phase.

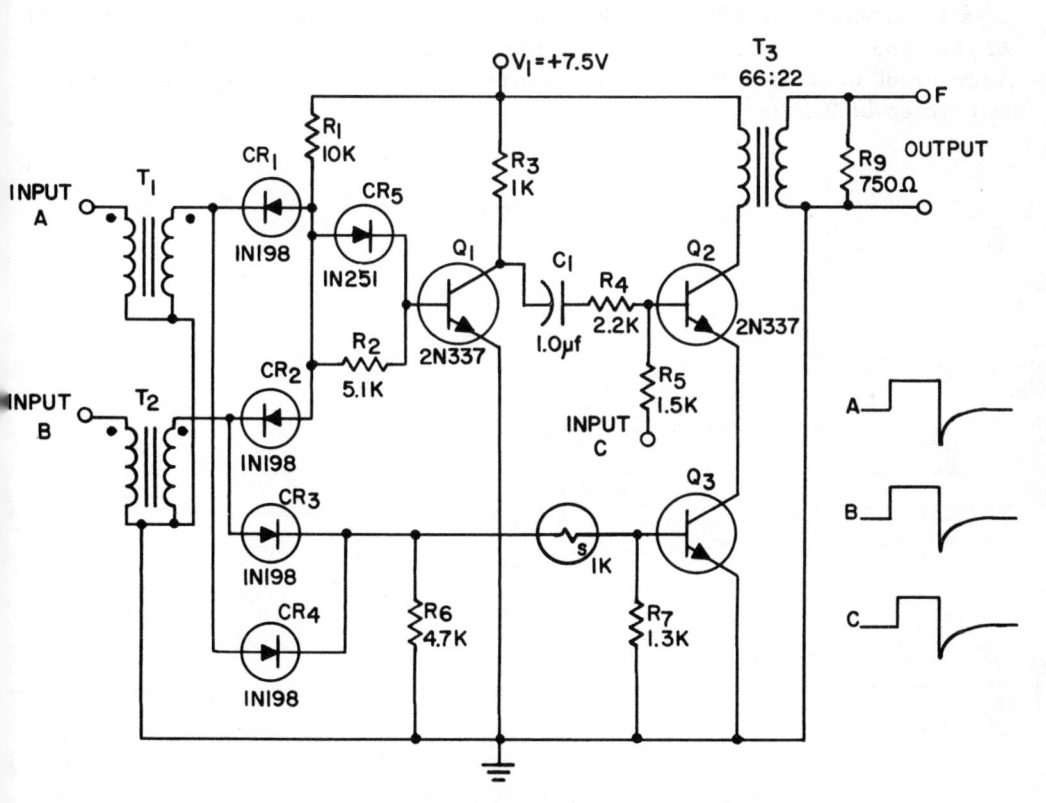

Circuit 7-3 - Diode-transistor gated amplifier, exclusive "or"

Circuit 7-3, Diode-Transistor Gated Amplifier, Exclusive "Or" (cont'd.)

The circuit operation may be described in the following manner. If either input A or B are present, a positive output from the transformer is coupled through either diode CR_3 or CR_4 to Q_3 turning Q_3 "on" for the duration of the pulse A or B. The "on" condition of Q_3 is required for any further positive action in the remaining part of the circuit. The diodes CR_1 and CR_2 are normally conducting, holding transistor Q_1 in an "off" condition. If A and B are both present, diodes CR_1 and CR_2 are opened by the positive output of the transformers, as described in Section 1. The current through R_1 is then diverted to Q_1, causing a negative output at the collector. The negative output of Q_1 and the input C oppose one another in the turn "on" of Q_2. In effect, Q_2 will conduct, providing an output if Q_3 is conducting, input clock C is present, and there is no negative output from Q_1. The overall operation may logically be expressed as $(A \cdot B)' \cdot (A+B) C$ as described above and this is equivalent to $(A \cdot B' + A' \cdot B) C$.

The circuit operates over a temperature range using silicon transistors from -65°C to +85°C with concurrent variations in supply voltage of ±20%. The minimum input trigger for A or B is 1.8 volts, while the minimum input for C is 1.5 volts. The output is a 2.5 volt pulse with rise and fall times of 0.1 μs.

CIRCUIT 7-4

TRANSISTOR RESISTOR GATE

Philco Corporation   Contributed by R. W. Carney

Circuit 7-4 is the Transistor Resistor Logic (TRL) gate and is designed for a medium number of inputs (5). The grounded resistors represent the four input circuits that are saturated and thereby give the worst input loading conditions. The circuit is designed to provide adequate margins for this case, and the general circuit characteristics are listed as follows:

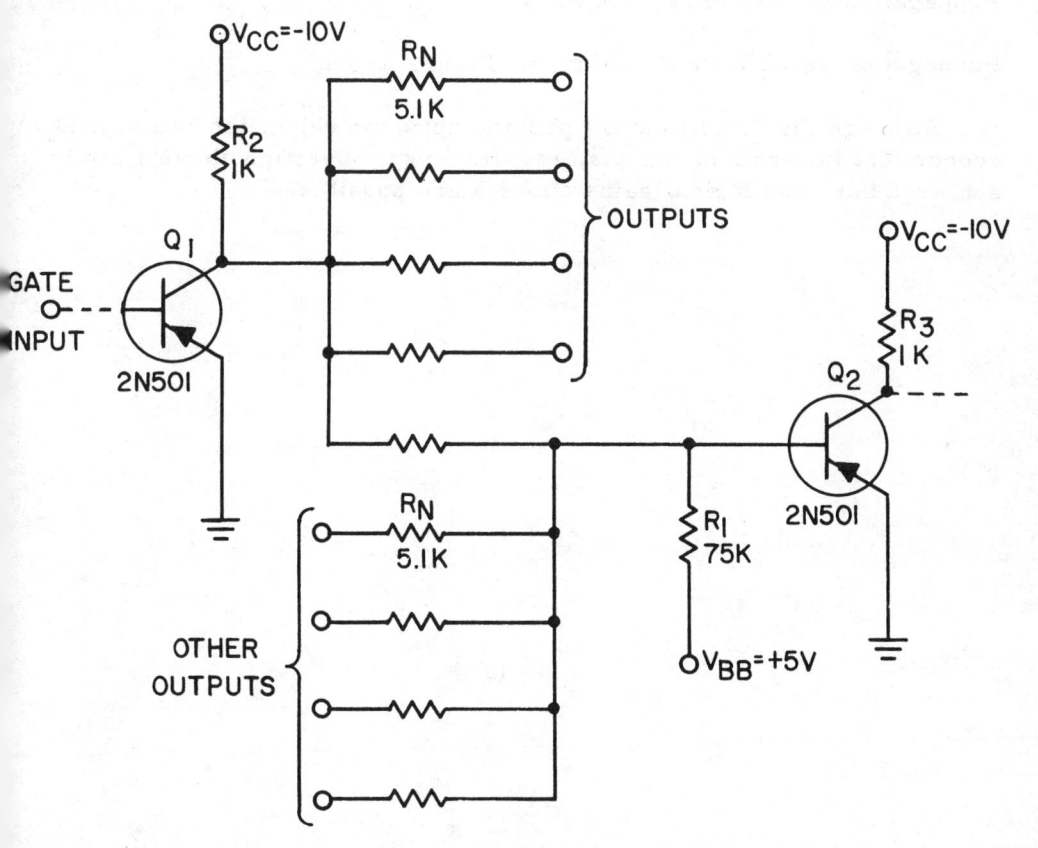

Circuit 7-4 - Transistor resistor gate

Circuit 7-4, Transistor Resistor Gate (cont'd.)

Time	Temperature 25°C	65°C
Delay time (0 to 10%)	50 mμs	50 mμs
Rise time (10% to 90%)	50 mμs	60 mμs
Saturation time (100% to 90%)	150 mμs	170 mμs
Fall time (90% to 10%)	65 mμs	110 mμs
Propagation A	40 mμs	40 mμs
Propagation B	110 mμs	160 mμs

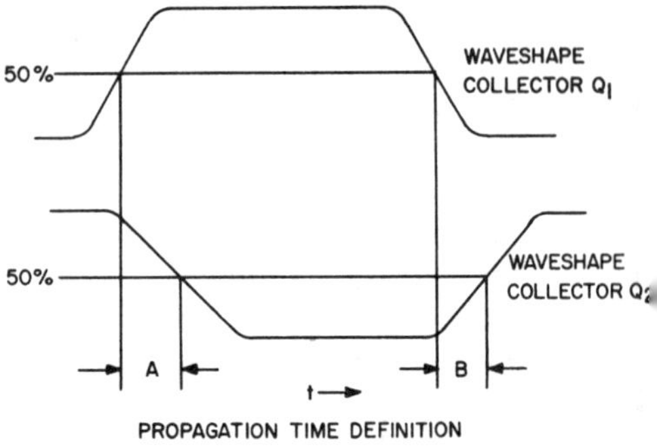

Fig. 7-4.1 - To Circuit 7-4 Propagation

Propagation is defined as shown in Figure 7-4.1.

Although the circuit cannot provide optimum repetition rates, it is economical in terms of transistors. However, operation above 1 mc is achieved here and logical gains above 5 are possible.

CIRCUIT 7-5

CURRENT TYPE DUAL GATE

Philips Research Laboratories 　　　Contributed by Dr. Nico C. de Troye
Eindhoven, Netherlands

Circuit 7-5 is a current type dual gate that illustrates effective use of the inherent current properties of transistors. The basic circuitry exhibits optimum response characteristics and excellent stability. There is also less circuit dependence on the transistor parameters.

The 6 mAdc current source, I_1, may be generated from a transistor, as can the current sinks I_2 and I_3. If any one of the inputs a, b, or c, is low (-0.6v) the current I_1 is diverted from Q_4, and the output current of Q_4 becomes zero. However, the load R_2 has 3 mAdc of current (I_2) remaining. When the inputs a, b, and c are all high (+0.6v), the current I_1 flows into the emitter of Q_4, providing an output current of 6 mAdc. The load current is 3 mAdc in a direction opposite to the previous case described. The output at the collector of Q_4, then, acts as an "and" gate. The collectors of Q_1, Q_2, and Q_3 are used to generate an "or" output at terminal E. The output current at this terminal remains constant even if all three inputs a, b, and c are low, since the current I_1 is divided between the three transistors.

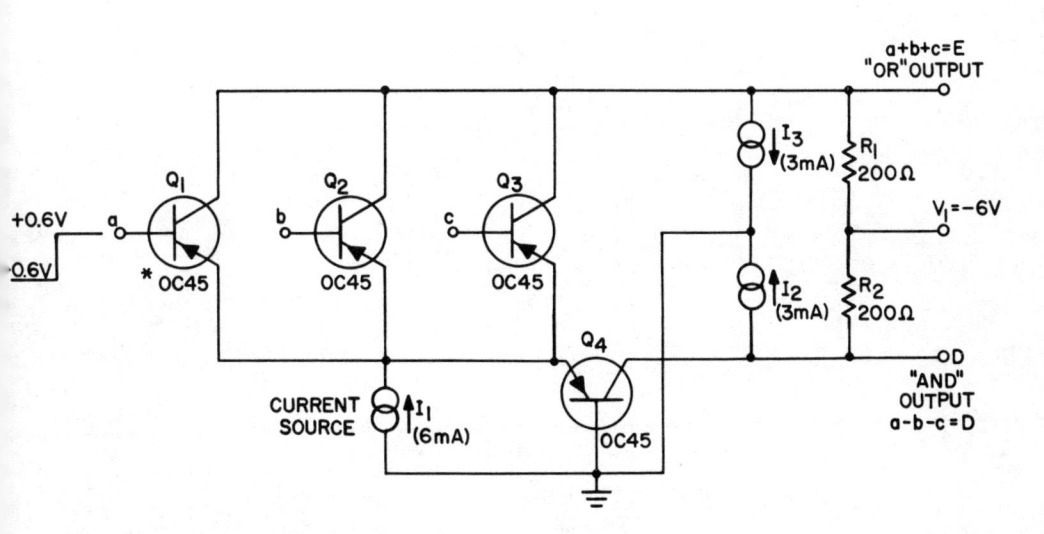

*2N404 NEAREST EQUIVALENT

Circuit 7-5 - Current type dual gate

Circuit 7-5, Current Type Dual Gate (cont'd.)

The voltage levels in this type of gate circuitry can be low, as illustrated here. In addition, the impedances are also low, minimizing the effect of stray capacitance.

The gate output rise time and fall time is about 150 mμs or less.[1] The delay time is low and the circuit will operate with minimum circuit variations over a very wide temperature range.

The circuit is primarily alpha dependent and a minimum alpha value is usually specified depending on the circuit application.

[1] If 2N404 transistors are used, the output rise and fall time are improved.

CIRCUIT 7-6

EXCLUSIVE "OR"

I.B.M.   Contributed by W. N. Carroll

Circuit 7-6 is an exclusive "or" circuit that uses RC coupled techniques and has been designed for reliable circuit operation and for improved margins. The circuit has a truth table as shown below.

Inputs		Output
A	B	
ground	ground	-3.5v
-3.5v	ground	ground
ground	-3.5v	ground
-3.5v	-3.5v	-3.5v

The Boolean expression for this circuit is $AB' + A'B$.

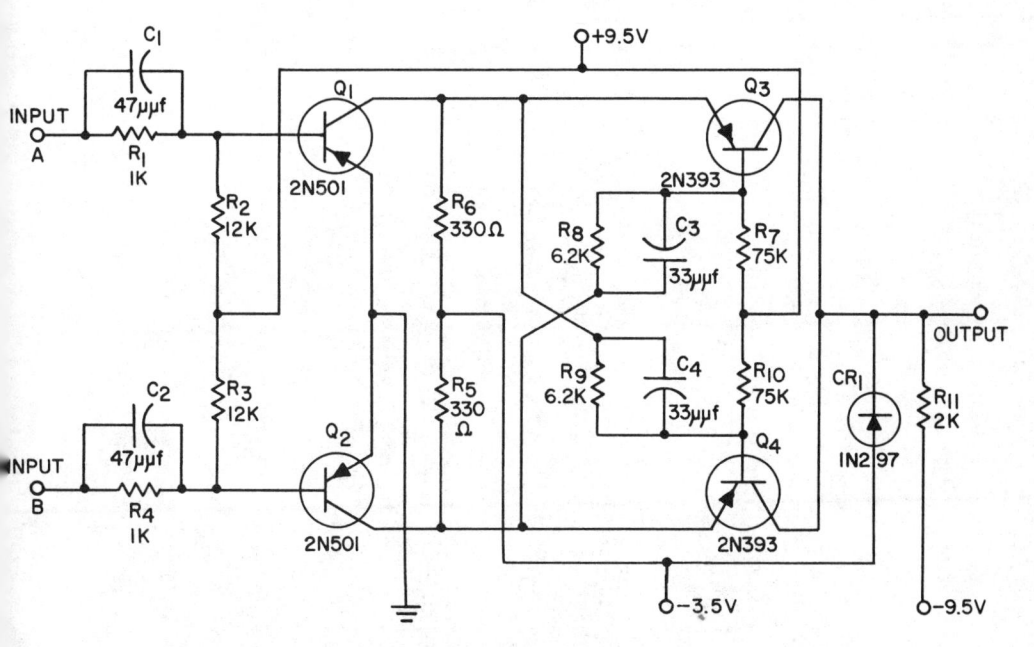

Circuit 7-6 - Exclusive "or"

Circuit 7-6, Exclusive "Or" (cont'd.)

The circuit operates as follows. With input A at ground, Q_1 will be cut-off and its collector will be at -3.5 volts with input B at ground, its collector will also be at -3.5 volts. Thus the input potential of Q_3 and Q_4 will be the same as their emitter potential and neither Q_3 or Q_4 will conduct. With no conduction, the output potential will be -3.5 volts. With both A and B inputs at -3.5 volts, the collectors of Q_1 and Q_2 will be at ground. However, the inputs to Q_3 and Q_4 will again be identical with their emitters and no conduction will take place. Thus the output will again be at -3.5 volts.

If input A is at -3.5 volts and input B is at ground, the collector of Q_1 will be at ground while the collector of Q_2 will be at -3.5 v. With this condition the emitter of Q_4 will be -3.5 volts while its base input will be at ground. Under this condition Q_4 will not conduct. However, the emitter of Q_3 will be at ground and its base input will be at -3.5 volts. This is the proper condition to Q_3 to conduct and bring the output to ground. With just the reverse condition, Q_3 will not conduct but Q_4 will conduct and since both use a common collector resistor (R_{11}) the output will also be at ground.

CIRCUIT 7-7

DIRECT-COUPLED COUNTER STAGE WITH GATING

Bell Telephone Laboratory

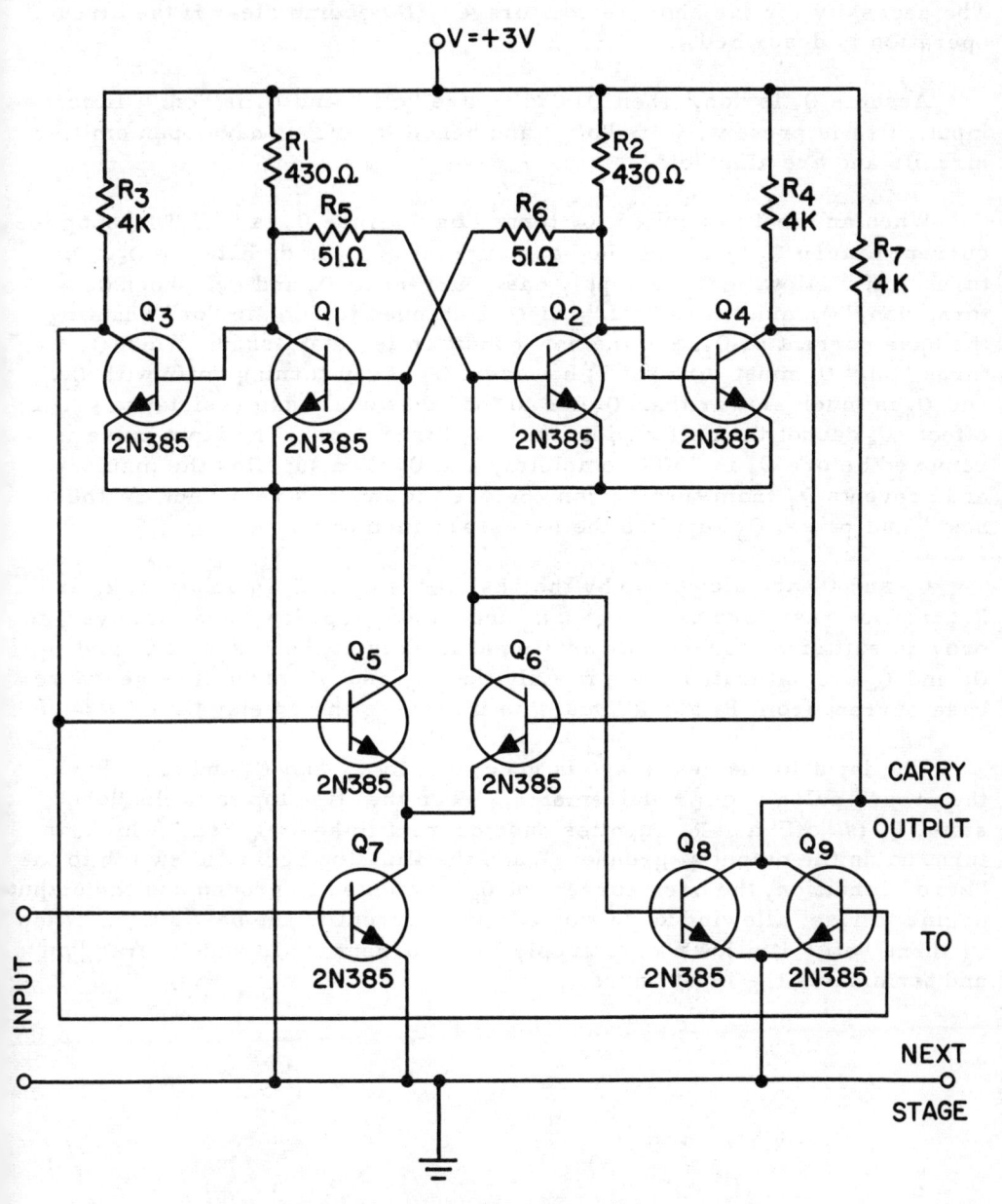

Circuit 7-7 - Direct-coupled counter stage with gating

Circuit 7-7, Direct-Coupled Counter Stage with Gating (cont'd.)

Circuit 7-7 is a single stage of a direct-coupled transistor type counter. The transistors Q_1 and Q_2 form the flip-flop, while Q_3 and Q_4 supply the short term storage. Q_5, Q_6, and Q_7 are steering networks for the input pulse. Q_8 and Q_9 generate a carry to be fed to the next stage. The necessity for the short term storage will become clear if the circuit operation is described.

Assume Q_1 is "on," then Q_2 and Q_3 are "off," and Q_4 is "on." If no input pulse is present, Q_7 is "off," and hence Q_5 and Q_6 have open emitter circuits and are also "off."

When an input is applied, Q_7 turns "on." Since Q_3 is "off," R_3 supplies current to turn Q_5 "on" and the base current of Q_1 is diverted to Q_5. Q_1 turns "off," allowing R_1 to supply base current to Q_2 and Q_3. When Q_3 turns "on," Q_5 must turn "off," but Q_2 continues to remain "on" shunting the base current of Q_1, and the new condition is established. When Q_2 turns "on," Q_4 must turn "off"; however, Q_4 began turning "off" with Q_1, and Q_4 is much slower than Q_2 due to the additional base resistance R_2. In effect, Q_4 cannot turn off rapidly until Q_2 turns "on." The input pulse is removed before Q_4 is "off" completely and Q_4 then supplies the memory and prevents Q_6 from turning "on" before the input is removed. At the next input pulse, Q_3 supplies the necessary memory.

Q_3 and Q_4 are slowed up by the resistance R_3 and, in addition, R_5 and R_6, if these resistors are large. R_5 and R_6 are kept low here, however, to provide sufficient current for additional loading at the bases of Q_1 and Q_2. Q_3 and Q_4 are saturated more heavily than Q_1 and Q_2, since they get more base current from R_1 and R_2, and this increases their delay time.

The input to the next stage is generated by Q_8 and Q_9 and must have the proper pulse width requirements. When the flip-flop is in the "one" state, Q_1 is "off" and R_1 supplies base current to keep Q_8 "on," which, in turn, holds the output at ground. When the flip-flop begins to switch to the "zero" condition, the base current of Q_8 is shunted to ground and the output begins to rise, allowing R_7 to supply input current to the next stage. When Q_3 turns "off," it allows R_3 to supply base current to Q_9, which turns "on" and terminates the input current.

CIRCUIT 7-8

RC-COUPLED FLIP-FLOP WITH DIRECT-COUPLED GATING

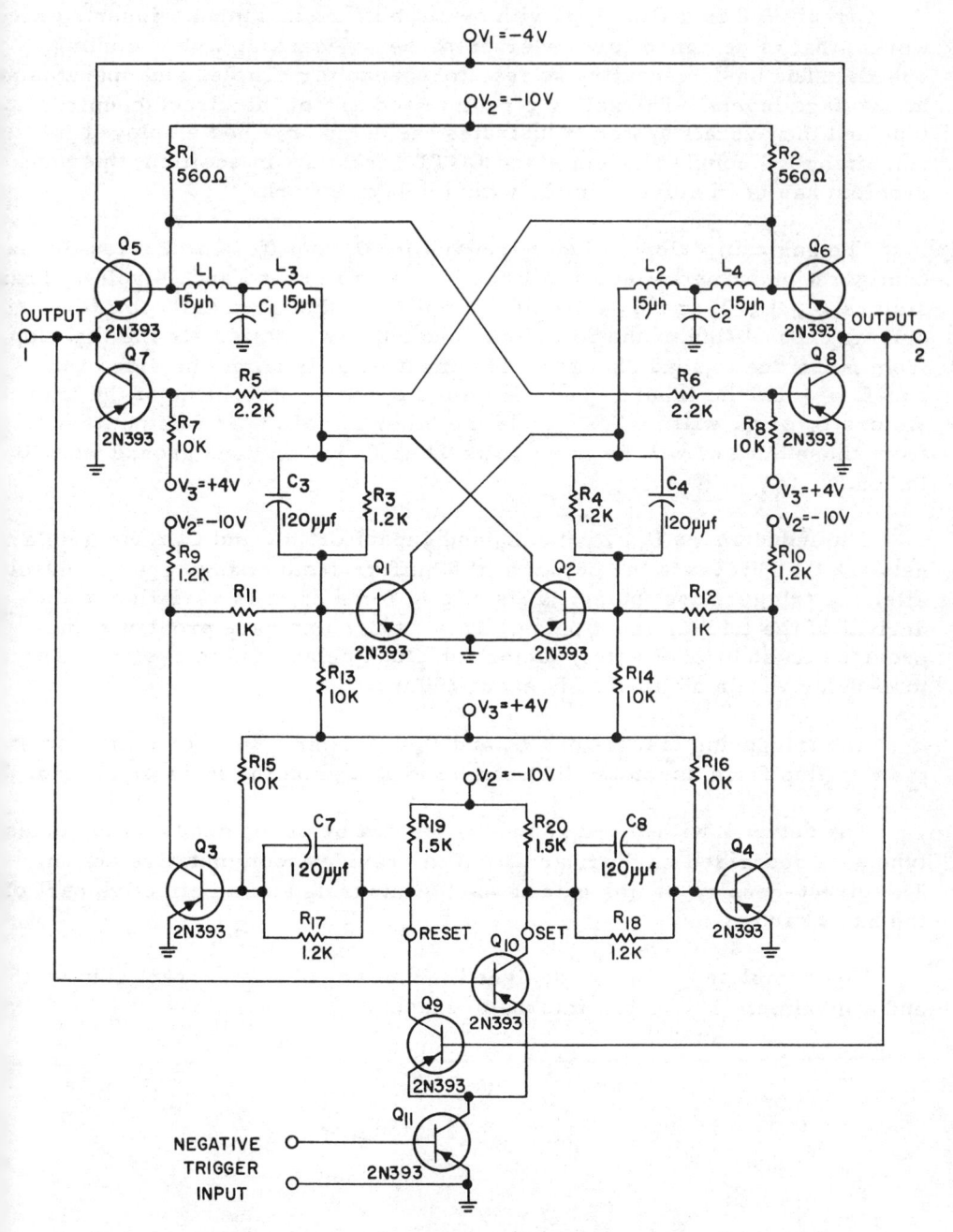

Circuit 7-8 - RC-coupled flip-flop with direct-coupled gating

Circuit 7-8, RC-Coupled Flip-Flop with Direct-Coupled Gating (cont'd.)

Sylvania Electric Products Inc.   Contributed by J. Monahan

Circuit 7-8 is a flip-flop, with output buffers and input triggering networks, that is designed for moderate-to-heavy loads at high switching speeds. The basic circuitry is resistor-capacitor coupled and operates a low voltage levels. The gating circuits used are of the direct-coupled type and the overall system illustrates the design method employed in combining RC coupled circuits and DCTL circuits. In addition, the race problem has been solved simply with a delay network.

The basic flip-flop includes transistors Q_1 and Q_2 in an Eccles-Jordan configuration. The collector voltage levels are ground and -4 volts. Transistors Q_5, Q_7, Q_6 and Q_8 are buffer amplifiers that primarily improve the driving capabilities of the flip-flop. The buffers also isolate the flip-flop from noise due to load changes. The input to Q_5 is from one side of the flip-flop while the input to Q_7 is from the opposite side. One of the transistors Q_5 or Q_7 will be "on" while the other is "off." The output then from these buffers will be at -4 volts when Q_5 is "on" and ground when Q_7 is "on."

The inductances L_1, L_2, L_3, L_4 and capacitors C_1 and C_2 form a delay network that prevents the output at the buffers from changing until 160 mμ after the trigger input pulse. This allows for a 25 mμs variation in the arrival of the trigger input pulse. This assures no race problems and provides an additional safety factor for clock delays in the system. The total delay within a flip-flop is about 200 mμs.

The triggering transistors Q_3 and Q_4 are simple inverters that isolate the flip-flop from any noise that can be expected on the input or ground.

The network with Q_9, Q_{10}, and Q_{11} should be recognized as the series type gate described earlier, arranged to provide a commutative action. The direct-coupled series gate is easily integrated as an effective part of the basic circuit.

The complete circuit is designed for an end of life current gain of 15 and a maximum I_{CO} of 500 microamperes.

CIRCUIT 7-9

DIRECT-COUPLED TYPE RING COUNTER

Philco Corporation   Contributed by R. W. Carney

Circuit 7-9 is a ring counter that uses "and" gates for trigger pulse steering. The capacitors (730 $\mu\mu$f) act as memory storage and eliminate race problems between the input trigger and a collector voltage change. In effect, the capacitor slows down the response of the gate transistor to avoid an improper transfer. The race problem is discussed in the design procedure, and this is an illustration of a practical solution to the problem. The use of the capacitor here is a simple design approach and gives effective results.

The basic circuit is the direct-coupled type, and the gates have been described earlier as compatible with DCTL. The collector rise and fall times for each binary stage are less than 20 mμs, and the circuit delay time is less than 15 mμs.

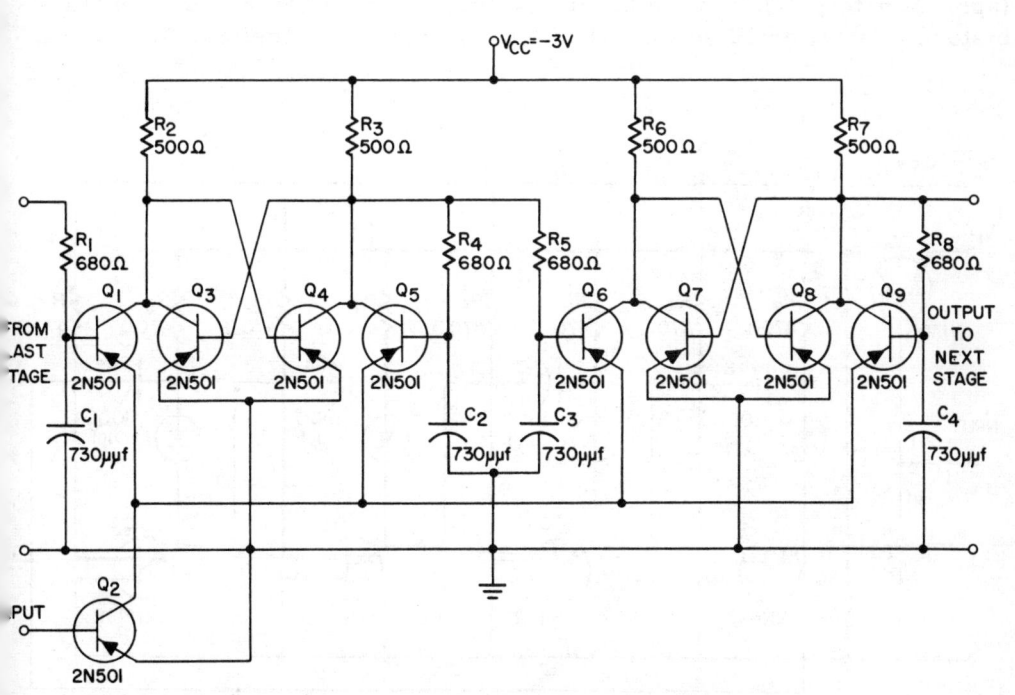

Circuit 7-9 - Direct-coupled type ring counter

CIRCUIT 7-10

UNIJUNCTION RING COUNTER

General Electric Company   Contributed by T. P. Sylvan

Circuit 7-10 is a unijunction ring counter that is characterized by a small number of components and an additional trigger input for setting and resetting control. The operation of the basic unijunction circuit (monostable) is described in Part 6 - Switching Circuits, Circuit 6-15. For bistable operation the load resistor R_3, R_5, R_7, etc. must intersect the "off" region below the peak to breakdown voltage and the "on" region beyond the valley point as shown in the voltage current characteristic.

To turn the circuit "on," a positive pulse is applied at the emitter that raises the emitter voltage above peak point. The circuit may be turned "off" by applying a negative pulse at the emitter that lowers the emitter voltage below the valley point (V_B).

In the ring counter, assume Q_2 is "on." The collector current of Q_1 flows through R_3, CR_1 and the emitter of Q_2. The voltage of Q_2 is at V_B (approximately 3v), and the emitter voltage of the other unijunction transistors is at V_1 or 10 volts. When a trigger pulse is applied, Q_1 cuts off,

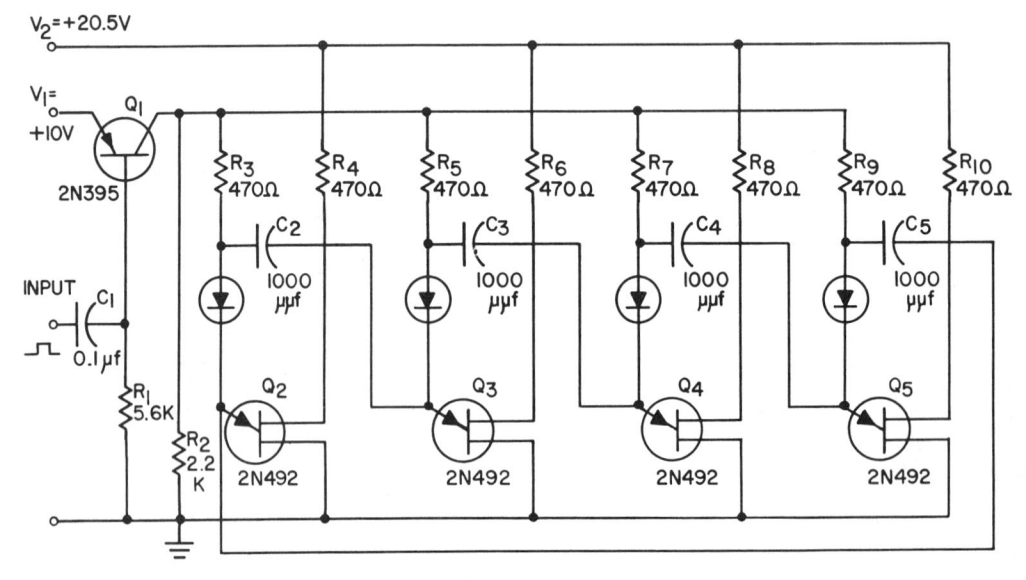

Circuit 7-10 - Unijunction ring counter

Circuit 7-10, Unijunction Ring Counter (cont'd.)

and the emitter current of Q_2 decreases to zero cutting off Q_2. During the trigger pulse, capacitor C_5 discharges through R_9, R_3 and CR_1. Capacitor C_2 remains charged to $(V_1 - V_B) = 7v$, since diode CR_2 and the emitter of Q_3 are off preventing it from discharging. At the end of the trigger, the voltage at the collector of Q_1 rises to V_1. Since there is no voltage on C_3, C_4 and C_5, the voltages at the emitters of Q_2, Q_4, Q_5 will rise to V_1; however, this is below their peak voltage, so that they remain off. The voltage on C_2 is equal to $V_1 - V_B$, so that the voltage at the emitter of Q_3 rises to $2V_1 - V_B$, or about 17 volts, and exceeds the peak voltage causing Q_3 to turn "on." In a similar manner, the "on" state is transferred to succeeding stages.

The circuit is designed to work with the normal range of unijunction parameters. The coupling capacitor cannot discharge during the trigger pulse, so that the trigger pulse width can vary over a wide range. Emitter triggering is used, since it requires less current, and operating margins are at a maximum. The resistors R_4, R_6, etc. are used for peak point stabilization, if desired, and should be low resistance values.

The counter operates at rates up to 80 kc, with a temperature range of -55° C to 120° C. (Q_1 not included.)

CIRCUIT 7-11

DIRECT-COUPLED TYPE RING COUNTER WITH CONDITIONAL STEERING

Burroughs Corp. Research Center   Contributed by E. G. Clark

Circuit 7-11 is a ring counter that employs direct-coupled circuitry and conditional type steering. This circuit employs an end-around-carry, and a preset which sets all stages except one in the "zero" state. The single "one" is made to advance around the ring, one stage per input pulse. Only one-half of the usual gating is required to reset the stage having the "one" and advance the "one" to the succeeding stage. The conditional steering allows operation from near static to multimegacycle rates. The conditional steering fundamentals are described in Section 1. The preset trigger is omitted in Circuit 7-10 for clarity, but the operational features may be described by assuming Q_1 is "on" and Q_2 "off," and each succeeding stage has been preset in an opposite fashion. Under these conditions Q_3 is conducting and its collector is held at ground. Q_7 and Q_8 are not conducting and their collector level is low. The gate composed of Q_3 and Q_4 is inactive when the advance input pulse is applied, while the gate with Q_7 and Q_8 provides an output that reverses the state of the second stage. The first stage, (Q_1 and Q_2) reverses due to the action of the gate associated with the last stage that completes the ring.

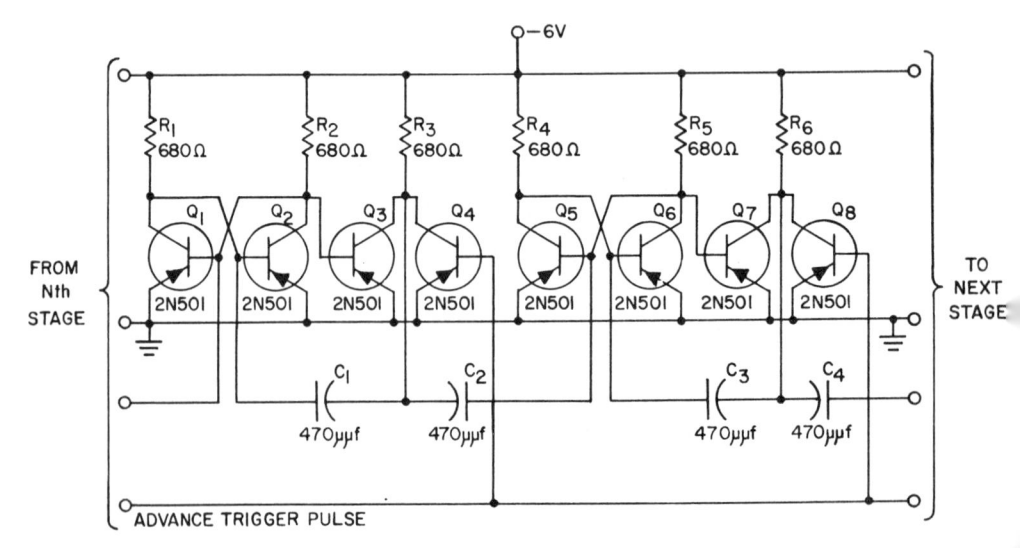

Circuit 7-11 - Direct-coupled type ring counter with conditional steering

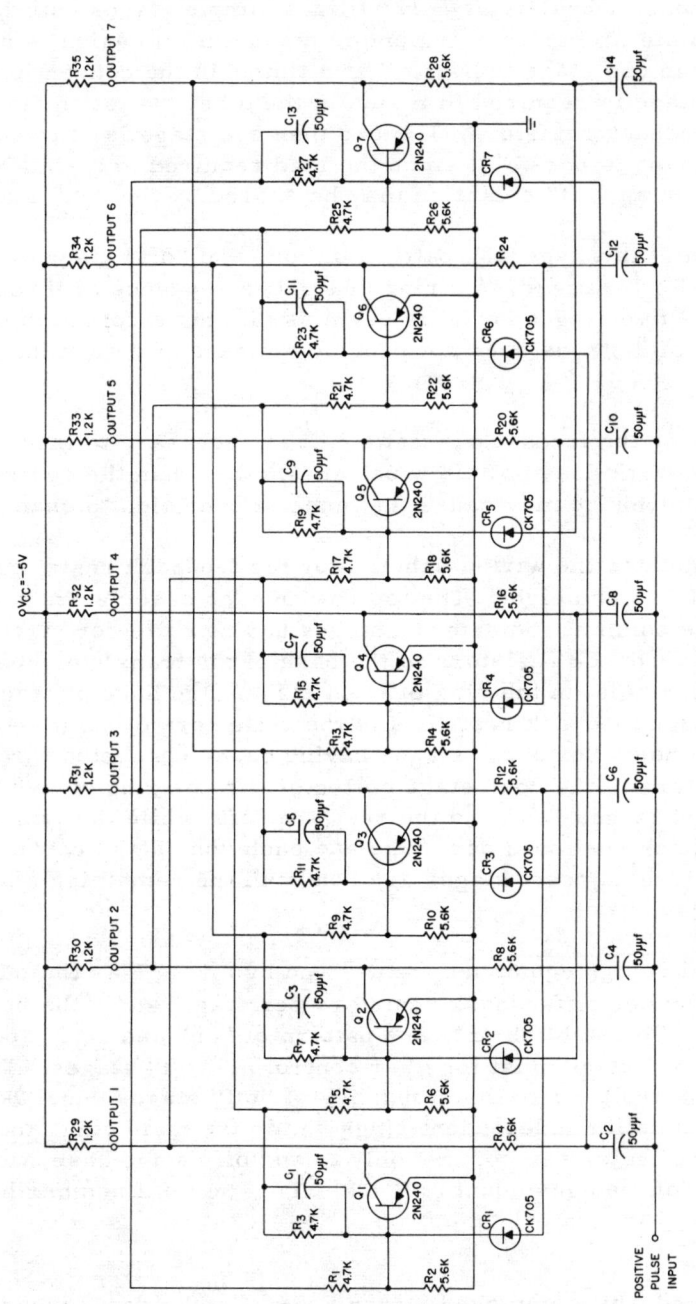

CIRCUIT 7-12

SEVEN-STAGE RING COUNTER

Circuit 7-12 - Seven-stage ring counter

Circuit 7-12, Seven-Stage Ring Counter (cont'd.)

Transistor Applications, Inc.*   Contributed by A. W. Carlson

The ring counter of Circuit 7-12 utilizes simple stages with feedback resistors to obtain proper counting and is unique in its design, since it allows more than one stage to be "on" at a time. In the conventional ring, a single "on" stage is required to maintain the other stages in an "off" condition by feedback resistors. If more than one stage is allowed to be "on," a longer ring is possible, since the load required to maintain the correct stages in an "off" condition may be shared by the "on" stages.

In a design of this type, the pattern of "on" and "off" stages must be recognizable. For example, in a ring of eight, a sequence of three adjacent stages "on" and five stages "off," is permissible, but a condition of alternate "on" and "off" stages does not provide a means of determining the count.

In positioning the feedback resistors, it is necessary to insure that the feedback network does not allow patterns, other than the desired one; otherwise a disturbance may cause the mode of operation to change.

Table I indicates the wiring scheme for the feedback resistors in a ring counter of five and eight. The top row in each case refers to the stage, while the columns indicate the stages having collector circuits connected through feedback resistors to the base of the transistor heading the column. For example, in the ring of eight stages, the base of stage 1 is connected through feedback resistors to the collectors of stages 4, 5 and 6. The "on" stage holds "off," the stages having bases connected through feedback resistors to the "on" stage collector. In the ring of five, there are two adjacent stages "on" and the rest are "off" while the ring of eight has three feedback resistors per stage and each "on" stage controls three "off" stages; three adjacent stages are "on" and the remaining stages are "off."

The logical design equations are derived by taking N as the number of stages, R the number of feedback resistors per stage, and S the number of "on" stages. To establish only one pattern of "on" and "off" stages, the first and last "on" stage must together control all "off" stages. This restriction requires the maximum number of "off" stages equal 2R. In order to avoid smaller independent rings within the main ring, and since an "on" stage collector can connect only to an "off" stage base, we provide the number of "off" stages equal to S + R - 1. The total number of "off" stages then is

*Work performed while employed at Air Force Cambridge Research Center

Circuit 7-12, Seven-Stage Ring Counter (cont'd.)

$$N = 2S + R - 1. \qquad 7\text{-}12.1$$

From this, the maximum number of stages for a given number of feedback resistors per stage is

$$N_{max} = 3R + 1 \qquad (R > 1). \qquad 7\text{-}12.2$$

In a ring of eight, an odd number of feedback resistors per stage are required. Equation 7-12.2 indicates that three feedback resistors make possible a ring of 10, so that three resistors are required for the ring of eight. Taking R equal to three, then, N as eight, Equation 7-12.1 gives the number of "on" stages, S, as three.

The collector of a particular stage is connected to the bases of three succeeding stages through the feedback resistors, starting with the Sth stage after the one in question. Table I is derived in this manner.

The seven-stage ring of Circuit 7-12 is designed using the techniques described above, and operates at megacycle rates. The gates are arranged to turn "off" the last stage in the group of "on" stages during the input trigger interval. The coupling capacitors assist in switching "on" the stage ahead of the group.

Ring of 5				
1	2	3	4	5
4	5	1	2	3
3	4	5	1	2

Ring of 8							
1	2	3	4	5	6	7	8
6	7	8	1	2	3	4	5
5	6	7	8	1	2	3	4
4	5	6	7	8	1	2	3

Table I -- Wiring Scheme
for Rings of 5 and 8.

CIRCUIT 7-13

SHIFT REGISTER WITH CONDITIONAL STEERING

Burroughs Corp. Research Center   Contributed by E. G. Clark

Circuit 7-13 is a shift register using conditional steering to avoid race problems. The gates operate as described in Section 1, and their outputs are connected to the adjacent stage to give shift register operation.

The capacitor recharging, as described earlier, generates a noise pulse at the termination of the input pulse; however it has negligible effect, since it operates into a saturated collector. The circuit is capable of operation in the megacycle range and exhibits excellent reliability.

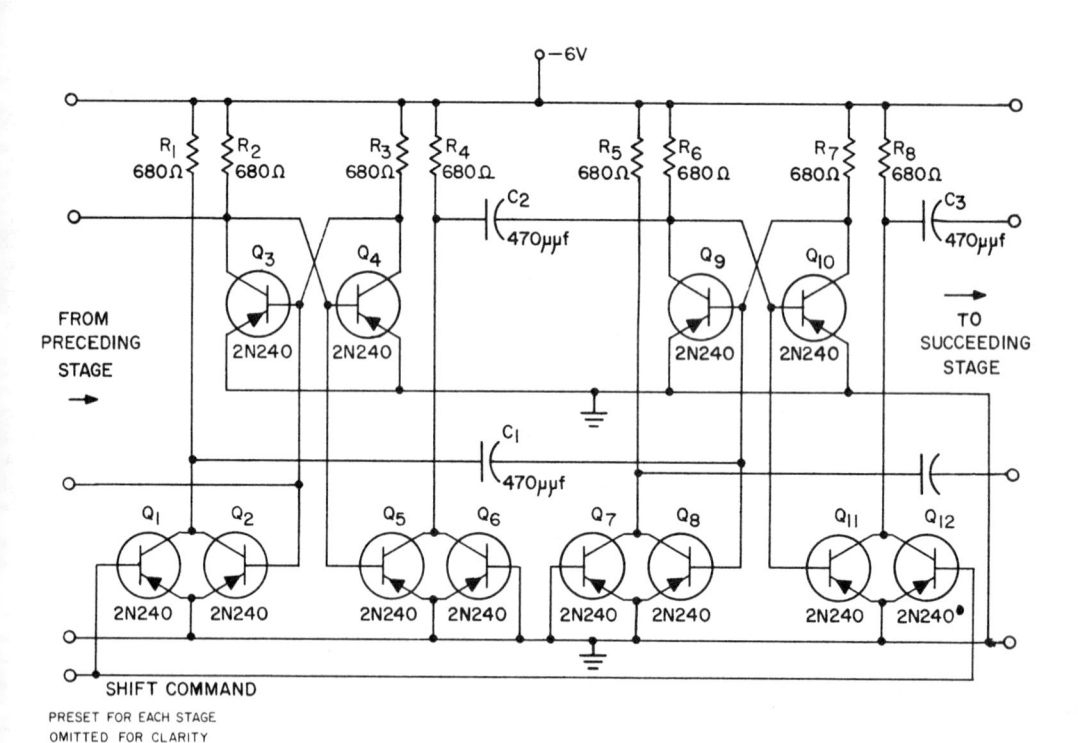

Circuit 7-13 - Shift register with conditional steering

CIRCUIT 7-14

DIRECT-COUPLED HALF ADDER

Philco Corporation   Contributed by Ralph Brown

Circuit 7-14 is a half adder that employs direct-coupled transistor logic. The unique simplicity of DCTL is evident here, and allows a relay type realization of logic designs. The half adder is a combination of the DCTL gates previously described such that the relay representation is achieved.

The Boolean expressions for the two input adder are:

$$\text{Sum} = ab' + a'b \qquad 7\text{-}14.1$$

$$\text{Carry} = a \cdot b. \qquad 7\text{-}14.2$$

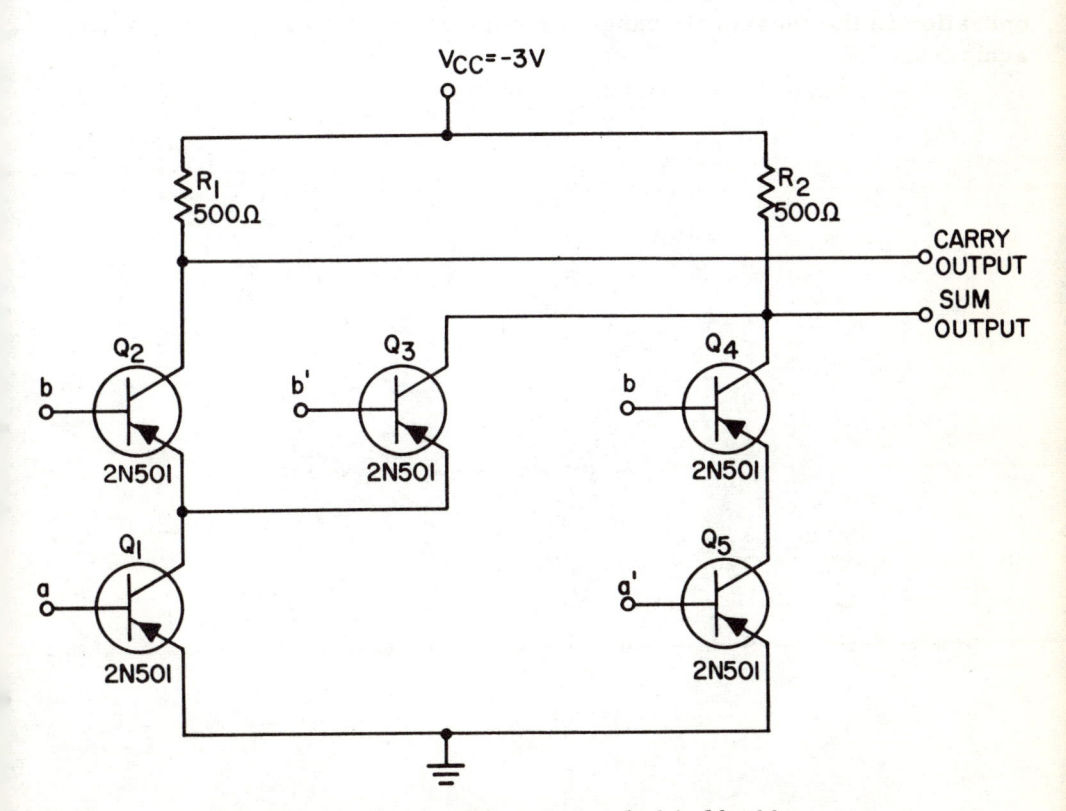

Circuit 7-14 - Direct-coupled half adder

Circuit 7-14, Direct-Coupled Half Adder (cont'd.)

If the inputs a and b are assumed to be from two flip-flops A and B, the circuit operation is described by first taking A in the "one" state. With A in the "one" state, the a' input will be near a ground level, while a will be negative. Under these conditions, Q_1 is "on" and Q_2 is "off." If A is in a "zero" state, the opposite condition holds and Q_1 will be "off" while Q_2 will be "on." If flip-flop B is in the "one" state, Q_3 is "off" while Q_2 and Q_4 are "on." If B is in the "zero" state, the opposite condition exists. The complete operation of the circuit may be analyzed by taking the possible input combinations and comparing these with the resulting sum and carry outputs. A sum and carry flip-flop are operated from the adder outputs.

The DCTL circuit limitations discussed previously must be considered in the adder design. However, the half adder shown here is small and does not require more than the normally accepted DCTL transistor characteristics.

The circuit speed depends on the transistor response primarily and operation in the megacycle range for current transistor types is easily achieved.

CIRCUIT 7-15

CURRENT SWITCH HALF ADDER CIRCUIT

Burroughs Corp. Research Center   Contributed by Carl M. Campbell, Jr.

    Circuit 7-15 is a half adder circuit that employs current switching type circuitry. The basic circuitry is nonsaturating and allows high speed operation. In addition, the circuits are largely independent of transistor characteristics and employ voltage swings of sufficient magnitude to be relatively noise insensitive.

    The basic circuits are current switches of either the p-n-p or n-p-n type. The input to the p-n-p switch is required to be +5v or +3v while the output level is -5v or -3v for an "off" or "on" condition respectively. The n-p-n switch, on the other hand, requires an input of -5v or -3v and provides an output of +5v or +3v for an "off" or "on" condition respectively.

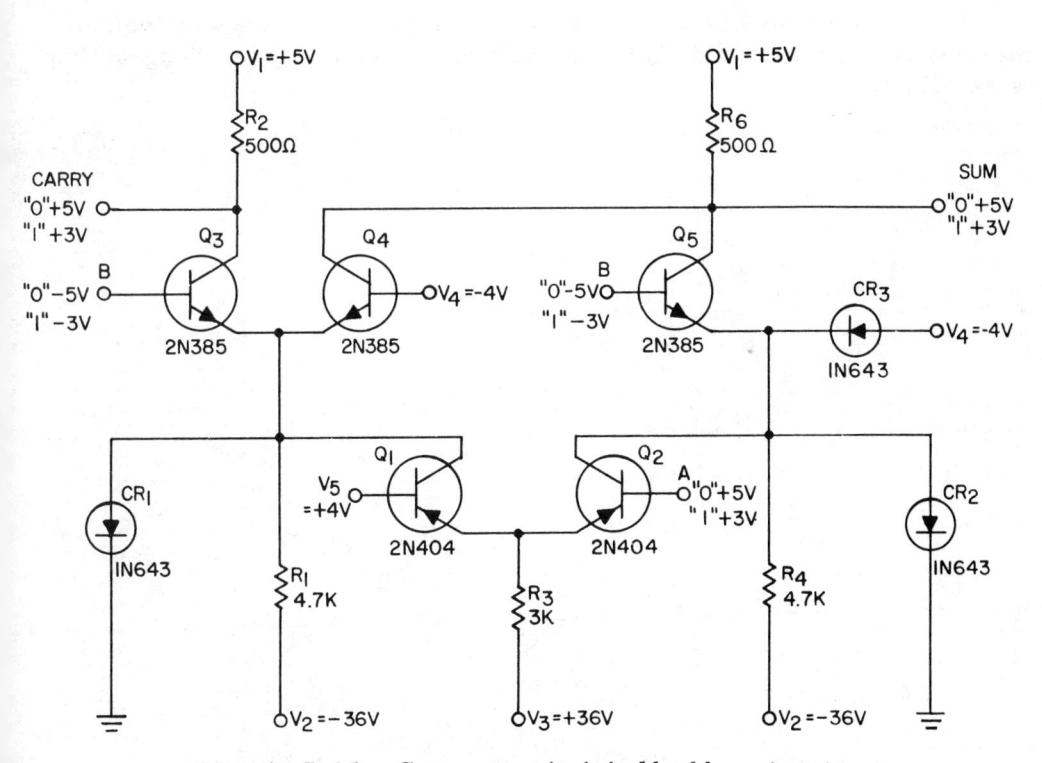

Circuit 7-15 - Current switch half adder circuit

Circuit 7-15, Current Switch Half Adder Circuit (cont'd.)

It should be clear that the two types of switches may be coupled to each other for extended circuitry.

In the half adder circuit shown, transistors Q_1, Q_2, Q_3, and Q_4 form what might be called an inhibit circuit. If Q_2 is "on" and Q_1 is "off," the current switch comprising Q_3 and Q_4 operate normally with input B determining the output condition of Q_3. However, if Q_2 is "off," and Q_1 is "on," then the collector current of Q_1 flows into R_1 and clamps the emitters of Q_3 and Q_4 at ground. It is then impossible for input B to affect the output Q_3 and an inhibiting condition exists.

The complete half adder is designed so that the current switch comprised of Q_1 and Q_2 inhibits either the switch composed of Q_3 and Q_4 or the switch with Q_5 and the diode CR_3, depending on the input at A. The input at B then determines where the current in the uninhibited switch will flow.

The current switch with Q_5 and the diode CR_3 is a simplification of the basic switch when the second output is unnecessary.

The circuits have been "worst case" designed and have wide voltage margins and excellent stability. In addition, transistor interchangeability is excellent.

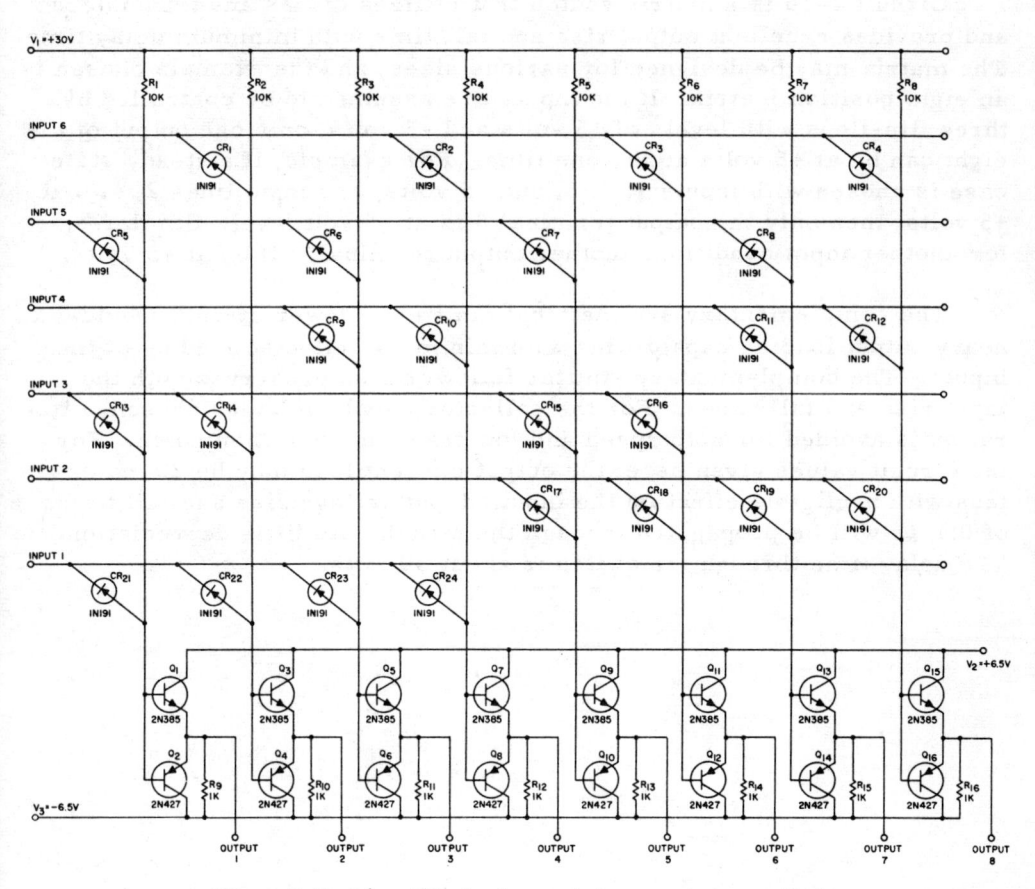

CIRCUIT 7-16

DIODE-TRANSISTOR MATRIX SWITCH

Circuit 7-16 - Diode-transistor matrix switch

Circuit 7-16, Diode-Transistor Matrix Switch (cont'd.)

Lincoln Laboratory, M.I.T. Contributed by Robert E. McMahon

Circuit 7-16 is a matrix switch that utilizes diodes and transistors and provides excellent output rise and fall time with minimum delay time. The matrix may be designed for various sizes, and the example chosen is an eight position matrix. If the inputs are assumed to be controlled by three flip-flops with levels of +5 volts and -5 volts, only one output of the eight can be at +5 volts at any one time. For example, if a steady state case is chosen with inputs 1, 3, 5, and -5 volts, and input lines 2, 4, 6 at +5 volts, then only the output terminal 8 is at +5 volt level. Similarly, for another input condition, another output terminal will be at +5 volts.

The complementary symmetrical emitter follower circuit provides heavy output loading capabilities with minimum reflected loading of the inputs. The complementary emitter follower also preserves both the input rise and fall times. For the collector supply voltages chosen, saturation is avoided for both speed and low transient condition noise. For the circuit values given here, the output current load may be 20 ma or less with negligible effect on the input. Input voltage rise and fall times of 0.1 μs will be propagated through the switch with little depreciation. The delay time through the switch is about 30 mμs.

CIRCUIT 7-17

SHIFT REGISTER WITH RESISTOR-CAPACITOR-DIODE GATES

Transistor Applications, Inc.   Contributed by A. W. Carlson

Circuit 7-17 is a shift register similar to the type described in Figure 7-0.12 and the corresponding design procedures. The basic building blocks are R-C coupled flip-flops that are designed for saturated operation while the gates are of an R-C diode type. Reliable operation is achieved with simple circuit techniques.

The gates are arranged such that the collector of each flip-flop control the gate action to provide proper pulse steering to the succeeding flip-flop inputs. The base triggering employed is simple and at the same time reliable provided the input shift clock does not exceed 7 volts in amplitude.

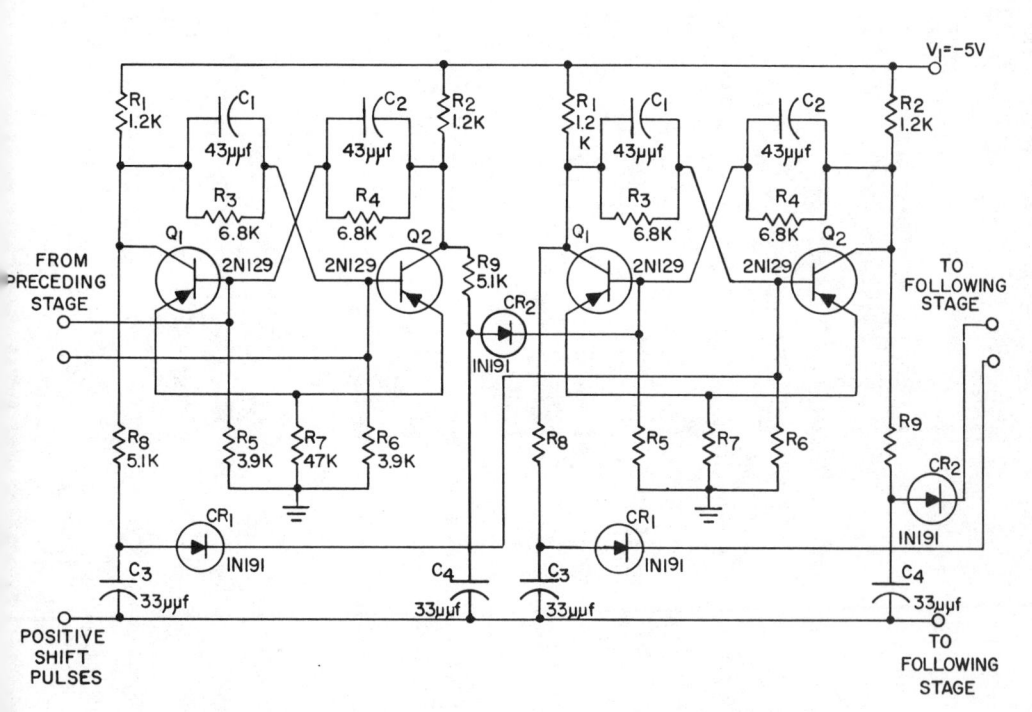

Circuit 7-17 - Shift register with resistor-capacitor-diode gates

Circuit 7-17, Shift Register with Resistor-Capacitor-Diode Gates (cont'd.)

The flip-flop resistance values are designed for suitable margins and moderate output loading. The capacitor coupling is designed to give adequate a-c gain without recovery problems at greater than 1 mc shift rates. A similar design procedure has been used in the choice of the input capacitors at the shift clock point.

The output voltage swing at the collectors is about 4.3 volts with a rise time of 30-50 mμs. The circuit operation is stable over a considerable temperature range and transistor interchangeability is good.

CIRCUIT 7-18

LOW POWER RING COUNTER

Army Ballistic Missile Agency　　　　　　　　Contributed by O. B. King

Circuit 7-18 is a ring counter that is designed for minimum power input and exhibits excellent stability.

Each stage of the counter shown employs an n-p-n and p-n-p transistor. The circuit resistors function in a manner similar to a conventional arrangement. The circuit has two stable states; one with both transistors on the second with both transistors off. Only one stage is normally on at any one time and the voltage developed across the common emitter resistor acts as the reverse bias for the other stages.

The off condition of the stages not activated reduces the current drain considerably below that of a conventional counter. The average power per stage is about 12 mw and may be as low as 4 mw per stage as the voltage output swing is lowered.

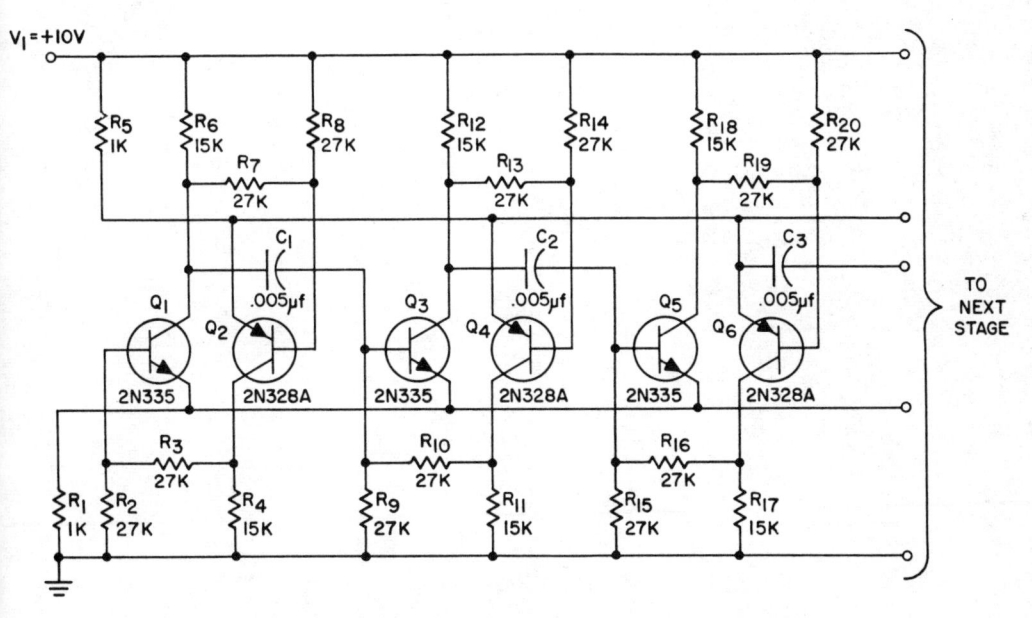

Circuit 7-18 - Low power ring counter

Circuit 7-18, Low Power Ring Counter (cont'd.)

The emitter "hold off" voltages permit reliable very high temperatures, approaching that of the junction rating temperature.

If the collectors of either type transistors are loaded heavily, it is necessary only to increase the size of the emitter resistor to regain the unloaded stability factor. In such a case, relatively equal loads should be placed on every stage to prevent unbalance.

The emitter buss is an ideal input for the count pulses. A negative input pulse appearing at the emitters of the p-n-p transistors will not affect the off stages but will turn off the on stage. The resultant positive step at the collector of the n-p-n transistor is differentiated and coupled to the base of the next n-p-n transistor. The time constant of the coupling network is such that this transfer pulse duration is longer than that of the count pulse. In turning the next stage on the cross coupling network also turns on the p-n-p transistor in the same stage. In effect, the single count pulse transfers the "on" stage to the right by one stage. The same action may be obtained by applying a positive count pulse to the emitters of the n-p-n transistors. The count operation is simple and does not require steering diodes as in the case of a conventional counter.

PART 8

A.C. TO D.C. POWER SUPPLIES

Section 1. Design Philosophy

A. Unregulated Power Supplies

1. Introduction

Circuit designers with experience in electron-tube circuitry frequently design unregulated power supplies, such as capacitor- and choke-input-filter rectifiers, in a surprisingly casual manner. The largest possible choke or capacitor is selected, because this produces the lowest ripple, and a rectifier type having the desired d.c. output current and inverse voltage ratings is chosen. Possibly the statement, "When filter condensers larger than 40 μf are used, it may be necessary to add additional plate supply impedance" may be pondered for a moment, but at this point the whole design of the power transformer is turned over to the transformer manufacturer with the statement that a certain d.c. output is required. Since electron-tubes will take tremendous abuse before catastrophic failure, this process often seems to be adequate, at least initially.

When semiconductor rectifier circuits are designed with this casual approach, the results can be disastrous because these rectifiers can fail in one cycle if sufficiently abused. However, it is possible to design semiconductor rectifier circuits which have excellent life and reliability, if care is taken to design the circuits properly.

Design information for rectifier circuits used in electronics is rather sparse, and there seems to be no completely comprehensive reference for use with semiconductors. An attempt is made here to present comprehensive design procedures for use with semiconductor rectifiers in capacitor- and choke-input-filter rectifier circuits. A comparison is made first among six possible rectifier configurations to aid in the selection of the optimum circuit. Given the output specifications of d.c. output voltage and current, voltage regulation, and ripple, the design procedures then specify:

(1) The input capacitor: its capacitance and rms current.

(2) The input choke: its inductance and resistance.

(3) The rectifier: its peak-inverse voltage, turn-on surge current, recurrent surge current, and rms current.

(4) The power transformer: its open circuit secondary rms voltage, secondary winding resistance, and secondary rms current.

2. Capacitor-Input-Filter Rectifiers

The capacitor input filter circuit is the type most widely used with rectifiers in electronic applications. Its design features include a high d.c. output voltage for a given a.c. input voltage, relatively low ripple output, and fair output voltage regulation. It can be built at low cost in a small volume and is relatively light in weight because it does not require a filter choke for its operation. Voltage-multiplying rectifiers which produce several times as much d.c. output voltage as the applied a.c. voltage are possible.

Its most serious weakness is the fact that it draws current from its source in the form of sharp pulses which have a high r.m.s. heating value. The high r.m.s. current heats the transformer winding and rectifier more than would be the case when the same output current is required from a choke input filter and rectifier. Thus the capacitor system is not able to produce as much d.c. output power with the same transformer and rectifier as the choke system.

A lesser weakness of the capacitor system is the surge of current which must be carried by the rectifier for about one cycle of the line frequency when the rectifier is turned on. This surge, which usually amounts to 30 to 100 times the c.c. output current, is frequently the determining parameter which governs the choice of a rectifier. The design of a reliable capacitor-input-filter rectifier requires careful attention to the amplitude and duration of this surge.

Four of the possible capacitor-input-filter rectifier circuits are presented in Figure 8-0.1 (A) to (D). The circuit in (A) is a half-wave rectifier; that in (B) is a full-wave center-tap rectifier; (C) is a full-wave bridge rectifier; and (D) is a full-wave voltage-doubler rectifier. Figure 8-0.1 (E) and (F), using a choke-input filter, will be discussed later. The performance data listed to the right of each circuit provides a means of comparing all six rectifier circuits.

The assumptions used for the purpose of comparison are:

(1) Identical transformer secondary windings are used in all six circuits. (Each of the center-tap circuits use two such windings.)

(2) For the capacitor systems, the product ωCR_S is 0.3, where ω is the frequency of the line voltage in radians per second.

NOTES:
  V_{OC}, R_S, C AND L ASSUMED EQUAL IN ALL CIRCUITS;
  $\omega C R_S = 0.3$; L IS TWICE CRITICAL INDUCTANCE AND HAS NO
  RESISTANCE; v_{OCP} = PEAK VALUE, V_{OC}; V_{OC} = OPEN CIRCUIT
  (NO LOAD) SECONDARY VOLTAGE, RMS; I_W = RMS WINDING CURRENT;
  I_R = DC OUTPUT CURRENT PER WINDING;
  v_{PI} = RECTIFIER PEAK INVERSE VOLTAGE.

Circuit	V_{DC}/v_{OCP} AT I_{DC} MAX.	I_{DC} MAX/WINDING	I_W/I_R AT I_{DC} MAX.	RELATIVE POWER $(V_{DC}/v_{OCP}) \times (I_{DC}$ MAX)	v_{PI}/v_{OCP}
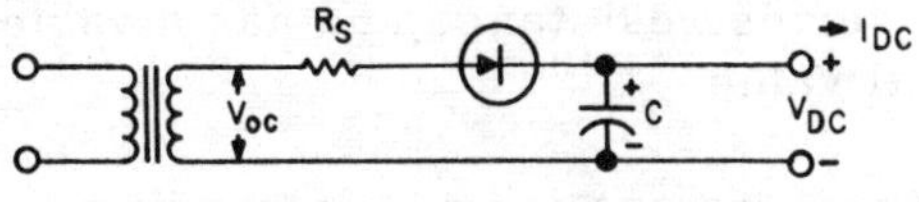 (A) HALF-WAVE, CAPACITOR-INPUT FILTER	0.83	1.0	2.65	0.83	2.0
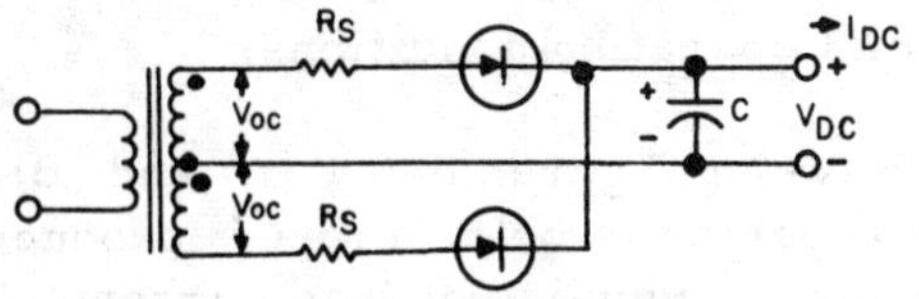 (B) FULL-WAVE, CENTER-TAP, CAPACITOR-INPUT	0.855	1.0	2.65	0.855	2.0
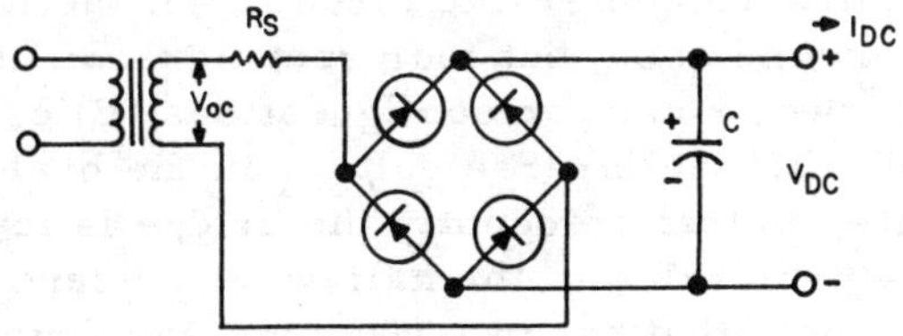 (C) FULL-WAVE BRIDGE, CAPACITOR-INPUT	0.80	1.56	1.70	1.25	1.0
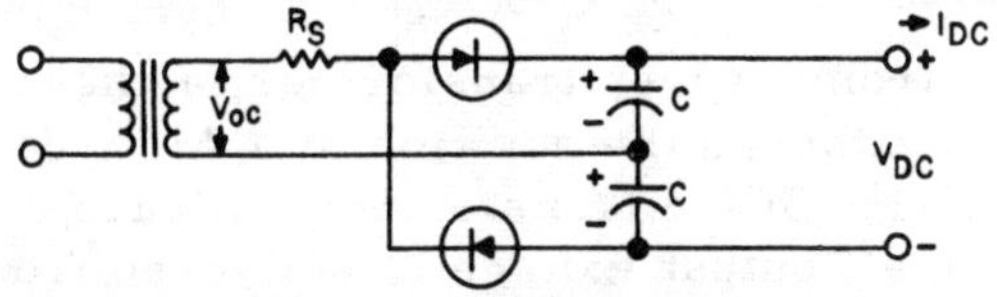 (D) FULL-WAVE, VOLTAGE DOUBLER, CAPACITOR-INPUT	1.57	0.78	3.40	1.22	2.0
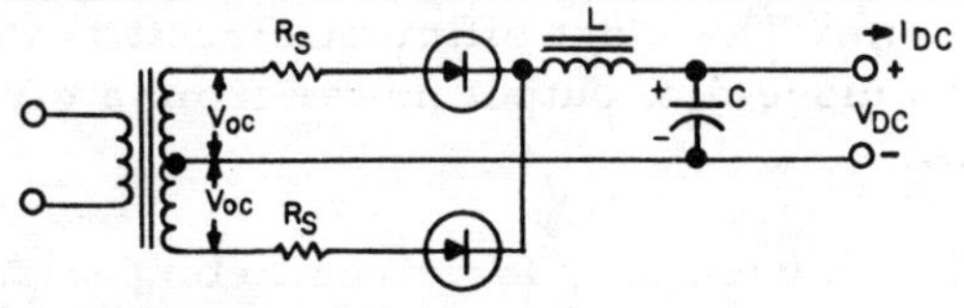 (E) FULL-WAVE, CENTER-TAP, CHOKE-INPUT	0.592	1.77	1.50	1.05	2.0
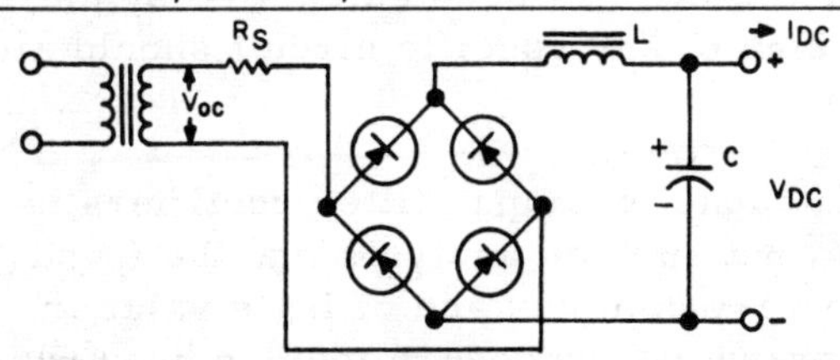 (F) FULL-WAVE BRIDGE, CHOKE-INPUT FILTER	0.573	2.50	1.06	1.43	1.0

Fig. 8-0.1 - Rectifier circuit comparison

(3) R_S is chosen at a value which will produce a turn-on rectifier surge current, i_S, which is equal to 40 I_{DC} in rectifier (B). R_S actually includes the transformer secondary winding resistance, the reflected primary winding resistance, and the dynamic resistance of the rectifier, and is assumed to be identical in all circuits.

(4) The inductance L has no series resistance and has a value equal to twice the critical value.

(5) Current and power comparisons are for single windings.

These assumptions are close to normal operating conditions.

While there is little difference between Figures 8-0.1 (A) and (B), the bridge rectifier in (C) is found to be superior to both in that it produces a smaller r.m.s. transformer current for a given output load current. This effect permits increasing the output load current by about 56% over that possible with the half wave and full wave center-tap circuits. If the load is fixed, a smaller transformer may be used, but four rectifiers will be required for the bridge instead of two or one for configurations (B) or (A). The rectifiers need withstand only half the inverse voltage in the bridge as compared to the other two circuits, so that frequently the bridge is found to be economically and physically preferable to the half-wave or center-tap rectifiers. It is interesting to note that the full-wave voltage doubler rectifier circuit Figure 8-0.1 (D) is also very efficient.

For the greatest power output from a given transformer, a choke-input-filter rectifier must be used, although the addition of a filter choke may nullify the saving made in the size of the transformer. Figure 8-0.1 indicates that, while a relatively small output voltage is delivered from a choke system, the output current which may be drawn is much greater than that possible with a capacitor system. The most efficient circuit is that of (F) which can produce about 67% more d.c. output power from a given transformer than the circuit in (B).

It is intended that Figure 8-0.1 be used only for circuit comparison and estimation. The data presented there has been calculated by the procedures which follow. The design of any specific circuit should also follow these complete procedures.

The mathematical analysis of capacitor-input-filter rectifiers is beyond the scope of this handbook, but such an analysis may be found elsewhere (1). The equations are very complex and of little value in designing a circuit. However, a graphical procedure may be used which will give quite accurate results with reasonable effort. The procedure which follows has been adapted from two sources (2, 3).

At the outset, a warning must be posted concerning the published ratings of rectifiers. Capacitor-input-filter circuits are relatively difficult to calculate and produce rectifier d.c. current ratings which are smaller than with any other circuit. Therefore, manufacturers frequently have rated semiconductor rectifiers in terms of resistive or inductive loads which may be easily calculated and which produce larger d.c. output currents with a given rectifier. The prospective user of a rectifier should realize that its d.c. output current rating must be derated by at least 20% and conceivably as much as 43% from its resistive-inductive load rating when the rectifier is to be used with a capacitor-input filter. Turn-on and repetitive surge current ratings are also important in capacitor systems.

In any of the systems (A) to (D) of Figure 8-0.1, capacitor C may be completely discharged at the instant that line a.c. power is applied. If it happens that the power is applied at the instant that the a.c. voltage is at its peak, a surge current will flow through the rectifier which is initially equal to the peak a.c. voltage divided by R_S. This turn-on surge current will decay to a normal repetitive surge current with a time constant approximately equal to the product CR_S. The turn-on surge current must be held within the limits established by the manufacturer, even at the extremes of line voltage. Some manufacturers state this limit as a "one-cycle surge current," while others provide an $I^2 t$ energy limit (4) for the surge, and still others express the limit graphically, as in Figure 8-0.2 (3). Since heating of the semiconductor is the effect being limited by this rating, the shorter the surge duration the larger the surge current may be.

When using semiconductor rectifiers, the surge duration is normally taken to be equal to the time constant CR_S which is a parameter that determines, in part, the d.c. output voltage regulation and ripple, as in Figures 8-0.3, 8-0.4, and 8-0.5 (3). These figures apply to any type of rectifier. When surge current ratings, such as those in Figure 8-0.2 for silicon rectifiers, are related graphically to the appropriate Figure 8-0.3, 8-0.4, or 8-0.5, it is found that the greatest d.c. output voltage from a given a.c. input voltage is obtained when ωCR_S is between 0.2 and 0.4. When related to specific line frequencies in radians ω, this means that a given rectifier will produce its best voltage regulation when operated within its turn-on surge current ratings if the time constant CR_S is between 0.5 and 1.0

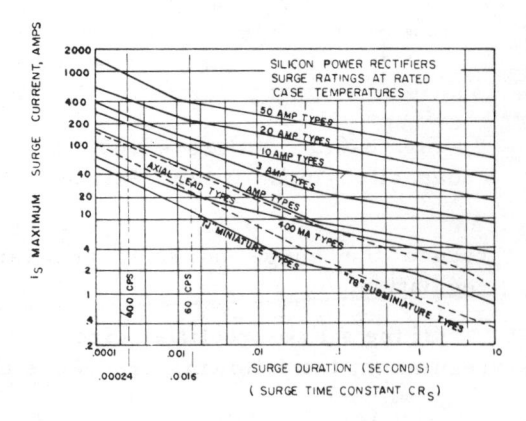

Fig. 8-0.2 - Typical silicon rectifier surge current ratings

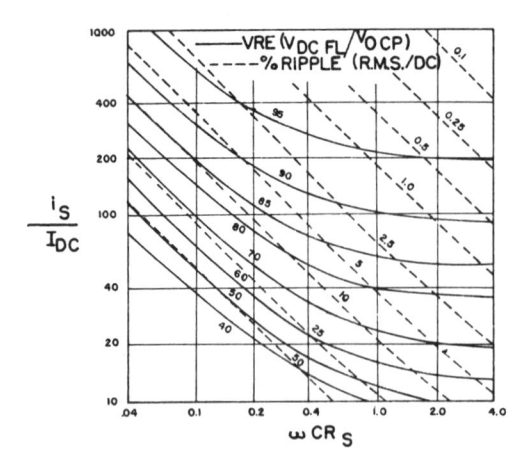

Fig. 8-0.3 - Half-wave capacitor-input-filter rectifier

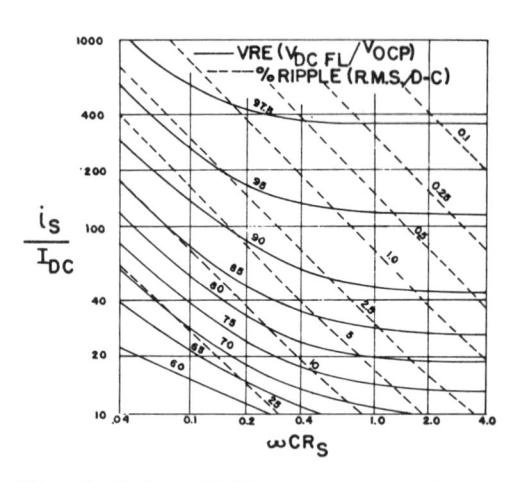

Fig. 8-0.4 - Full-wave capacitor-input-filter rectifier

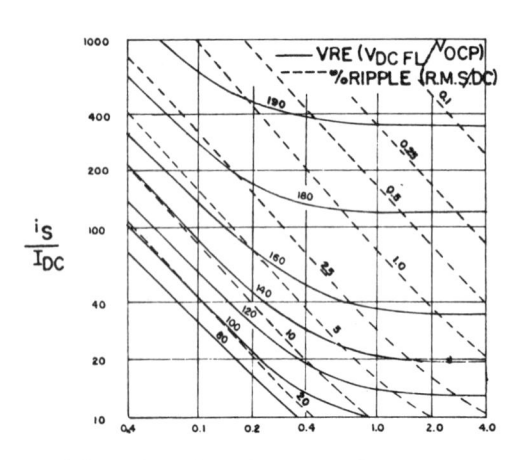

Fig. 8-0.5 - Voltage-doubler capacitor-input-filter rectifier

milliseconds for 60 c.p.s. operation and between 75 and 150 microseconds for 400 c.p.s. duty. This is true for silicon rectifiers only. Surge-current data is not available for germanium and selenium in a form which will permit this type of analysis.

This method of operation, however, does not produce the lowest ripple content in the d.c. output. If desired, the time constant CR_S may be increased to any value for the reduction of ripple at the expense of poorer d.c. regulation or a larger rectifier. According to Figures 8-0.2 to 8-0.5, an increase in the time constant CR_S must be accompanied by an increase in R_S to reduce the surge current, and this in turn will degrade the d.c. regulation.

If the silicon rectifier circuit is to be designed for the best voltage regulation, the following procedure may be followed (3):

(1) Determine the cuircuit requirements for the maximum d.c. output current I_{DC} full-load output voltage $V_{DC\ FL}$, line radian frequency ω, and percentage voltage regulation. Calculate VRE as 100% less the percentage voltage regulation required, for the half-wave and full-wave circuits. This is merely the percentage $V_{DC\ FL}/V_{DC\ NL}$, where "FL" and "NL" indicate "full load" and "no load" voltages. The VRE for voltage-doubler circuits will be twice the value so calculated.

(2) Using the appropriate Figure 8-0.3, 8-0.4, or 8-0.5, determine i_S/I_{DC} at $\omega CR_S = 0.3$ and the VRE calculated above. Multiply the ratio i_S/I_{DC} by the maximum d.c. output current I_{DC} to obtain the turn-on surge current i_S.

(3) Select a rectifier type able to carry the surge i_S for at least $2CR_S$. If ωCR_S is 0.3, this will amount to 1.6 milliseconds at 60 c.p.s. line frequency and 0.24 milliseconds at 400 c.p.s. Normally a rectifier able to carry this turn-on surge will also be able to handle the steady-state rectification of I_{DC}.

(4) Determine the open-circuit r.m.s. voltage required from the transformer as $V_{oc} = 0.707\ (V_{DC\ FL}/VRE + 0.5)$. The 0.5-volt term allows for the forward voltage threshold of a silicon rectifier. The selected rectifier type should have a peak-inverse-voltage rating of about $3.7\ V_{oc\ max}$, which includes a voltage derating factor of about 25%. If a bridge rectifier circuit is used, the peak-inverse-voltage rating need only be $1.85\ V_{oc\ max}$.

(5) Calculate the surge-limiting resistance $R_S = 1.414\ V_{oc}/i_S$. This value of R_S includes the transformer secondary winding resistance, the reflected primary resistance, and the rectifier dynamic resistance. The smallest possible transformer will result when its windings provide the entire value of R_S less the rectifier dynamic resistance.

(6) Determine the input capacitor, $C = 0.3/\omega R_S$, or $C = 800/R_S$ microfarads at 60 cps and $C = 120/R_S$ microfarads at 400 cps.

(7) Determine the percentage r.m.s. ripple, $V_r/V_{DC\ FL}$, from the appropriate Figure 8-0.3, 8-0.4, or 8-0.5. Calculate the r.m.s. current, I_c, in capacitor C from the ripple percentage:

$I_c = (V_r/V_{DC\ FL})\ V_{DC\ FL}\ \omega C$ for Figure 8-0.1 (A),

$I_c = 2(V_r/V_{DC\ FL})\ V_{DC\ FL}\ \omega C$ for Figure 8-0.1 (B) and (C), and

$I_c = 1/2(V_r/V_{DC\ FL})\ V_{DC\ FL}\ \omega C$ for Figure 8-0.1 (D).

Select a capacitor with appropriate capacitance voltage, and r.m.s. current ratings.

(8) Calculate the value

$$k = n \frac{i_S}{I_{DC}} (VRE),$$

where n is 1.0 for half-wave rectifiers, 2.0 for full-wave center-tap or bridge rectifiers, and 1/2 for voltage doublers. Then calculate the values of $k \omega CR_S$ and the percentage 100/k, and with these values enter Figure 8-0.6 to find the ratio I_r/I_R. This is the ratio between the r.m.s. current, I_r, and d.c. current, I_R, flowing in a single rectifier of the circuit being analyzed. The r.m.s. current is then calculated as $I_r = I_R(I_r/I_R)$. The current I_R is equal to $I_{DC}/2$ for the full-wave center-tap or bridge rectifiers, and is equal to I_{DC} for the half-wave and voltage-doubler rectifiers.

(9) The transformer winding r.m.s. current, I_w, may be calculated from the rectifier r.m.s. current, I_r. A single, untapped secondary winding is normally specified for all rectifier configurations except the full-wave center-tap rectifier, which requires two windings connected in series-aiding with the connection brought out as a center tap. The r.m.s. current flowing in each winding, I_w, is then equal to the rectifier r.m.s. current, I_r, in the half-wave and full-wave center-tap rectifiers, but it increases to $1.414\,I_r$ in the bridge and voltage-doubler configurations.

(10) Taking the values for $k \omega CR_S$ and 100/k from step (8) above, enter Figure 8-0.7 and find the ratio i_{RP}/I_R. The recurrent

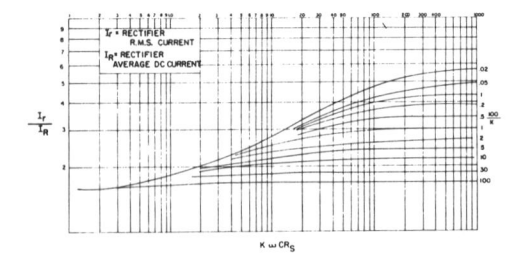

Fig. 8-0.6 - Ratio of r.m.s. to d.c. rectifier currents

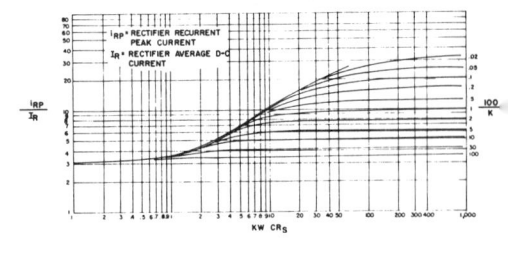

Fig. 8-0.7 - Ratio of recurrent peak current to d.c. rectifier currents

peak rectifier current, i_{RP}, is then equal to: $i_{RP} = I_R(i_{RP}/I_R)$, where I_R is the same as defined in step (8) above. At this point the rectifier currents can be expressed in four ways, in terms of the d.c. output, I_R, the r.m.s. value, I_r, the recurrent peak value, i_{RP}, and the turn-on surge current, i_S. Each of these calculated current values should now be checked against the specifications of the rectifier type selected in step (3) above to ensure that its current ratings are not exceeded at its maximum ambient or case temperature.

(11) Specify the transformer as follows:

   (a) Open-circuit secondary voltage, V_{oc}, from step (4).

   (b) Secondary winding resistance, including that reflected from the primary, from step (5). This resistance is equal to R_S less the rectifier dynamic resistance and any surge limiting resistance added to the circuit external to the transformer. Since the center-tap circuit has two secondary windings, the calculated value of winding resistance is valid for each of the two. Primary winding resistance reflects into each one of these two secondaries individually.

   (c) The r.m.s. current in each winding, from step (9).

Occasionally, a standard rectifier circuit constructed with semiconductor rectifiers, notably silicon, will be found to be unreliable because of rectifier failure during input or output switching. Whenever currents are changed suddenly, the magnetizing and leakage inductances of the power transformer will develop large voltage transients which may carry rectifiers into large inverse voltages. Germanium and selenium rectifiers have sufficient reverse leakage to damp these transients somewhat, but a good silicon rectifier has extremely low reverse current until it reaches its avalanche voltage breakdown. Hence, the magnetic energy of the transformer may be dissipated in the rectifier during these transients. If the energy is large, the rectifier may be destroyed.

Usually, a full-wave rectifier with a capacitor-input filter, operating from a power transformer having close coupling between its secondary windings, is inherently self-protecting, since a voltage transient tending to bias one rectifier inversely will also bias the other rectifier into conduction. The conducting rectifier then transfers the magnetic energy into the filter capacitor whose voltage normally will not rise appreciably. Hence, configurations (B), (C), and (D) of Figure 8-0.1 are normally self-protecting.

The half-wave rectifier in (A) of Figure 8-0.1, however, has no means of absorbing a transient tending to bias its rectifier inversely. Hence, such a rectifier circuit should be examined closely for switching transients. Protection may be obtained by shunting the transformer secondary with a silicon or selenium voltage-limiting device of suitable energy capacity, or with a series resistance-capacitance network across the secondary which is designed to resonate the inductance in question at a low Q so that its transient voltage peak is held to a low value.

When designing a capacitor-input-filter rectifier circuit, it is good practice to derate the turn-on surge current by about half. The possibility that the capacitor may have a +100% tolerance has already been allowed for by selecting a rectifier type capable of carrying the surge current for a time $2CR_S$, even though the design time is only CR_S.

High temperature is perhaps the major cause of rectifier deterioration, but it is difficult to express this fact as a recommended current and voltage derating on a rectifier. Conservative design requires the lowest possible rectifying junction temperature produced by derating the manufacturer's current rating and adequate cooling of the rectifier case. Probably a satisfactory derating method is to design for a maximum junction temperature less than the manufacturer's rating by a derating margin of possibly 50°C for silicon and 25°C for germanium rectifiers.

The reverse leakage current in a well-made rectifier is small enough to contribute very little to its heating. Germanium and silicon rectifiers normally break down suddenly at their avalanche voltages, and each type carries a guaranteed minimum breakdown voltage. Since the avalanche is relatively constant with time and actually increases with temperature, it seems that a peak-inverse voltage derating of only 20% to 30% (permitting use of 70% to 80% of rating) is all that is required for conservative design. However, it is essential that this derated voltage not be exceeded at high line voltage or during switching transients.

3. Choke-Input-Filter Rectifiers

The design of choke-input-filter rectifiers using semiconductor rectifiers is quite straightforward and is similar to the procedure used with electron tubes (5). Figure 8-0.1 (E) and (F) are the most frequently-used configurations. If, as is the normal case, the inductance of the choke L is large enough to maintain current flow through one of the pair of rectifiers during the entire a.c. cycle, the design of the circuit becomes simple.

The following procedure may be followed:

(1) Determine the circuit requirements for the maximum d.c. output current I_{DC}, no load d.c. output voltage $V_{DC\ NL}$, line radian frequency ω, and percentage voltage regulation.

(2) Calculate the total voltage drop in the transformer, rectifier, and choke as: V_D = (percentage voltage regulation) $\times V_{DC\ NL}$. Then,

$$R_S + R_R + R_L = V_D/I_{DC},$$

where R_S is the transformer resistance in each side of the center-tap winding or in the entire bridge winding, and includes reflected primary resistance,

R_R is the rectifier dynamic resistance, and

R_L is the resistance of the choke inductance L.

(3) Calculate the open circuit (no-load) transformer r.m.s. output voltage as $V_{oc} = 1.11\ (V_{DC\ NL} + 0.5)$, where $V_{DC\ NL}$ is the no-load d.c. output voltage. The 0.5-volt term represents the forward-bias threshold of silicon rectifiers. It will be a slightly different value with other types of rectifiers.

(4) Select a rectifier type capable of supplying the required I_{DC} to an inductive load at the maximum ambient temperature. The rectifier peak-inverse-voltage rating should be about 3.7 $V_{oc\ max}$ with a center-tap circuit, and about 1.85 $V_{oc\ max}$ with the bridge circuit. Determine the forward dynamic resistance, R_R, of this rectifier type.

(5) In order that one of the rectifiers conduct at all times, the inductance of the input choke must be greater than a minimum value, called the critical value, L_c. This relationship may be expressed as follows (5):

$$L \geq L_c = \frac{\dfrac{V_{DC\ NL}}{I_{DC\ min}} + R_S + R_R + R_L}{3\omega}$$

If the value of I_{DC} becomes very small in operation and thereby requires a large value of inductance L, a bleeder resistance across the output will maintain a minimum value of I_{DC}. In addition, a "swinging choke" may be used for L. This type of choke has little or no air gap in its magnetic path so that inductance is high at low d.c. currents but drops rather quickly at higher d.c. currents. Such a unit will provide so much low-current inductance that bleeder current requirements will be minimized. Current I_{DC} includes the bleeder current in all equations, and $V_{DC\ NL}$ is the output voltage obtained when I_{DC} is barely sufficient to produce rectifier conduction during the entire cycle of power frequency.

(6) Resistance $(R_S + R_L)$ may be found by subtracting R_R, from step (4), from the total $(R_S + R_R + R_L)$ found in step (2). This resistance $(R_S + R_L)$ must then be divided between the choke and transformer in such a manner as to provide the smallest and least expensive pair of units. Consultation with transformer manufacturers is recommended on this point.

(7) The ratio of rectifier r.m.s. to d.c. current I_r/I_R, is much smaller than is the case with capacitor systems. With an infinite L, the ratio is only 1.414 as compared to a typical 2.8 with a capacitor system. If L is a normal value about twice L_c, the ratio I_r/I_R may be taken as 1.50. Since the center-tap circuit of Figure 8-0.1 (E) provides $I_R = I_{DC}/2$, then I_r is 1.50 $I_{DC}/2$ or $I_r = I_w = 0.75\ I_{DC}$ r.m.s. The bridge circuit of Figure 8-0.1 (F) is even more efficient and can deliver 67% more power than the circuit in Figure 8-0.1 (B) from the same components because $I_w = 1.06\ I_{DC}$ in a single secondary winding.

(8) Specify the transformer as follows:

   (a) Open circuit secondary voltage, r.m.s., V_{oc}, from step (3).

   (b) Secondary winding resistance, R_S, from step (6) which includes the reflected primary resistance. In the center-tap circuit, R_S is for each side of the secondary winding and primary resistance reflects into each side individually.

   (c) The r.m.s. current, I_w, in the secondary winding, from step (7).

(9) Select a value for the capacitor C which provides the desired output ripple by reference to Figure 8-0.8.

(10) Calculate the turn-on resonant surge current in the rectifier. This surge is caused by the sudden application of what amounts to a d.c. voltage step to a series L-C resonant circuit. The current surge will be (6):

$$i_S = \frac{0.9\, V_{oc}}{\sqrt{\dfrac{L}{C}}} \text{ amperes.}$$

Determine whether or not the rectifier selected in step (4) will carry this surge current.

The resonance effect is frequently of importance to the load, since sudden changes in load current will also excite this resonance. It may or may not be adequately damped by the output load resistance and choke resistance. The load will receive a resonant voltage instead of the resonant current experienced by the rectifiers. The resonance is difficult to damp and may be sufficient to cause a voltage regulator supplied from it to become inoperative occasionally during load current transients.

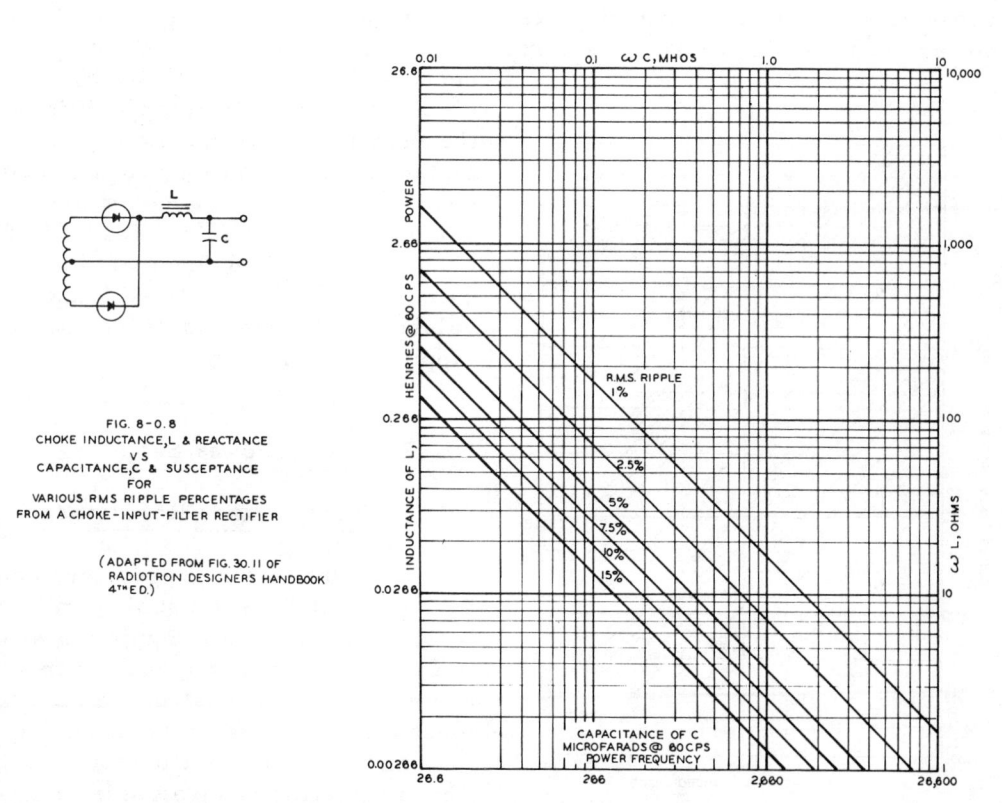

FIG. 8-0.8
CHOKE INDUCTANCE, L & REACTANCE
VS
CAPACITANCE, C & SUSCEPTANCE
FOR
VARIOUS RMS RIPPLE PERCENTAGES
FROM A CHOKE-INPUT-FILTER RECTIFIER

(ADAPTED FROM FIG. 30.11 OF RADIOTRON DESIGNERS HANDBOOK 4TH ED.)

Fig. 8-0.8 - Choke inductance, L & reactance vs capacitance, C & susceptance for various r.m.s. ripple percentages from a choke-input-filter rectifier

While most of the same comments on transient voltage surges made in the section on capacitor systems apply here, the choke must now also be taken into consideration. Fortunately, it normally will not produce damaging transients. When the rectifier is turned on, the choke has no energy to release; when it is turned off, the choke surges the rectifiers into normal forward conduction and discharges its energy into the filter capacitor and load. Load current switching transients are largely absorbed by the filter capacitor so that the current in the choke need not change rapidly and voltage surges remain small.

The application of semiconductor rectifiers to choke-input-filter rectifiers is quite simple. In general, the resistive-inductive load current rating is used if the turn-on resonant surge is not too large. Derating procedures similar to that described for capacitor systems are adequate.

4. Filtering

In the event that the output ripple of one of the capacitor-input or choke-input-filter rectifier circuits of Figure 8-0.1 is excessive and cannot be reduced with reasonable components, a series-L, shunt-C filter section may be added to the rectifier circuit to reduce the ripple. Since the inductance will have a d.c. resistance, the added d.c. voltage drop should be taken into account in the design of the rectifier circuit. Reference to Figure 8-0.9 will permit calculation of the required L and C values for the desired reduction of ripple.

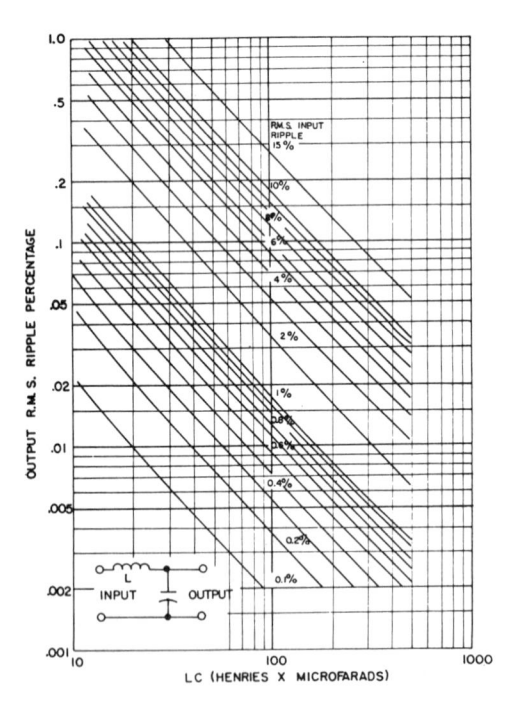

Fig. 8-0.9 - Output ripple vs LC product for various r.m.s. input ripple percentages 120 cps ripple

B. Regulated Power Supplies

1. Shunt Voltage Regulators

Shunt voltage regulators can be the simplest, and usually are the least efficient, of all regulating systems. The prime reasons for their use are their low cost in the simplest form and their inherent overload and short-circuit protection. This protection is invaluable, especially in experimental work where accidental short-circuits are relatively frequent. In contrast, a poorly-protected series regulator may have every transistor destroyed by a short-circuit lasting for a small fraction of a second.

Figure 8-0.10 illustrates the three basic types of shunt voltage regulators. The first and simplest, (A), is identical in configuration to the familiar VR (voltage regulator) electron tube circuit. Its design is somewhat simpler than that of the VR circuit since no "starting" voltage surge is required to ignite the breakdown diode, CR_1. It is probable that both the input voltage and output load current will vary in use, so that this regulator must be designed to operate at the extremes to be encountered.

For instance, R_1 will be determined by considering simultaneously the minimum instantaneous unregulated input voltage (including the effect of ripple) V_U, the maximum output load current, the minimum current desired in CR_1, and the highest voltage expected from any diode CR_1 which might be used. By then considering simultaneously the effect of maximum possible V_U, minimum load current, and minimum voltage from CR_1 with the value of R_1 just calculated, the maximum power dissipation which may occur in CR_1 can be determined.

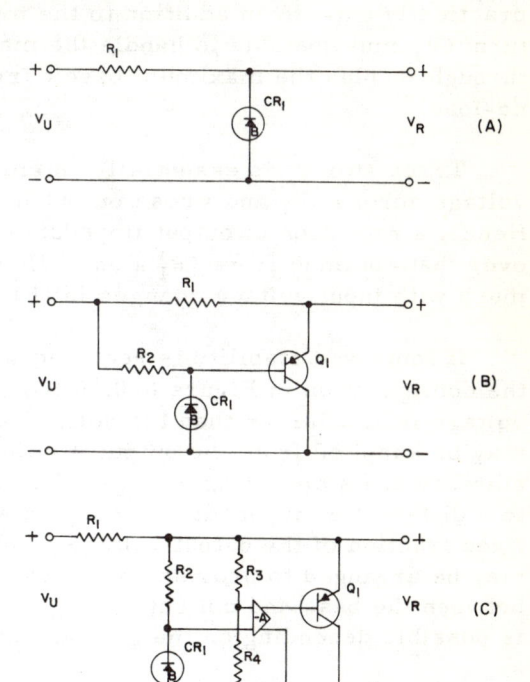

Fig. 8-0.10 - Shunt voltage regulators

At the present time, breakdown (or "Zener") diodes are available in 5%, 10% and 20% tolerances between 2 and 200 volts and with power dissipation ratings between 0.2 and 50 watts. Voltage ratings near six volts produce the lowest regulating impedance and approximately zero voltage temperature coefficient. At other voltages, either higher or lower, both the impedance and temperature coefficient are degraded. Present diode types carry impedance ratings between 0.8 ohms and 300 ohms.

If the impedance of the breakdown diode is insufficient for the application, or if it is not able to carry the required current, an arrangement such as shown in Figure 8-0.10 (B) may be used. Resistor R_1 must be determined as before. Resistor R_2 must be able to provide the maximum I_{cbo} of Q_1 under conditions of maximum output current, when Q_1 should be practically cut-off, in addition to the minimum current desired in CR_1. In turn, CR_1 must be able to handle the maximum current which might flow through R_2 plus the maximum base current which might flow from Q_1 at no-load.

Transistor Q_1 is essentially an emitter follower whose input is the voltage across CR_1 and whose output is the regulator load resistance. Hence, a reduction of output impedance by a factor of about a_{fe} is obtained over that obtained from CR_1 alone. However, the output will change just as much with input voltage changes in (B) as in (A).

If improved stability is required with respect to line and load changes, the configuration in Figure 8-0.10 (C) is a possibility. Since the output voltage is now larger than the voltage across CR_1, this breakdown diode may be supplied from the output it helps regulate with much improved stability of its operating voltage. The voltage from CR_1 is then one input to a differential-input, direct-coupled amplifier whose other input is a fixed fraction of the output voltage provided by R_3 and R_4. The amplifier may be arranged to provide an inverting gain with its output voltage placed between the base and collector of Q_1. Almost any degree of performance is possible depending on the gain and stability of the differential amplifier.

It will be noted that in all the regulators of Figure 8-0.10 that an overload or short circuit on the output merely decreases the voltage across and the current through Q_1 toward zero so that there is no possibility of destroying Q_1 or its driving amplifier by such means. However, it is also important to note that at no-load, Q_1 must dissipate the full output power. For this reason, this type of regulator has an inherently low efficiency and is most often used in fixed-voltage, fixed-load applications.

2. Series Voltage Regulators

The series voltage regulator, as its name implies, places the transistor regulating element in series with the load. Regulation occurs as the result of varying the voltage drop across the series transistor. No output current is wasted through a path shunting the output, as with the shunt regulator, so that the small-load efficiency of the series regulator can be made much greater than that of the shunt regulator. Full-load efficiencies can be made practically the same with both series and shunt regulators. Despite its fail-safe operation, the shunt-regulator's poor small-load efficiency usually causes a series regulator to be used instead.

The series regulator does not have inherent overload protection, and can be utterly destroyed by a momentary short circuit unless suitable protection is employed. Protective schemes will be mentioned later in Section B.6. They may be added to any regulator circuit, but in order to avoid confusion no protective circuits will be added to any of the basic regulators to be discussed.

The simplest series regulator is shown in Figure 8-0.11 (A). Transistor Q_1 is merely an emitter follower whose input is the voltage across the breakdown diode CR_1. Resistor R_1 provides the base drive current for Q_1 and the breakdown current for CR_1. The regulation of this circuit is only fair, but adequate for many applications. Changes in the unregulated input voltage will cause a change in the breakdown voltage of CR_1 and thereby cause a small change in the regulated output. The output load will "see" an output resistance slightly higher than the emitter input resistance of transistor Q_1 operating in the common-base connection. This output resistance will be about 30 ohms with a 1 mAdc load and will drop to perhaps one ohm at 1 A load and possibly 0.2 ohms at 10 A load.

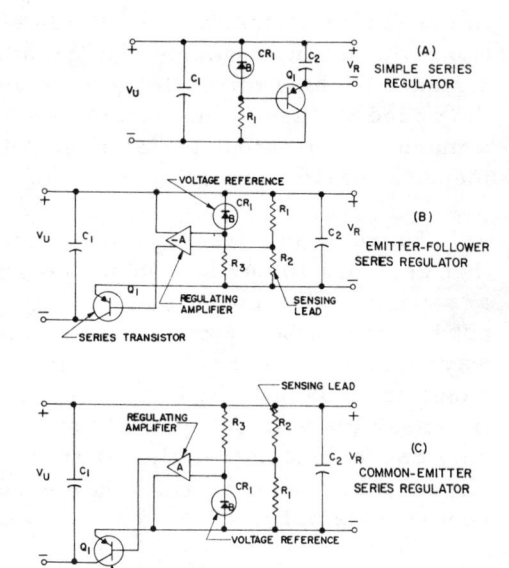

Fig. 8-0.11 - Series voltage regulator

This simple circuit has no voltage amplifier and yet it can provide an output resistance of an ohm at a one-ampere load. If voltage amplification is added, this output resistance is divided by the voltage gain. For instance, to achieve a one milliohm output resistance at one ampere load, a voltage gain of only 1,000 is required with practically any power transistor.

There are two general methods of applying voltage gain to a series-regulating transistor. Both have about the same performance but they differ in the method of operating the series transistor. The first method, illustrated in Figure 8-0.11 (B), uses the series transistor as an emitter follower, since the input drive to the series transistor is placed between its base and collector while the output is taken between collector and emitter. Thus, the collector is common to input and output, and the circuit is an emitter follower.

The second method, shown in Figure 8-0.11 (C), utilizes the series transistor as a common-emitter amplifier stage, since its input signal is applied between base and emitter, while its output is taken between collector and emitter. Although voltage gain is obtained between base and collector, there is no gain advantage to the common-emitter configuration in the series element, since the basic output resistance of this configuration, with its input shorted, is greater than the shorted-input output resistance of the emitter-follower configuration by a factor closely equal to the added voltage gain. Thus the additional gain has to be used to reduce a much higher output resistance so that both configurations have nearly identical performance.

There is an advantage to the use of the common-emitter configuration that becomes important when designing a regulator for a wide output voltage range. This configuration requires that the regulating amplifier output be applied between base and emitter where the voltage levels are always small, regardless of the ultimate output voltage delivered by the regulator. Usually, it is convenient to build the entire regulating amplifier near the voltage level of the emitter of the series transistor, and the only connection to the other output terminal is a voltage "sensing" lead. Hence, there is no restriction placed on regulator output voltage by the regulating amplifier.

In contrast, the emitter-follower configuration which is normally used with electron tubes places the regulating amplifier near the potential of the output lead which does not involve the series transistor. The regulating amplifier then senses the voltage across the output terminals and drives the base of the series transistor. However, the base may be many volts away from the regulating amplifier in a high-voltage regulator so that a dangerously large collector-emitter voltage is required in the last transistor stage of the regulating amplifier.

If the regulator is designed to produce a non-adjustable output, this collector-emitter voltage will be relatively constant. In this event, a large portion of the voltage may be absorbed by a breakdown diode in series with the collector. However, a wide-range regulator will not permit the use of such a scheme and the necessarily large voltage on the last amplifying transistor becomes a real design hazard. At this point, the common-emitter configuration may become quite attractive.

Both configurations in Figure 8-0.11 (B) and (C) are similar in other respects. Resistor R_2 "senses" the output voltage and provides, with R_1, an input voltage to the regulating amplifier which will be adjusted by feedback action to equal the voltage on the voltage reference, CR_1. Resistor R_3 passes current through CR_1 in order to establish its reference voltage. The regulating amplifier will be discussed in Section B.5.

If the regulator is required to produce an output that necessitates a possible excessive voltage drop across a single series transistor, the series stack of Figure 8-0.12 may be used. Resistors R_4, R_5, and R_6 are equal and divide the total voltage across Q_1, Q_2, and Q_3 equally. Emitter followers Q_4, Q_5, and Q_6 transfer the voltage levels from the resistor divider to the stack, Q_1, Q_2, and Q_3 in such a way that each of these transistors will assume one third of the total voltage drop.

Such a stack may be increased in length as desired. However, each level must be provided with a turn-off bias current, such as that provided by R_1, R_2, and R_3, in order that the emitter-follower drivers Q_4, Q_5, and Q_6 can remain in control at high temperatures. Also, the resistance divider must be of a suitably low resistance level so that the base currents of the emitter followers will not alter the voltage division appreciably.

3. Constant-Current Regulators

The emphasis in voltage regulators is given to the reduction of output impedance. In contrast, the current regulator is constructed to have the highest possible output impedance. Instead of sensing the voltage at the output terminals, the regulated current is passed through a resistor and the voltage developed across the resistor is compared with a voltage reference. Any voltage difference is then amplified and applied to a series regulating transistor which completes the regulating feedback connection by controlling the regulated current flow.

Three representative current-regulator circuits are illustrated in Figure 8-0.13. The first regulator, (A), utilizes R_1 as the current-measuring resistor whose voltage is applied to the emitter of Q_1. The voltage from the reference breakdown diode, CR_1, is applied to the base of Q_1 so that Q_1 is itself the comparison element. The result is that Q_1 attempts to pass a current which will keep the voltage across R_1 closely equal to the voltage across CR_1.

Performance of this simple circuit is limited because the current

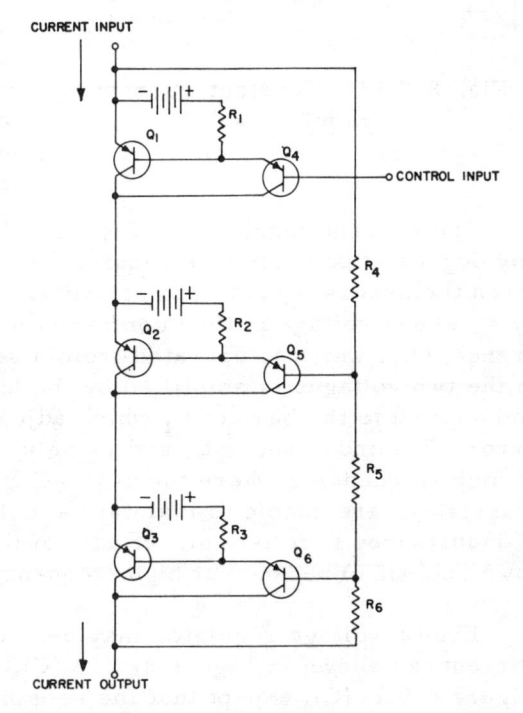

Fig. 8-0.12 - High voltage transistor stack

measured by R_1 is not exactly the output current, since it also contains the base current of Q_1 and does not contain I_{CO}. The only regulating gain is that provided by Q_1 itself. Performance improves as R_1 and the voltage across CR_1 are increased, but the output resistance can never surpass the collector-base resistance of Q_1 which is effectively shunting the output.

The output current, I_R, will be regulated over a voltage range extending from a positive limit of about $(V_U - V_{CR_1})$ to a lower limit which can even become negative. This assumes, of course, that Q_1 has suitable collector-base voltage and dissipation ratings. In this case, the transistor is acting as a common-base amplifier so that the collector-base voltage rating, which is higher than the collector-emitter rating, may be used as long as the total base circuit resistance is held low enough to prevent instability from an alpha greater than one.

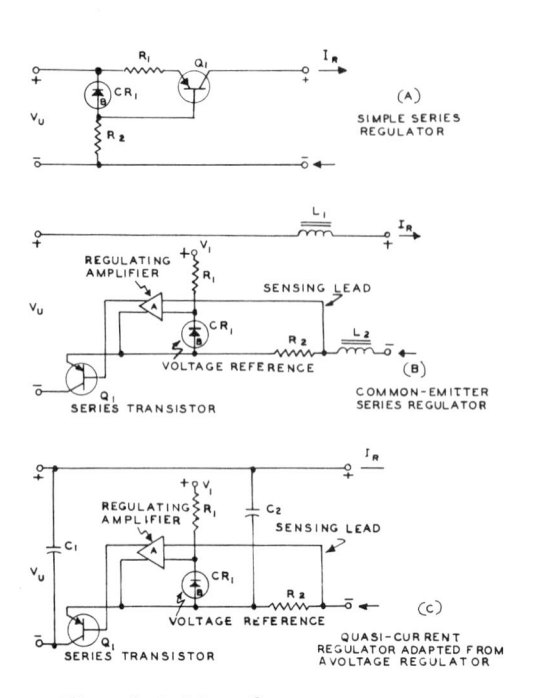

Fig. 8-0.13 - Constant-current regulators

The current regulator in Figure 8-0.13 (B) is capable of practically any degree of performance required, providing the regulating amplifier is given the necessary gain and stability. The actual output current is sensed by R_2 whose voltage is then compared with that of the stable voltage reference, CR_1, which is operated from a separate voltage source. Any error in the two voltages is amplified by the high-gain, direct-coupled amplifier and applied to the base of Q_1 which adjusts the current I_R to minimize the error. The inductances L_1 and L_2 help to maintain a high output impedance at high frequencies where the regulating gain inevitably decreases and the transistors are unable to maintain a high impedance themselves. The use of inductances here is a direct dual of the use of a shunting capacitor to lower output impedance at high frequencies in a voltage regulator.

Even a voltage regulator may be used to provide a constant output current, as shown in Figure 8-0.13 (C). This is the identical circuit from Figure 8-0.11 (C), except that the sensing lead now measures the voltage across a current-monitoring resistor, R_2, instead of the output voltage. The voltage reference is now supplied from a separate voltage source because the output voltage may fluctuate so widely during current regulation.' The capacitors C_1 and C_2 do not assist in current regulation and in fact prevent the regulator from producing a true constant current except at d.c. At any higher frequency, the capacitors lower the output impedance toward the level expected from a voltage regulator. Hence, this might be called a "quasi-current regulator."

4. Switching Regulators

It was stated previously that series voltage regulators have relatively high efficiency. This is true in relation to the shunt voltage regulator, but the series regulator still depends on a controllable loss of power to perform its function of regulation. There is one additional type of voltage regulator which can be made almost lossless in operation and achieve efficiencies approaching 100%. Such a regulator utilizes a series semiconductor which is rapidly switched between its limiting conducting and non-conducting states.

Regulation control is obtained by variation of the conducting duration during each cycle of switching. Semiconductor switches can be made to have very small dissipation in both the conducting and non-conducting states. The only appreciable dissipation occurs during the transitions between states, and this is minimized by rapid switching and the use of a suitably low switching frequency. The net result is a very high efficiency.

The bursts of power passed by the semiconductor must then be filtered to produce usable d.c. power. Because of the relatively low switching frequencies involved, ranging from about 50 cps to 4 kc, and the low-pass filter which follows the switch, the response speed of the switching regulator is rather slow in comparison to the shunt or series voltage regulator previously discussed.

Three possible forms of switching regulators are illustrated in Figure 8-0.14. The first, the Silicon-Controlled-Rectifier Voltage Regulator in (A), accomplishes regulation while rectifying. The two silicon controlled rectifiers, SCR_1 and SCR_2, form a bridge rectifier, with CR_3 and CR_4. When either of the controlled rectifiers is pulsed on, it will remain conducting until its terminal voltage reverses, in the manner of a thyratron. It is the function of the unijunction transistor pulser to pulse the two controlled rectifiers twice each line cycle so that the one which is biased in a forward direction will be brought into conduction at the point in the a.c. cycle which is appropriate to the d.c. output required.

The d.c. output voltage is sensed by R_4 and R_5 and compared with the voltage reference from CR_6. The error is amplified and applied to the unijunction transistor along with a waveform derived from the a.c. line by CR_1 and CR_2. The pulser than produces a trigger in each half-cycle of a.c.

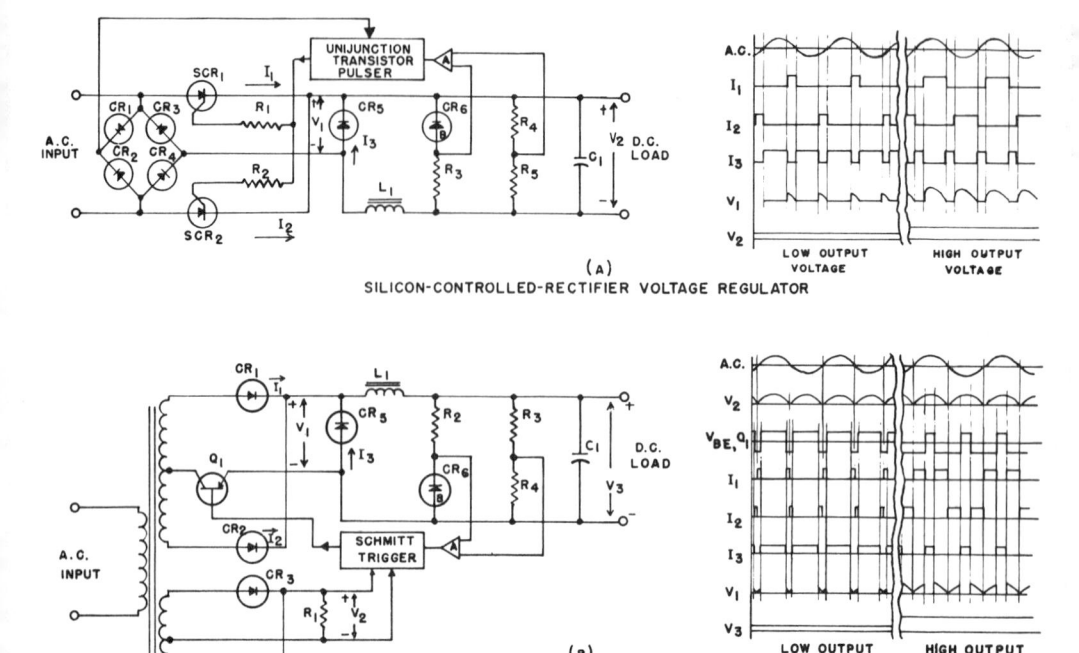

Fig. 8-0.14 - (Parts A & B) Switching voltage regulators

power, earlier in the cycle if more d.c. output is required, or later if less is needed. Each controlled rectifier is then brought into conduction once each cycle, a half-cycle out-of-phase with its companion.

The waveforms in Figure 8-0.14(A) explain this action further. It is assumed that L_1 is very large so that the load current is constant. This current must then flow through one of three paths, SCR_1, SCR_2, or CR_5. The current waveforms I_1, I_2, and I_3 illustrate the manner in which this constant load current switches between the three possible paths.

This circuit is very similar to the ordinary choke-input-filter rectifier and also has a critical inductance value for L_1 which is a function of d.c. output current. If L_1 is greater than this critical value, then the circuit will produce the same good output regulation typical of a choke-input-filter rectifier, even at a fixed conduction time in the controlled rectifiers. For this reason, the conduction time controls the output d.c. voltage rather than the current. The two waveforms for V_1 show how the conduction time affects this voltage. The inductance, L_1, and capacitance, C_1, take the average value of the voltage waveform V_1 so that the d.c. output voltage, V_2, increases with an increase in conduction time.

Silicon controlled rectifiers are still quite expensive. It is possible to duplicate the performance of the controlled rectifier with the transistor-phased rectifier, which uses a single power transistor and two conventional rectifiers, as outlined in Figure 8-0.14(B). The d.c. output sensing and comparison with a voltage reference is accomplished as before, and the error is amplified and applied to a Schmitt trigger circuit along with a full-wave rectified waveform, V_2. The Schmitt trigger produces two output states which can drive transistor Q_1 into hard saturation conduction or into cut-off.

The Schmitt trigger will produce intervals of conduction in Q_1 which are centered in time about the zero-crossings of the a.c. power sinusoid. As more output voltage is required, the conduction time (or angle) is increased until it reaches a maximum of 100%, when the circuit becomes merely a conventional choke-input-filter rectifier. At smaller conduction times, the voltage V_1 will consist of segments of a full-wave-rectified sinusoid resembling a sequence of letters "M". Again, the inductance, L_1, and capacitance, C_1, produce V_3, the d.c. output, by taking the average value of the "M" waveform.

The load current, again assumed constant because of the larger-than-critical inductance, L_1, also finds three possible paths as before, through CR_1, CR_2, or CR_5. When Q_1 is cut-off, the load current is blocked from flowing through CR_1 and CR_2, so that it is driven through CR_5 by the discharge of energy from L_1. When Q_1 conducts, the load current is free to flow through CR_1 or CR_2, whichever the a.c. power has forward-biased.

In cases where the voltage peaks across Q_1, which are equal to no more than the peak value of the voltage on either half of the winding supplying CR_1 and CR_2, exceed the voltage limit on a single transistor, a high-voltage transistor stack may be used to replace Q_1. Note that the load line for Q_1, or the stack replacing it, is rectangular; that is, the current through Q_1 will not begin to decrease until the full voltage is developed across Q_1 during turn-off, nor will the voltage begin to decrease until the full current is passing through Q_1 during turn-on. This produces an instant of simultaneous maximum current and maximum voltage during each switching transient. Not only must switching be rapid to keep the dissipation low, but also there is the possibility of collector-emitter avalanche breakdown during turn-off. However, it is readily possible to produce efficient and useful regulators using the transistor-phased-rectifier scheme.

Both the silicon-controlled-rectifier and transistor-phased-rectifier systems must operate at power frequency because they include the power rectification function. Because of the low ripple frequency, the inductance, L_1, must be quite large to be greater than the critical value, and C_1 must be large to keep the ripple in the output low. These large values of L_1 and C_1 are included within the regulating feedback loop so that the speed at which the regulator can respond to a load transient is very slow.

When the load current is suddenly increased, C_1 must provide the increased current until the current through L_1 can rise to the increased value. To increase the current through L_1, the average voltage value of the sinusoid segments at its input must exceed the d.c. voltage at its output for a length of time proportional to its inductance. The current will change at a rate equal to the voltage difference divided by the inductance. For this reason, no switching regulator may be used to provide a continuous output voltage corresponding to 100% conduction at the lowest input voltage; there must always be a surplus of conduction time available to be used during load transients to increase the current through L_1 quickly. Also for this same reason, it is desirable to reduce the value of L_1 so that both recovery time and transient voltage difference across the inductance may be minimized.

Since there is a large minimum value for L_1 in both of these circuits, the transient performance of both is only fair. Another approach to the problem, called the Transistor-Variable-Chopper Voltage Regulator, is presented in Figure 8-0.14(C). Here, rectification of the a.c. power is accomplished by ordinary rectifiers and the transistor switch operates on an unregulated d.c. input. There is no longer any requirement to switch at line frequency, so that a better compromise between switch transistor dissipation, size of L_1, and transient performance may be made by increasing the chopping frequency to the region between 400 c.p.s. to possibly 4 kc.

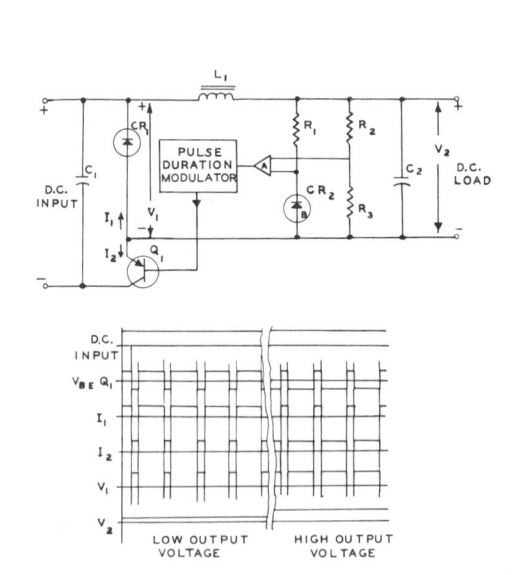

Fig. 8-0.14 - (Part C) Transistor-variable-chopper voltage regulator

This chopping frequency is relatively constant and is generated by the "pulse duration modulator." This device receives the amplified error signal in the usual way and generates a waveform which drives Q_1 between hard, saturated conduction and cut-off. If more output voltage is required, the conduction pulse generated by the modulator is made a greater portion of the chopping period, as shown in the waveforms in Figure 8-0.14(C). The modulator can be made easily to have a reasonably linear error-voltage-to-duration characteristic. Since the output voltage varies linearly with pulse duration, the regulating gain is nearly constant over the entire operating range and it will produce good regulation at any output voltage from about 5% to 100% of maximum output. The regulating gain of the previous two switching regulators involves trigonometric terms and is not uniform over the operating range.

The critical value of L_1 is much smaller now, so that an improved transient recovery can be obtained with a much smaller inductance. Since the ripple frequency is higher, a smaller C_2 than before will suffice.

The load current has only two possible paths in this regulator, one being through CR_1 and the other going through Q_1. When Q_1 is conducting, the load current passes through it and L_1 is charged somewhat. When Q_1 is switched off, the load current switches to CR_1 and continues to flow by discharging L_1. During this interval, the energy which flows to the load is coming entirely from L_1 and is part of the energy which is stored while Q_1 conducted. In a sense, it performs like a d.c. autotransformer in that it can transform a high-voltage, low-current input into a lower-voltage, higher-current output at high efficiency.

5. Regulating Amplifiers

Basically, regulating amplifiers are merely high-gain, direct-coupled amplifiers with low noise and drift and good stability when used within a feedback loop. Any of the conventional techniques used in the design of direct-coupled amplifiers are applicable.

A method of circuit analysis is necessary, and the question arises concerning the use of small-signal or large-signal transistor parameters. Much of the published work on regulators contains a small-signal analysis of operation, but this does not yield accurate results when some of the transistor elements, especially those near the end of the amplifier, are driven over a wide range of their characteristics, possibly into cut-off. Hence, it is necessary to use a large-signal, or "d.c.," analysis in all stages where the operating point varies sufficiently to cause a change in the conventional small-signal, or "a.c.," parameters. It will be found that the series, or shunt, regulating transistor must then be handled on a large-signal basis, while the first stage of the regulating amplifier may be analyzed on a small-signal basis.

If the permissible change in output voltage with a certain change in load is known, reference to the transistor characteristic will determine the change in regulating transistor input voltage and current necessary to permit the known change of load current. If no regulating amplifier is used, the input voltage thereby determined will be the change in the regulator output voltage caused by the load change.

If a regulating amplifier is to be used, its required voltage gain will be the regulating transistor input voltage change divided by the desired regulator output voltage change. For instance, if the series regulating transistor requires a bias change from -0.2 volts to -0.8 volts to produce the change in output load current, and if an output voltage change of 600 microvolts is the design objective, then the regulating amplifier must provide a d.c. gain of 0.6 volts/600 microvolts, or 1,000. This voltage gain, incidentally, includes the attenuation of the sensing voltage divider, represented by R_1 and R_2 in Figure 8-0.11(B) and (C). Therefore, the amplifier itself may have to provide as much as thirty times the value calculated, depending on the attenuation of the divider.

In an electron-tube regulator, this voltage gain requirement is the only gain requirement. However, since transistors have both current and voltage input requirements, it is also necessary to specify an overall current gain as well as an overall voltage gain. An equivalent specification might refer to a certain voltage gain operating at specified input and output impedances, and this might also be referred to as a required power gain between specified input and output impedances. Regardless of the manner in which it is specified, a certain power gain, or current gain, must be produced by the transistor regulating amplifier in order to achieve the required voltage gain.

In an ideal direct-coupled amplifier, where all load resistances are current sources, and collector output resistances are infinite, the amplifier voltage gain (not including the sensing voltage divider) is merely the product of all the individual stage common-emitter current-gains times the final output load resistance divided by the amplifier input resistance. The actual amplifier, of course, uses actual resistors instead of current sources as load resistances, so that the actual voltage gain is less than this ideal value. However, this ideal situation serves to emphasize certain design facts.

One is that amplifier gain is controlled primarily by the common-emitter current gains rather than the audio power gains. Normally, the optimum collector load resistance of one stage is much larger than the input resistance of the following stage. No matching transformers can be used, since it is a direct-coupled amplifier. Hence, there is a fundamental loss in power gain due to the impedance mismatch. The input resistance of one stage is then the predominant load resistance on the collector of the preceding stage. As a result, almost all of the available collector current passes into the base of the following stage regardless of the base resistance of that stage, within reason. Whereas base input resistance is an important parameter in determining audio amplifier power gain, it becomes of almost no importance in a direct-coupled amplifier when compared to the importance of common-emitter current gain.

Another point that the ideal direct-coupled amplifier stresses is that any circuit change which does not alter the current gains or the input and output load resistances has little effect on amplifier gain. This means that a change of operating point in one stage, which doubles its voltage gain, will not affect the overall voltage gain, since the input impedance of that same stage will become half of its former value so that the stage preceding it will experience a drop in gain to half its former value.

A further fact that can be gleaned from the ideal case is that emitter-follower stages, which do not have a voltage gain greater than one can be used in a direct-coupled amplifier to increase the overall voltage gain significantly. This is accomplished by an increase in the current-gain product alone. In an actual amplifier, the effect of the emitter follower is to increase the apparent input resistance of a voltage amplifier so that the preceding stage will see a larger load resistance and develop a larger gain. It is possible to obtain d.c. voltage gains of as much as 10,000 in one voltage-amplifying stage which is followed by emitter followers.

While the point might be argued that two stages with gains of 100 each would provide the same performance with possibly fewer transistors, it must be remembered that the regulating amplifier must operate within a closed feedback loop where high-frequency phase shift must be minimized and controlled. The two-stage scheme will develop almost twice the ultimate phase shift that would be produced by the single stage with its emitter followers. It is possible to build a common-emitter series transistor regulator which provides critically-damped transient recovery from a regulating amplifier having 94 db of voltage gain, including the input sensing voltage divider attenuation, by using this emitter follower principle.

The stage voltage gain in an actual direct-coupled amplifier may be calculated as equal to the current gain times the ratio of the output load resistance to the input resistance. Large or small-signal parameters may be used as is appropriate.

The output load resistance is comprised of the input resistance of the following stage shunted by the collector output resistance and any load or bias resistors. The input resistance is that resistance produced by the transistor at its base terminal and does not include any other circuit resistances. This method is valid for either the common-emitter or emitter-follower amplifiers.

When designing a regulating amplifier, it is imperative that sufficient turn-off bias current be provided for each stage so that its I_{CO} at high temperatures will not cause instability. The expected I_{CO} should be determined at the maximum junction temperature for the worst new transistor to be used in a given stage. A degradation factor of five over this maximum expected I_{CO} is not excessive. It will be found that the usual small germanium transistors in a 50°C ambient may require about 0.5 mAdc turn-off base bias, and the usual 3-ampere power transistor will require about 50 mAdc. It is also important to use a transistor that is not too large for its job, because the larger a transistor becomes the more severe is the problem of coping with its high-temperature I_{CO}.

A great deal of time can be spent trying to calculate the feedback stabilization components for a desired amplitude and phase margin, even if one is competent to undertake the task. It is fortunate that empirical procedures, coupled with a general analysis of the feedback problem (7), can be used to stabilize an amplifier quite satisfactorily. A useful procedure is to load the amplifying stages with capacitance until oscillation has ceased. Then switch load current with a continuously-running chopper or relay and observe the voltage transient response at the output. The phase-lead networks for each stage may then be experimentally determined for the best transient recovery.

Drift caused by temperature changes is always a problem with direct-coupled amplifiers. It is normally determined primarily by the input amplifier stage and is not influenced by negative feedback. In transistor amplifiers, it is caused by temperature variation of the base-emitter voltage, the common-emitter current gain, and I_{CO}. In a balanced differential amplifier, these temperature variations in the three parameters will tend to balance out.

Matching the two transistors in a differential amplifier for similar base-emitter voltage and current-gain temperature coefficients will help to keep the drift low. Matching the I_{CO} in the two transistors is usually not worth while, since it is the least predictable characteristic with time. The best way of handling I_{CO} is to keep it as small as possible. At high temperatures, I_{CO} will eventually become the predominant drift component. The low I_{CO} of silicon transistors makes them especially attractive for low-temperature drift differential amplifiers.

Associated with the problem of first-stage temperature drift is the problem of voltage reference stability with temperature changes. Several compensated breakdown diodes are currently available with coefficients of less than 0.001% per °C over a temperature range from -55°C to +100°C. Their reference voltage is usually 8.4 volts. A reference of this sort must be supplied with a stable and precise current in order to obtain the best temperature characteristic.

Another approach is the combination of a silicon transistor with a breakdown diode in such a way that the negative base-emitter voltage temperature coefficient helps to compensate for the positive breakdown diode voltage coefficient. These units, called Ref-Amps (8), are available in temperature coefficients as low as 0.002% per °C and will provide the same functions as a stable voltage reference and a two-silicon-transistor differential amplifier.

The design of a regulating amplifier should always be studied for possible transistor overloads during supply turn-on, turn-off, output overload, output short-circuit, release of output short-circuit, and so on. Electron tubes do not require such care, but since transistors can be destroyed so quickly by surges it is necessary to examine all possible surges for possible damaging effects.

If the d.c. output power from the regulator is to be delivered over a considerable distance to its load, it will probably be advantageous to sense the output voltage at the remote load. The regulating amplifier sensing connections to both output lines are then connected directly to the load. In this way, the load voltage itself is regulated and the wire resistance between the regulator and load is effectively divided by the regulating amplifier gain. It may be necessary for reasons of feedback stability to bypass the regulator output and sensing leads for high frequencies so that they do not have to travel to the load and back.

If the regulator is a narrow-voltage-range type, it may be possible for the regulating amplifier to operate directly from the output power of the regulator. However, a wide-range regulator, and especially one which operates down to zero output voltage, will require a separate regulated supply for its regulating amplifier. Fortunately, this separate regulator need not be as involved as the main regulator, and often is merely a power breakdown diode.

6. Overload Protection

As has been mentioned, overload protection is very important in transistor regulator design. The regulating amplifier must not be damaged by any expected or unexpected overload or transient. However, adequate protection for the amplifier is relatively easy to provide in comparison to protection for the series regulating power transistor. It stands directly in the path of a short circuit and can be destroyed in milliseconds.

Fuses are frequently used for protection, but it is always a race between the fuse and transistor to be first to melt during a heavy overload. Fuses are probably not the best protection and possibly are not even adequate protection in many cases.

The output lead may be interrupted by a magnetic circuit breaker which may be faster than a fuse and consequently safer. An even faster breaker may be made from a transistor flip-flop which is triggered by an excessive load current flowing through a monitoring resistor. The flip-flop can be made to bias off the series transistor in a matter of microseconds so that protection is excellent. A manual reset, or automatic reset after a pause, may be used to restore the flip-flop.

There is a question as to the suitability of circuit interrupters for power supply protection. If a supply with such protection is momentarily overloaded or shorted, its output will drop to zero and remain there until the fuse is replaced or the breaker is reset. If this supply had been supplying a bias voltage for a large transistor installation, conceivably hundreds of transistors could have failed during this brief interval without bias.

Possibly better protection can be afforded the equipment upplied by using a current-limiter method of protection. In this method, constant-voltage operation is maintained up to 100% of rated current, but beyond that limit the voltage regulation may turn into a form of current regulation so that even a short circuit will not draw more than 120% of rated current. A momentary short will then drop the voltage, but it will restore as quickly as the short is removed. In this way, the load is given the best continuity of power and the least chance of catastrophic damage.

Current-limiting may be done in a number of ways. One obvious way is to place a regulating transformer at the input to the power supply. When this type of transformer is overloaded, its output voltage will drop toward zero and limit the current it will supply. Another method limits the base drive current available for the series power transistor and thereby limits the current it will pass. One of the best current-limiting schemes places a monitoring resistor in series with the output current in such a way that an excessive current will give rise to a voltage across it which exceeds a present threshold. When the threshold is exceeded, the monitoring resistor voltage assumes control of the regulating amplifier so that the output current, rather than the voltage, is regulated. When the current demand at the output drops to normal values, the regulating amplifier is again returned to its normal voltage regulation function.

Still another method of obtaining current limiting is by placing a current regulator of the type illustrated in Figure 8-0.13(A) in series with one of the output leads. As long as the load current is less than the regulating level, the power transistor will remain saturated and have little effect on the voltage regulator. If the load current ever becomes excessive, the current regulator will unsaturate and begin to function, thereby protecting the series voltage regulator.

Overloads having nothing to do with output circuit malfunction can occur during turn-on when the output capacitor must be charged to the operating output voltage. This surge current will be controlled by a current-limiter, but if this means of protection is not used, some method of either charging the capacitor slowly so that its surge current is small, or of paralyzing the overload interrupting circuit when this turn-on surge is expected, will be required.

REFERENCES

1. E. E. Staff, M.I.T., "Applied Electronics," John Wiley & Sons, Inc., New York, 1947, pp 272-287.

2. Langford-Smith, F., "Radiation Designer's Handbook," Fourth Edition, Published by Wireless Press for Amalgamated Wireless Valve Co. Pty. Ltd., Sydney, Australia, and distributed by RCA Victor Division, RCA, Harrison, N.J., pp 1170-1182.

3. "Application Notes," AN 1351C, Transition Electronic Corp., Wakefield, Mass.

4. "Selection of Surge Resistor for Capacitive Input Filter Circuits," Semiconductor Rectifier Handbook, Semiconductor Products Dept., General Electric, Syracuse, N.Y., "Rectifier Applications" section, pp 5-8, July 1, 1957.

5. E. E. Staff, M.I.T., "Applied Electronics," John Wiley & Sons, Inc., New York, 1947, pp 289-300.

6. Langford-Smith, F., "Radiotron Designer's Handbook," Fourth Edition, et al, p. 1185.

7. Flood, J. E., "Negative Feedback Amplifiers--Conditions for Critical Damping," Wireless Engineer, Vol. 27, #322, July 1950, p. 201.

8. "Ref-Amp" Data Sheet, TE-1352E, Transition Electronic Corp., Wakefield, Mass.

BIBLIOGRAPHY

Chase, F. H., Hamilton, B. H., Smith, D. H., "Transistors and Junction Diodes in Telephone Power Plants," BSTJ, 33, pp 827-858, July 1954.

Chase, F. H., "Power Regulation by Semiconductors," Elec. Eng., New York, 75, pp 818-822, September 1956.

Deuitch, D. E., Paz, H. J., "A Phase-Regulated Transistor Power Supply," IRE Trans. on Circuit Theory, CT-4, pp 279-284, September 1957.

Foss, R. N., "Electrically-Rugged Power Supply," Electronic Design, 6, pp 48-49, December 10, 1958.

Hanson, G., "High Temperature Voltage Regulator Design," Products Bulletin, Texas Instruments, Inc.

Jensen, J. L., "Voltage and Current Regulation," U.S. Patent #2,776,382.

Lowry, H. R., "Transistorized Regulated Power Supplies," Electronic Design, 4, pp 38-41, February 15, 1956, and pp 32-35, March 1, 1956.

Mohler, R. D., Curtis, R. H., "Engineering Investigation of Transistor Voltage and Current Regulators," Final Report, Motorola, Inc., Communications and Electronics Division, July 18, 1958, ASTIA AD204900.

Sherr, S., Levy, P., Kwap, T., "Design Procedures for Semiconductor Regulated Power Supplies," Electronic Design, 5, pp 22-25, April 15, 1957.

Thompson, E. E., "A Transistorized D-C Power Supply Employing a Novel Method of Voltage Regulation," Program of 1958, IRE Canadian Convention.

"Strain Gage Power Supply Has 0.1 Percent Regulation," Electrical Design News, 4, pp 28-29, February 1959.

"Circuit Applications for Diffused Silicon Regulators," Texas Instruments, Inc., August 1958.

"Silicon Voltage Regulators," AN-1352A, Transitron Electronic Corp., October 1958.

"Notes on the Application of the Silicon Controlled Rectifier," G.E. Semiconductor Products Dept., December 1958.

"Stabilized Supply Contains Transistor and Magnetic Regulators," Electrical Design News, 4, pp 50-51, March 1959.

CIRCUIT 8-1

BREAKDOWN-DIODE REGULATED POWER SUPPLY

Texas Instruments, Inc.

Circuit 8-1 is an example of a two-stage shunt regulator using double-anode breakdown diodes.

The voltage provided at C_1 by the bridge rectifier, comprised of CR_1 through CR_4, is 150 volts d.c. with 11.5 volts peak to peak ripple. The resistor, R_2, and the first breakdown diode regulator, CR_5, provides additional filtering, reducing the ripple by a factor of 180, while reducing the d.c. voltage to 47 volts. The value of the series resistor, R_2, was calculated to allow a current of approximately 50 mAdc to pass through the first diode, at a nominal input of 120 volts r.m.s.

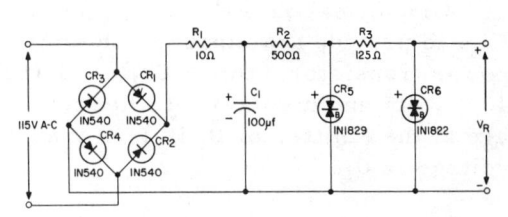

Circuit 8-1 - Breakdown-diode regulated power supply

The second regulator provides further ripple reduction by a factor of 16 at full load while regulating the output at 24 volts for load variations from zero to 150 mAdc. The output voltage drops about 2% when the load is changed from zero to 150 mAdc, resulting in an output resistance of approximately 3.3 ohms.

The circuit will take input fluctuations from 100 to 140 volts with little effect on the output. If lower voltages are expected, the circuit may be designed so that CR_5 will carry a larger current at 155 volts r.m.s. input. The upper limit is imposed by the rating of the 1N1829.

CIRCUIT 8-2

EMITTER-FOLLOWER REGULATOR

Texas Instruments, Inc.

Circuit 8-2 is a simple regulator which utilizes a breakdown diode, CR_1, to provide the input for an emitter follower, Q_1, which is a 2N457 power transistor. Diode CR_1, a 1N1827, breaks down at approximately 40.3 volts and displays an impedance of about five ohms. The output voltage at the emitter of Q_1 is less than 40.3 volts by the emitter-base bias voltage in Q_1.

A typical 2N457 will carry an emitter current range of 0.5 to 1.5 amperes with an emitter-base bias voltage range of 0.25 to 0.45 volts. This transistor also has a common-emitter current gain of about 50, so that the base current will change about 20 mAdc over the load current range of 0.5 to 1.5 amperes. This 20 mAdc change will pass into the breakdown diode, CR_1, and cause a change of 100 mVdc across its 5-ohm impedance.

If the unregulated source, V_U, has a regulation of 20%, corresponding to an impedance of 13 ohms, the change of load current over a one-ampere range will cause V_U to change by 13 volts. This will change the current in

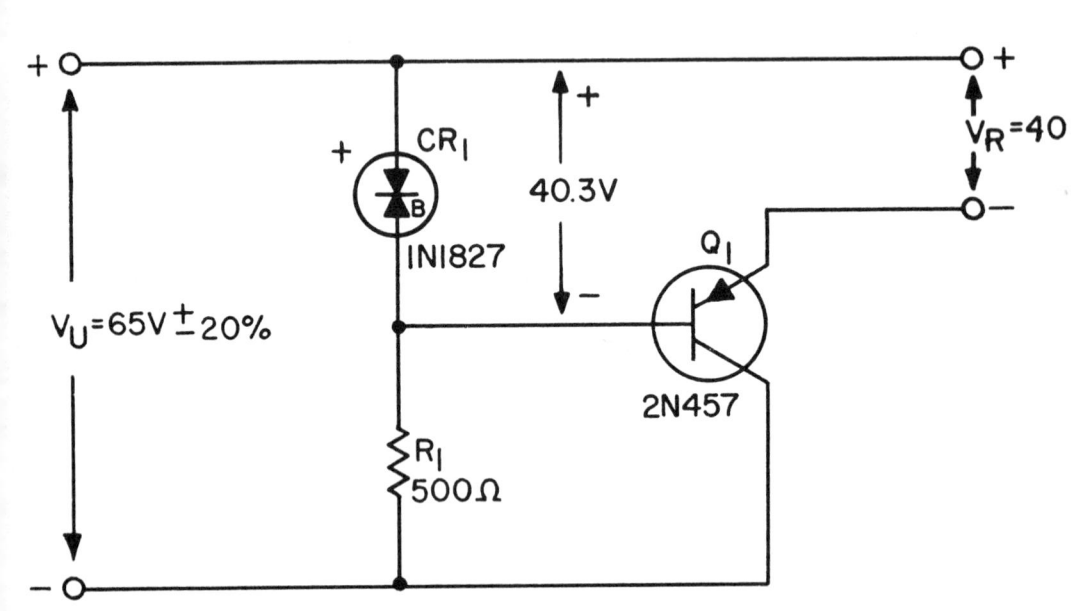

Circuit 8-2 - Emitter-follower regulator

Circuit 8-2, Emitter-Follower Regulator (cont'd.)

R_1 by 26 mAdc and cause an added change in the voltage of CR_1 of (5 x .026) or 130 mVdc. Thus, the total change in voltage across CR_1 will be 100 + 130 = 230 mVdc.

This change of 230 mVdc is added to the emitter-base voltage change of 200 mVdc so that the total output voltage change at V_R is 430 mVdc for a one-ampere change. Consequently, the output resistance is 0.43 ohms.

With an input d.c. voltage of 65 volts, ±20%, with 200 mVdc peak to peak ripple, this circuit will provide an output current between zero and 1.5 amperes at a voltage of 40 volts and a ripple of 4 mVdc peak to peak. At a case temperature of 60°C, the 2N457 will permit a maximum V_U of 60 volts at full 1.5-ampere load without carrying its junction temperature above 95°C. At the same case temperature, the 1N1827 may dissipate as much as 9 watts without exceeding its junction temperature limit. Since the circuit will dissipate no more than 1.6 watts in CR_1, the 1N1827 is well derated.

CIRCUIT 8-3

SIX-VOLT SILICON SHUNT REGULATOR

Texas Instruments, Inc.   Contributed by Gary Hanson

    This circuit is intended to deliver a fixed voltage output equal to 6.0 volts, ±3%, at load currents between 50 and 300 mAdc. It uses the shunt regulation principle with a high-gain voltage amplifier and shunt regulator built from silicon transistors.

    The voltage reference, CR_5, is a Texas Instruments Type 651C5, which operates at 5.0 volts, ±5%. It is supplied from an 8.0-volt breakdown diode, CR_2. The output voltage, V_R, of 6.0 volts is sensed by the voltage divider R_{10} and R_{11} which produces a voltage, equal to that across CR_5, at the base of Q_5. The differential amplifier, comprised of Q_4 and Q_5, compares the voltage from CR_5 with the voltage from R_{10} and R_{11} and drives the load resistor R_8. The collectors of Q_4 and Q_5 are supplied from a breakdown diode regulator, R_7, CR_3, CR_4 and C_2, which produces a constant 10.0 volts, ±5%.

    The output signal from R_8 drives the base of Q_3, a p-n-p 2N369, which is the second voltage amplifier. The emitter of Q_3 is held within 5% of 8.0 volts by CR_1 at an impedance of about 8 ohms. This small impedance in the emitter of Q_3 causes some degeneration and loss of gain to the

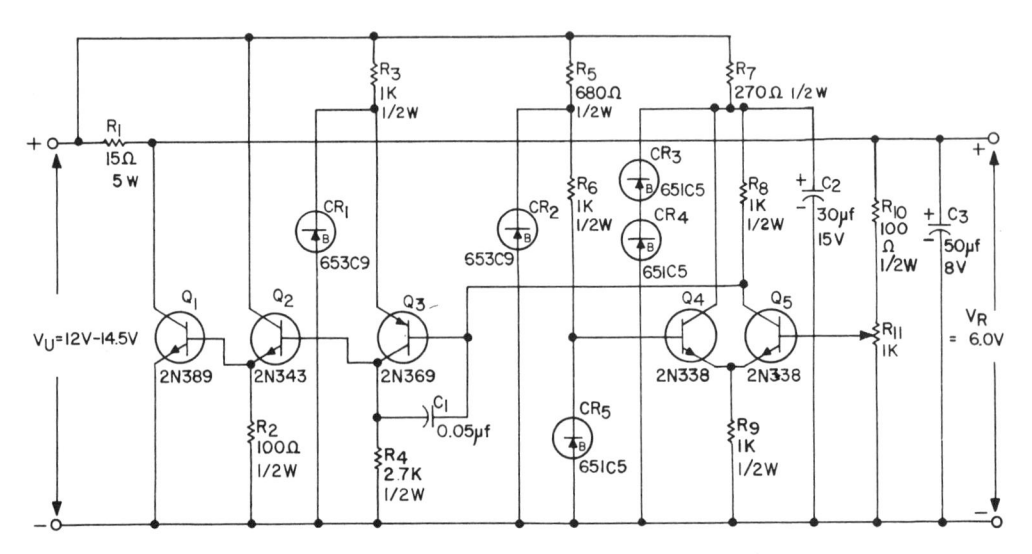

Circuit 8-3 - Six-volt silicon shunt regulator

Circuit 8-3, Six-Volt Silicon Shunt Regulator (cont'd.)

collector load, R_4, but the loss is neither serious nor easily reduced. Capacitor C_1 permits stability of the closed loop. The collector output from Q_3 drives Q_2 as an emitter follower which in turn drives the silicon regulating power transistor, Q_1.

If the output voltage, V_R, increases, the base of Q_1 will be driven positively, causing Q_1 to conduct more and increasing the voltage drop across R_1 and the source impedance of V_U. This action tends to reduce the output, V_R, and opposes the tendency for an increase.

The output ripple is less than 50 mVdc peak to peak and the dynamic output impedance is less than 4 ohms. The regulator will operate in ambient temperatures as high as 55°C.

CIRCUIT 8-4

CURRENT-LIMITED SERIES VOLTAGE REGULATOR

Applied Physics Lab., University of Washington

Contributed by R. N. Foss

This regulated power supply features a current-sensing circuit which takes over the regulating amplifier whenever the output current exceeds a threshold value and converts the voltage-regulator into a current-regulator capable of limiting overload currents to safe values. The voltage regulator is capable of providing output voltages from 0.5 to 30 volts at currents as high as 3.5 amperes. The regulating amplifier provides a modest voltage gain from a straightforward circuit which reduces the output resistance to about 0.15 ohms. While not a high-performance regulator, it will nevertheless perform well for a great many applications.

Power from T_1 is full-wave, center-tap rectified by CR_1 and CR_3, to provide the d.c. voltage across C_1 which is regulated by Q_4 to Q_8. It is also rectified by another full-wave, center-tap circuit comprised of CR_2 and CR_4 which develops about 37 volts across C_2 and supplies the 10-volt reference, CR_7, a base return resistor, R_{11}, and an output bleeder, R_{15}.

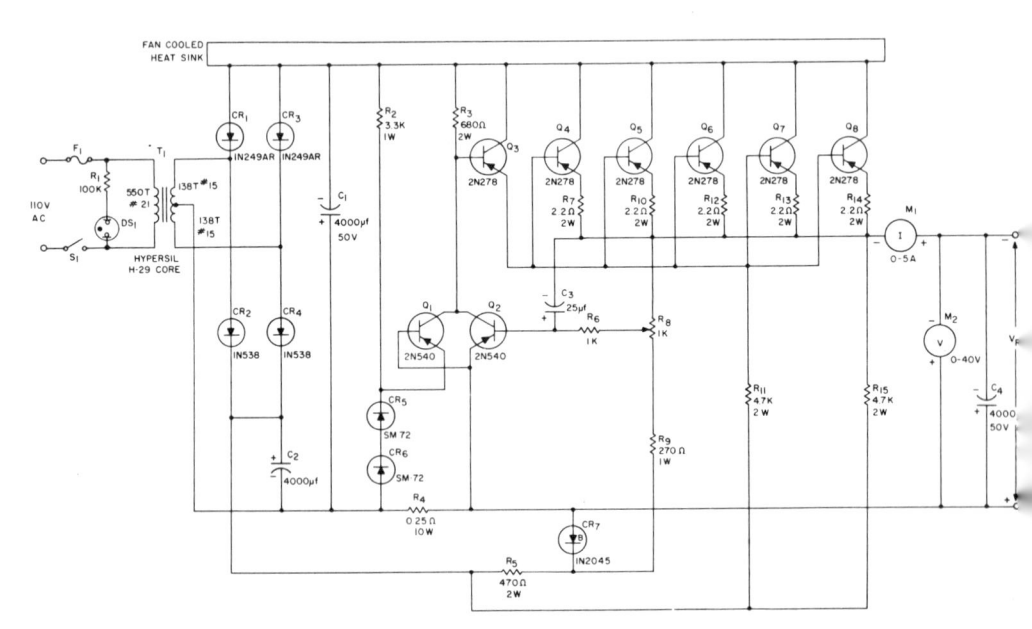

Circuit 8-4 - Current-limited series voltage regulator

Circuit 8-4, Current-Limited Series Voltage Regulator (cont'd.)

Current from CR_1 and CR_3 passes through the fan-cooled heat sink to the collectors of the paralleled series transistors, Q_4 to Q_8. Accurate division of the current through the five series transistors is produced by the 2.2-ohm emitter resistances.

The voltage developed across the load, V_R, is sensed by R_8 which also provides a comparison with the voltage reference, CR_7, through R_9. At any setting of the voltage-setting control, R_8, circuit regulating action will adjust the output voltage so that the slider of R_8 is maintained a few tenths of a volt negative with respect to the positive output terminal. The voltage sensed by R_8 is then amplified by Q_2 and applied to the base of emitter-follower Q_3 which in turn drives the paralleled bases of the series transistors. Since the collector-emitter output of Q_2 is applied to the bases and collectors of the series transistors, this is classed as an emitter-follower type of series regulator. The voltage gain of Q_2 varies from about 100 to 400, but the attenuation caused by R_6 and R_8 reduces this to an effective regulating gain of about five at a 30-volt output.

Output current flows through R_4 where it develops a positive voltage at the anode of CR_6. Both CR_5 and CR_6 are merely silicon power junctions which are biased in the forward direction by current from R_2 so that each produces a relatively constant 0.6-volt drop. Hence, there is normally a -1.2 volt bias developed by CR_5 and CR_6 on the emitter of Q_1 with respect to its base. However, when the positive voltage developed by load current passing through R_4 reaches this same level, Q_1 will be brought into conduction through CR_5 and CR_6, and will take control of Q_3 and the series transistors. As the load current tends to increase Q_1 will be brought into heavier conduction and the series transistors will be biased off further. Hence, the current is limited at about 4.6 amperes by a current-regulating circuit. As soon as the overload is removed, Q_1 is cut off and returns control to Q_2.

Current-limiting overload protection of this type requires that the series transistors carry the full 4.6 amperes if the output is shorted from a normal 30-volt output. Thus, a heat sink capable of dissipating about 160 watts is required. Circuit components are arranged so that all critical heat-producing semiconductors can be mounted on the same heat sink without mica electrical insulators. By using reverse-polarity 1N249A rectifiers for CR_1 and CR_3, even they can be mounted on the heat sink without mica.

Resistor R_6 reduces the regulating gain but is included as a means of protecting Q_2 against output voltage surges. Capacitor C_3 reduces the attenuation of the sensing network at high frequencies and thereby improves output impedance at these frequencies by utilizing the entire voltage gain of Q_2.

Circuit 8-4, Current-Limited Series Voltage Regulator (cont'd.)

Since this circuit was designed, transistors somewhat more suitable for the job than those shown in Circuit 8-4 have become available For instance, Q_1 and Q_2 handle a maximum current of about 50 mAdc at a maximum dissipation of about 0.5 watt. The power transistor type 2N540 is able to pass three amperes and dissipate over ten watts so that it is much too big for this application. Because it is so large, its I_{CO} will be much larger than is necessary and this will cause a serious high-temperature voltage drift. A better voltage amplifier to replace Q_2 would consist of a pair of smaller transistors connected as a differential amplifier.

Each of the series transistors must pass about an ampere, maximum, at a voltage drop of possibly 35 volts. The smallest possible junction capable of dissipating 35 watts and carrying an ampere should be used instead of the 2N278, which is a 15-ampere type. The 2N278 can have an I_{CO} as large as 8 mAdc at room temperature, so that the five together could conceivably require 40 mAdc. Since I_{CO} increases drastically with temperature, a turn-off bias supply of at least several hundred milliamperes should be available from R_{11}. However, only 9 to 15 mAdc pass through R_{11} so that the output voltage may frequently rise out of control with light output loads. The 2N278 should be replaced with a "3-ampere" type with a low thermal gradient and a low I_{CO}, and R_{11} should be reduced drastically to permit operation at higher temperatures.

CIRCUIT 8-5

150-VOLT SILICON SERIES REGULATOR

Raytheon Manufacturing Company

Contributed by J. Harmer
and J. Perkins

A regulator capable of operating at temperatures as high as 100°C while delivering 100 mAdc to a 150-volt load is shown in Circuit 8-5. Operation at temperatures as high as 100°C is made feasible through the use of silicon transistors and diodes throughout.

This regulator was designed for use with choke-input-filter rectifier system which is not shown but which produces about 170 volts to the input of the regulator. Since the output is fixed at 150 volts, the emitter-follower type of series regulator can be operated from its own output, and no auxiliary power supplies are necessary.

The output voltage is sensed by the voltage divider comprised of R_5, R_6, and R_7, and is compared with the voltage reference by Q_4. The voltage

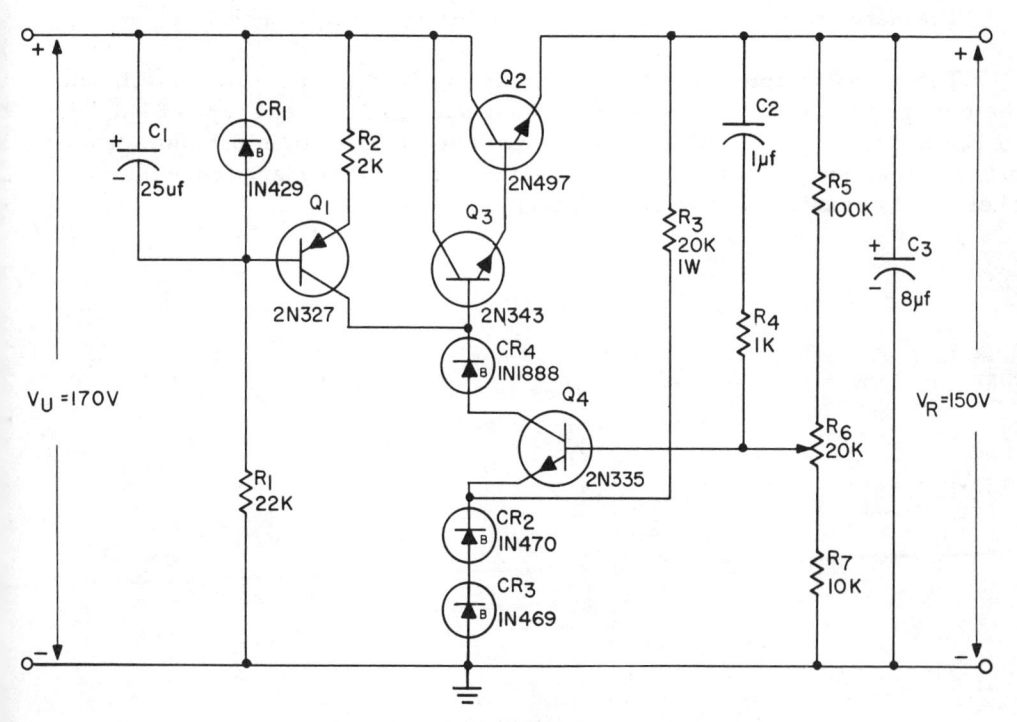

Circuit 8-5 - 150-volt silicon series regulator

Circuit 8-5, 150-Volt Silicon Series Regulator (cont'd.)

reference of about 12.9 volts is developed by two breakdown diodes in series, CR_2 and CR_3, which were chosen to produce a voltage temperature coefficient which would tend to compensate for the change in V_{BE} in Q_4 with temperature. This compensation results in a change in output voltage of less than $\pm 1\%$ over the 0°C to 100°C range.

The output load for the collector of Q_4 is about 137 volts from the emitter of Q_4 so that voltage breakdown and high dissipation could be a severe problem with Q_4. Both problems are solved by the use of CR_4, a 100-volt breakdown diode which reduces both the voltage and dissipation in Q_4 to manageable levels. This trick is only possible because the output voltage level is fixed.

Transistor Q_1 is used as a constant-current source of about 2.8 mAdc in conjunction with R_1, R_2, CR_1 and C_1. It provides a high-impedance source of current for both Q_3 and Q_4 and maximizes the gain available from Q_4. Transistor Q_3 acts as an emitter follower in transferring the regulating signal from Q_4 to the series-regulating transistor, Q_2, which also operates as an emitter follower.

The network C_2 and R_4 is used to stabilize the feedback loop.

The output ripple from this regulator is less than one millivolt, and the change in output voltage with a change in load current from 25 mAdc to 100 mAdc is 0.55%. The output impedance is less than 0.5 ohm from d.c. to 80 kc except for a peak of 1.5 ohms at the resonant frequency of the choke and capacitor in the unregulated rectifier source.

CIRCUIT 8-6

6.3-VOLT, 5-AMPERE SERIES REGULATOR

Motorola, Inc.

Suitable for supplying filament loads, this regulator features a current-limiter transient overload protector and positive current feedback to reduce output resistance.

The circuit configuration is basically an emitter-follower type of series regulator. Changes in the output voltage, V_R, are sensed by breakdown diode CR_3 which is operated by current from R_3 and drives the base of Q_5. Since the emitter of Q_5 is connected to the positive output terminal, Q_5 performs the comparison between the output voltage and the reference

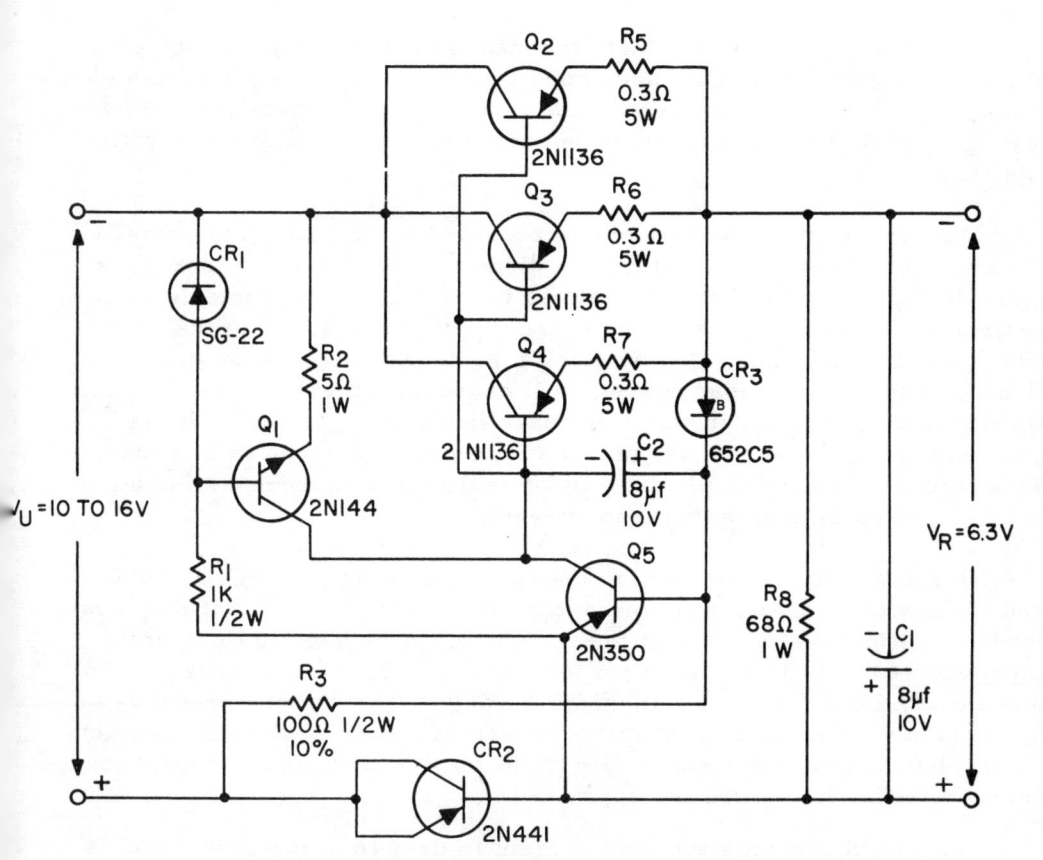

Circuit 8-6 - 5-ampere series regulator

Circuit 8-6, 6.3-Volt, 5-Ampere Series Regulator

voltage from CR_3. The temperature stability of the regulator depends largely on the fact that CR_3, which is a 6.1 volt-breakdown diode, has a positive temperature coefficient which very closely matches the negative temperature coefficient of the base-emitter voltage of Q_5. Hence, the sum of these two voltages, which is also equal to the output voltage, tends to be invariant with temperature.

Transistor Q_1, in conjunction with CR_1, R_1, and R_2, forms a constant-current regulator which provides base current for Q_2, Q_3, and Q_4 and collector current for Q_5. It is designed to provide just slightly more base current to the paralleled series transistors than they require to deliver the five-ampere output. Then, if the output is subjected to a transient overload or a short circuit, diode CR_3 will bias Q_5 off and the full current to Q_1 will pass from the series transistor bases. This will cause only slightly more than five amperes to flow, so that the circuit is protected by virtue of a limited current flow.

In normal operation, the high impedance presented by Q_1 to Q_5 also serves to maximize the gain developed by Q_5. Capacitor C_2 prevents oscillation of the feedback loop. Equal division of the regulated load current through transistors Q_2, Q_3, and Q_4 is encouraged by the 0.3-ohm emitter resistors, R_5 to R_7.

The positive current feedback is provided by CR_2, which is actually a 10-ampere transistor connected as a diode. As load current increases through CR_2, its voltage drop increases in the manner characteristic of a germanium diode. This small increase in voltage is added to the source supplying the reference, CR_3, by placing R_3 on the unregulated side of CR_2. Hence, an increase in load current will cause an increase in the current flowing through CR_3 and a small increase in its voltage. Since the reference voltage is increased, the output voltage will also tend to increase. This tendency is used to partially balance the normal decrease in output voltage caused by increasing load current.

This regulator is designed to operate in ambients as high as 60°C and transistor mounting base temperatures as high as 70°C. It displays load regulation of about ±1% at the limiting conditions of 10 to 16 volts input and -40°C to +55°C ambient temperature. Its voltage changes about ±1.3% over the same temperature range. The output impedance of the regulator is in the tens of milliohms from d.c. through audio frequencies. Since it is designed for filament loads, its high-frequency impedance characteristics have not been optimized.

Resistor R_8 permits the load current to drop to zero. If the load is always greater than 0.1 ampere, R_8 may be omitted.

CIRCUIT 8-7

WIDE-RANGE REGULATOR WITH 0.005% REGULATION

Owen Laboratories, Inc. Contributed by R. P. Owen

This is a complete a.c. to d.c. regulated power supply now in commercial production. It is a well-engineered circuit capable of much better than average performance. It is able to deliver up to 2.5 amperes at voltages from nearly zero to 40 volts with regulation against 10% variations in line voltage of 0.005% at high output voltage and 1 mVdc at low output voltage. Its regulation to load current variations between 0.25 amperes and 2.25 amperes is ±0.005% at high output voltage and ±2 mVdc at low output voltage.

No forced air cooling of the power supply is required because transistor dissipation is reduced to the point where convection cooling is adequate up to 50°C. A circuit breaker is used for protection against output overloads. Excellent stability results from the use of a

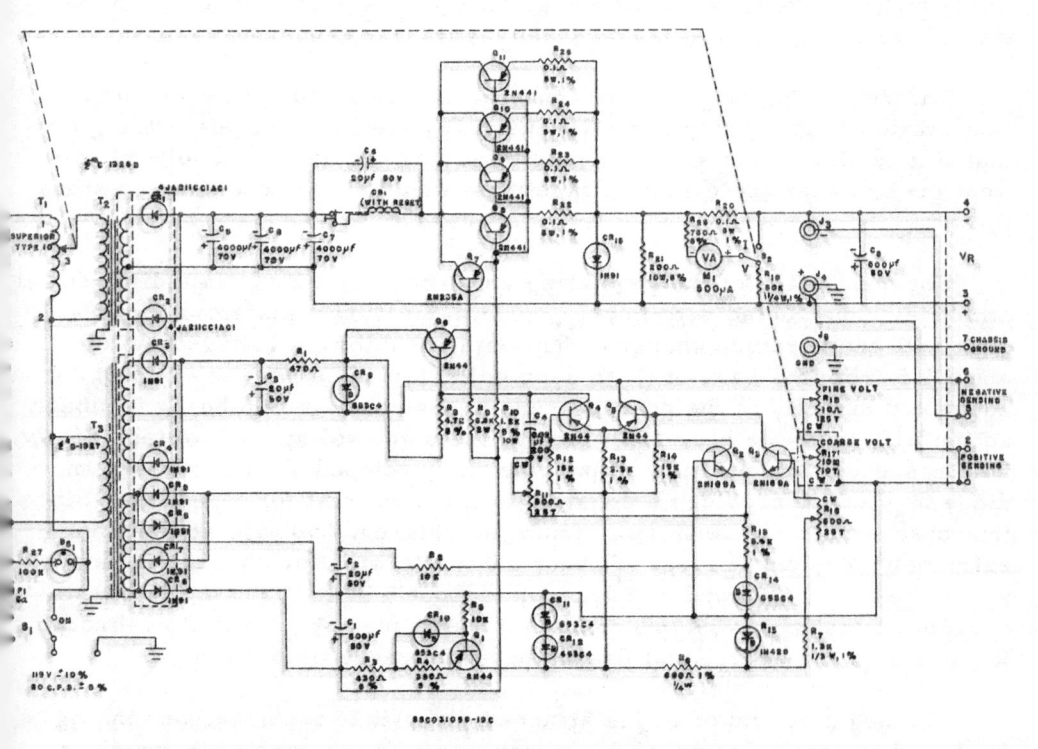

Circuit 8-7 - Wide-range regulator with 0.005% regulation

Circuit 8-7, Wide-Range Regulator with 0.005% Regulation (cont'd.)

temperature-compensated breakdown diode reference and two stages of differential amplification. Remote load sensing is possible.

The two windings of transformer T_3 operate three breakdown-diode voltage regulators attached to the positive and negative sensing leads. The d.c. voltage developed across C_1 by CR_6 and CR_8 passes through R_3 and a constant-current regulator comprised of Q_1, CR_{10}, R_4, and R_5. The constant 7 volts across CR_{10} is maintained by Q_1 across R_4 so that the current through R_4, and through Q_1 as well, is maintained relatively constant at about 21 mAdc regardless of line voltage changes.

This "constant" current then passes to CR_{11} and CR_{12} and their load which requires a constant 15.5 mAdc. The difference between 21 mAdc and 15.5 mAdc, or 5.5 mAdc, then flows through the breakdown diodes, CR_{11} and CR_{12}, causing them to develop a regulated 14 volt potential. The resulting +14 volts, measured from the positive output terminal, is used as a base and emitter return for Q_4 and Q_5 and as a source for the temperature-compensated voltage reference breakdown diode, CR_{13}, a type 1N429. The low breakdown impedance of CR_{13} in conjunction with the relatively high resistance of R_6 produces another stage of regulation so that the 6.2 volts across CR_{13} has only 200 μVac of ripple, peak to peak.

The regulating amplifier is constructed around the positive output line, while the regulating series transistors are in the negative line, so that this is an emitter-follower type of series regulator. A voltage somewhat greater than the output voltage must then appear across transistors Q_4 and Q_5.

The first stage of the regulating amplifier, Q_2 and Q_3, is a differential amplifier constructed with low-I_{co} grown-junction transistors for low drift with temperature changes. This stage exhibits a voltage gain of about 7.5 from the base of Q_3 to each collector. The base input to Q_2 is connected directly to the positive sensing line so that regulating feedback action will drive the base of Q_3 toward the same voltage. A voltage of -7v, with respect to the positive output line, is developed across breakdown diode CR_{14} which provides a constant emitter current for Q_2 and Q_3. Since this base is then 6.2 volts away from the reference voltage at CR_{13}, an extremely constant current of about 4.0 mAdc will flow through R_7 and R_{16}. The base of Q_3 will withdraw only about 0.01 to 0.015 mAdc of this current so that about 99.75% of the 4.0 mAdc from R_{16} must flow through R_{17} and R_{18}, the coarse and fine output-voltage-setting controls.

The negative end of R_{18} is attached directly to the negative sensing lead, and the positive end of R_{17} is connected to the base of Q_3 which is

Circuit 8-7, Wide-Range Regulator with 0.005% Regulation (cont'd.)

maintained by feedback at the voltage of the positive sensing lead. Hence, the voltage across R_{17} and R_{18}, which is 4.0 mAdc times the total resistance of R_{17} and R_{18}, is also the d.c. output voltage from the supply. Thus, when R_{17} is set to its maximum value of 10 kilohms, an output of 40 volts results. A calibrating resistor, R_{16}, permits the "4.0 mAdc" current to be adjusted for the tolerances in R_{17}, R_7, and CR_{13} so that precisely 40 volts is produced at the maximum setting of R_{17}.

This method of voltage control develops an output voltage which varies directly with shaft rotation when R_{17} has a linear taper. It is then possible to couple R_{17} mechanically to T_1, which is a variable autotransformer. By this means, the d.c. voltage developed across C_7, C_8, and C_9 is maintained at a level only slightly larger than the d.c. output voltage V_R which is set by R_{17}. Hence, the voltage drop across the series transistors, Q_8 to Q_{11}, is minimized at any output voltage so that the heat dissipation of these transistors is small enough to permit cooling by convection only. This technique also increases the efficiency of the supply at the lower output voltages.

The differential output from the collectors of Q_2 and Q_3 is applied to the bases of Q_4 and Q_5 which forms a second differential amplifier stage. The gain from the base of Q_4 to the base of Q_6 is about 375. The network C_4 and R_{11} stabilizes the feedback loop. An adjustment of R_{11} is provided to permit accurate stabilization. It is important that Q_4 and Q_5 have adequate collector-emitter voltage ratings since the circuit will place over 40 volts across each one.

Output is taken from the collector of Q_4 which has R_8 and the base of Q_6 as its load. Breakdown diode CR_9 is operated by one of the windings of T_3 and provides a -7 volt potential, measured from the negative sensing line, for the load resistor R_8. Transistor Q_6 operates as an emitter follower with a voltage gain of nearly one and a current gain of about 25. It, in turn, drives the base of power transistor Q_7 which operates as another emitter follower in the "Darlington connection" with power transistors Q_8 to Q_{11}. Transistor Q_7 again provides nearly unity voltage gain and a current gain of about 50. Resistor R_{10} provides about 40 to 80 mAdc of turn-off base bias current for Q_8 to Q_{11} so that Q_7 can remain in control at high temperatures.

The total voltage gain, excluding the attenuation caused by R_{17} and R_{18}, is about 2,800 or 69 db. When this attenuation is included, the closed loop gain is found to be this same 69 db at zero output voltage and about 200, or 46 db, at 40-volt output. If Q_8 through Q_{11} require a base-emitter voltage change of 200 mVdc to vary the load current between 250 mAdc

Circuit 8-7, Wide-Range Regulator with 0.005% Regulation (cont'd.)

and 2.25 amperes, then the regulating amplifier will cause this to occur with a change across the voltage sensing terminals of 200 mVdc/200 or 1.0 mVdc at a 40-volt output. This corresponds to a change of 0.0025% and an output resistance of 0.5 milliohms. At output voltages near zero, the same 200 mVdc will be required to drive the series transistors, but since the voltage gain is now 2,800, the output voltage need change by only 200 mvDc/2,800, or 72 μVdc and the output resistance will drop to only 36 microohms.

Circuit breaker CB_1, rated at 2.5 amperes, is directly in the path of the output current where it can open the power circuit when an overload occurs. By placing it in the collector of the series transistor, instead of its emitter, the impedance of the coil of CB_1 has no great effect on output resistance.

When local sensing is used, jumpers between output terminals 1 and 2, and 5 and 6 connect the sensing lines to the output lines at J_3 and J_4 which are the local output terminals. When remote sensing is used, these jumpers are removed and the sensing lines are connected directly to the load. As a result, the voltage at the load is regulated and the resistance of the connecting cables is effectively reduced by a factor between 200 and 2,800, depending on the output voltage used.

Careful shielding of the power transformer and its rectifiers isolate the power supply from the power line so that it can be "floated" across a strain gauge bridge and introduce only 2×10^{-7} volts of noise per ohm of load resistance to ground. Its normal noise and ripple is less than 1 mVdc peak to peak. Its output voltage is stable to within ±0.02% or 2 mVdc, whichever is greater, when operated for eight hours at constant line voltage and temperature. Within the range of 0°C to 50°C, it displays a temperature coefficient less than 0.02%/°C. Recovery from a load transient is complete, to within regulation limits, in less than 100 μs.

CIRCUIT 8-8

CHOPPER-STABILIZED STRAIN-GAUGE SUPPLY

Neff Instrument Corporation

This power supply, Neff model 2-300, is intended to be used as an airborne power source for strain gauges. A high degree of voltage stability is required in this application and it is provided in this unit by a regulating amplifier which uses an electromechanical chopper modulator as its input stage. It is capable of providing up to 1.0 ampere at a 5, 10, or 15 volts output which is constant to within ±0.1% despite a ±10 volt change in the 115 volt, 400 cps line, or a change from no-load to full-load output current, or a temperature change from -55°C to +85°C. Its drift is about ±0.05% in any 24-hour period in a constant environment.

The output voltage is selected by a screwdriver-set switch, S_2, which alters both the reference voltage and the unregulated d.c. voltage provided by transformer T_1 and rectifiers CR_9 and CR_{10} which operate into L_1 as a full-wave, choke-input-filter circuit. Rectifiers CR_5 and CR_6 also operate from the same winding and provide a d.c. output across C_1 which supplies breakdown diodes CR_1 through CR_4.

Diodes CR_1 and CR_2 in series act as a stage of regulation for the voltage reference, CR_3, with a temperature coefficient of 20 parts per million per degree C. The breakdown diode voltage reference, CR_3, has a

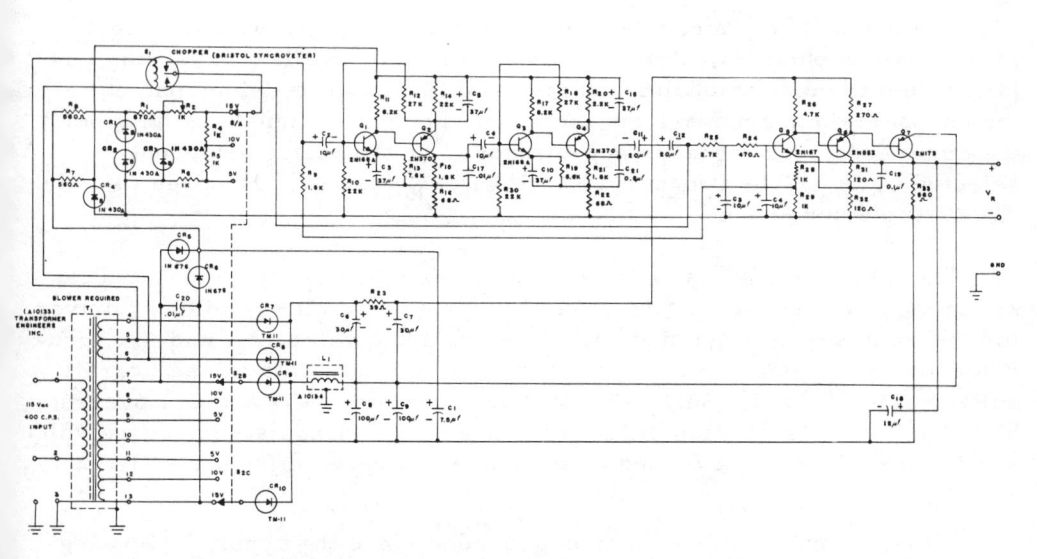

Circuit 8-8 - Chopper-stabilized strain-gauge supply

Circuit 8-8, Chopper-Stabilized Strain-Gauge Supply (cont'd.)

coefficient of only 10 ppm/°C, and establishes a stable voltage across the voltage divider comprised of R_2, R_4, R_5, and R_6 whose output is selected by the voltage selector switch, S_2. Resistor R_2 is a multi-turn potentiometer which sets the voltage at the junction of R_2 and R_4 to precisely 7.5 volts. Breakdown diode CR_4 provides a regulated positive voltage which is used to supply the collectors of the a.c. amplifier.

The voltage divider R_{28} and R_{29} delivers one-half of the supply output voltage through R_9 to the chopper modulator. Here it is connected to the selected output voltage from the voltage reference at a frequency of 400 cps. Any deviation between these two voltages will result in a voltage fluctuation at the base of Q_1 which has an amplitude proportional to the deviation and a phase relative to the 400 cps line which corresponds to the polarity of the deviation. This process then converts a very small d.c. error voltage into a 400 c.p.s. a.c. signal which can be amplified easily and without drift in an a.c. amplifier.

The a.c. amplification is accomplished by transistors Q_1, Q_2, Q_3, and Q_4 arranged in two direct-coupled pairs which are practically identical. Each of the pairs incorporates local feedback provided by R_{16} and R_{22} which stabilizes the gain of each pair at about 27.5. The emitter capacitors bypass the 400 c.p.s. signals on the emitters and permit maximum 400 c.p.s. gain from each transistor. Each pair is d.c. stabilized by resistors R_{10} and R_{12} in the first pair and resistors R_{30} and R_{18} in the second. Capacitors C_{17} and C_{21} prevent high-frequency oscillation within each pair.

The output of the a.c. amplifier is coupled through two opposing electrolytic capacitors, which are so arranged to handle a reversing polarity, into the d.c. amplifier and the synchronous-rectifying contact on the chopper which recovers the amplified d.c. error signal. At this point, the amplified error signal is superimposed on the voltage reference selected by S_{2_A}. The two-stage RC filter, R_{25}, and C_{14}, removes the 400 c.p.s. ripple component.

The d.c. amplifier is comprised of transistors Q_5, Q_6, and Q_7. Transistor Q_5 provides the voltage gain while Q_6 is an emitter-follower driver for the main series regulator, Q_7. The voltage divider, R_{31} and R_{32}, provides a feedback signal to the emitter of Q_5 which maintains the overall voltage gain of the d.c. amplifier at 2.0. Capacitors C_{18} and C_{19} prevent high-frequency oscillation of this amplifier. An unregulated positive voltage for the collector of Q_5 and emitter of Q_6 is derived from rectifiers CR_7 and CR_8.

Either output terminal may be grounded since the supply is floating with respect to its chassis ground. The output contains less than 1 millivo

Circuit 8-8, Chopper-Stabilized Strain-Gauge Supply

r.m.s. ripple and noise, and the output impedance is less than 10 milliohms at d.c. and less than 50 milliohms at any frequency up to 10 kc. It is ruggedly constructed for use up to 70,000 feet. A vibration of one inch peak to peak displacement from 0 to 20 c.p.s., or 20 g's from 20 to 2000 c.p.s. will not cause damage or operational error. It can also operate under 20 g acceleration in any axis and will not be damaged by a 50-g shock. It may operate in humidity as high as 100%.

CIRCUIT 8-9

SMALL LAB SUPPLY WITH CURRENT LIMITING

Hewlett-Packard Company   P. N. Sherrill

    Circuit 8-9 is a compact, hand-sized power supply capable of as much as 150 mAdc output at voltages between zero and 30 volts. It features a novel adjustable current limiter as protection, not only for the power supply but also for a semi-conductor load. The current limiter can be set to limits of 25, 50, 100 and 225 mAdc. The load regulation is less than 0.3% or 30 mVdc, whichever is greater at any output. The output ripple and noise is less than 150 μVac r.m.s.

    Winding 1-2-3 of transformer T_1 supplies power which is rectified by CR_1 and CR_2 and applied to the series power transistor, Q_1, for regulation. Winding 4-5-6 of T_1 and rectifiers CR_3 and CR_4 develop about 20 volts for the operation of the regulating amplifier, comprised of Q_2, Q_3, and Q_4.

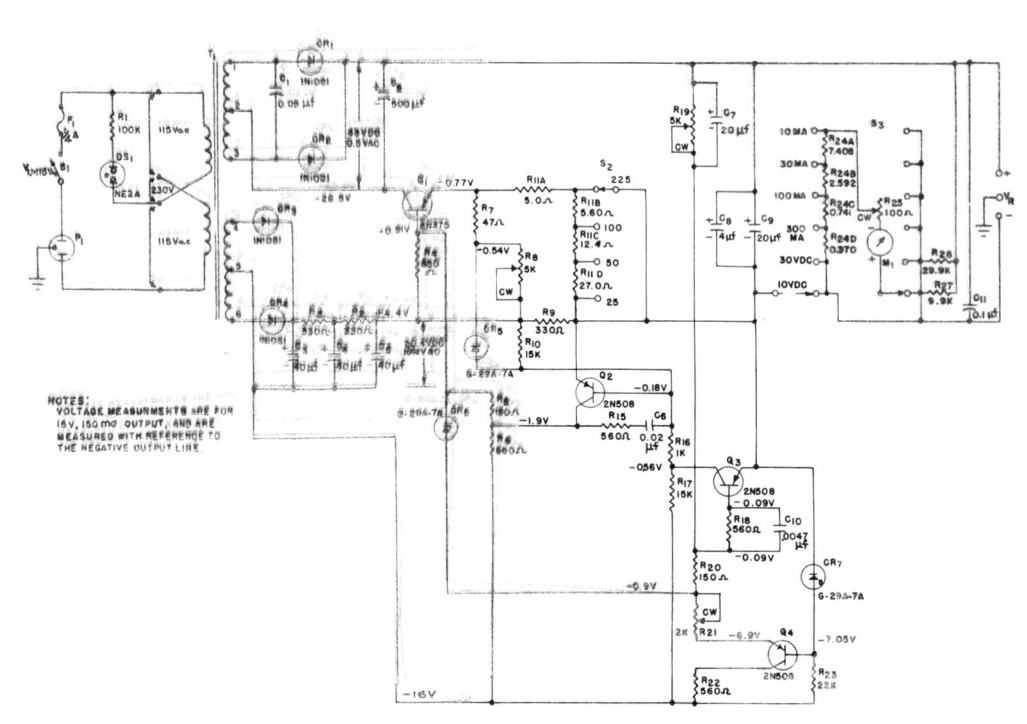

Circuit 8-9 - Small lab supply with current limiting

Circuit 8-9, Small Lab Supply with Current Limiting (cont'd.)

The voltage reference breakdown diode, CR_7, operates at about -7.05 volts with current from R_{23}. It is able to operate at a low current, since emitter follower Q_4 reduces the 6 mAdc reference load from R_{21} to a load of about 60 μAdc at CR_7. Hence, CR_7 is able to operate at about 400 μAdc which permits R_{23} to be quite large and provide considerable regulation in conjunction with the low impedance of CR_7.

This circuit is a good example of a common-emitter type of series regulator. Transistor Q_3 is the first stage of the regulating amplifier. Resistors R_{19}, R_{20}, and R_{21} comprise a voltage divider running between the positive output line and the reference output from the emitter of Q_4. Since CR_6 is normally non-conducting and the base current from Q_3 is very small (less than 10 μAdc), essentially the same current flows through all three resistors. Regulating feedback action will cause the positive output to assume a value which permits the base of Q_3 to be about 90 mVdc negative with respect to its emitter which is tied to the negative output line.

A constant voltage of about 6.8 volts then appears across R_{20} and R_{21} which causes a constant current to flow through R_{19}. The voltage across R_{19}, which is only 90 mVdc larger than the supply output voltage, is then this constant current times the resistance of R_{19}. By varying R_{19} then, control of the output voltage is obtained. Adjustment of R_{21} is made so that the constant current has a value of about 6.2 mAdc which will cause the output to be 31 volts when R_{19} has its maximum value.

Resistor R_{17} is the collector load for Q_3. The collector signal is raised slightly in d.c. level and is transferred to the base of Q_2, the second voltage amplifier, by means of the voltage divider R_{16} and R_{10} which operates against a positive 4.4 volts created by current from CR_3 and CR_4 flowing through R_9. The output from Q_2 taken across its collector load, R_6, is again raised in d.c. level and applied to the base of the series power transistor, Q_1, by the voltage divider R_5 and R_4. Since Q_1 is in series with the negative lead of the main power rectifiers, CR_1 and CR_2, it is able to regulate the power supply output.

All of the output current reaches the emitter of Q_1 by flowing through resistor R_{11_A} and any of the resistors R_{11_B}, R_{11_C}, and R_{11_D} which are not shorted out by "Short Circuit Current" switch S_2. The voltage built up across these resistors produces a negative voltage at the emitter of Q_1 which will tend to drive CR_5 into forward conduction. At a particular value of output current, the diode CR_5 will be brought into conduction so that the base of Q_2 will be driven negatively with any further increase in output current.

Circuit 8-9, Small Lab Supply with Current Limiting (cont'd.)

This action tends to override the signal from Q_3 which attempts to hold the output voltage constant since a negative base drive will raise the collector of Q_2 positively and tend to cut off series transistor Q_1. Hence, while CR_5 conducts the circuit becomes a current regulator which limits the overload current which the supply can produce. Adjustment of R_8 will cause the limiting action to occur at the output current levels of 25, 50, 100, and 225 mAdc as selected by switch S_2.

The feedback loop is stabilized by components R_{18}, C_{10}, R_{15}, and C_6. Capacitor C_7 improves the loop gain at higher frequencies by reducing the attenuation of the divider R_{19}, R_{20}, and R_{21}.

A fine example of the sort of hazards experienced in designing a complete regulated power supply is provided by CR_6. This diode has no function during normal operation of the supply. However, when the supply is set to a low output voltage and is lightly loaded, turning the power switch off will cause the internal power supply voltage across C_5 to collapse much faster than that across C_2. The result is that, without protection, the regulator would become inoperative long before the energy of C_2 had been dissipated and a dangerous surge of voltage would occur at the output.

Protection from this voltage surge is provided by CR_6. Normally there is no forward voltage across CR_6, but when the -16 volt line collapses at turn-off, the anode of CR_6 is pulled positively by current which continues to flow through R_{19}. This same current is conducted by CR_6 to the base of Q_1 where it is effective in biasing Q_1 off and preventing an output surge.

CIRCUIT 8-10

CONVECTION-COOLED, WIDE-RANGE POWER SUPPLY

Lambda Electronics Corp.

This commercial power supply features better than 0.15% or 20 mVdc line and load regulation over the wide output voltage range of zero to 32 volts at currents as high as 2 amperes. By minimizing their power loss, it is possible to dissipate the heat of the series transistors with an efficient convection-cooled radiator at ambient temperatures as high as 50°C. Since no blower or fan is required, the power supply may be operated at any power line frequency between 50 and 400 c.p.s. without alteration. A magnetic circuit breaker is employed for overload protection.

The upper two tapped windings on transformer T_1 provide the power which is regulated and delivered to the output, while the lower winding provides a positive voltage from CR_5 for use as a base-bias source for the series regulators and a negative voltage from CR_6 which operates the voltage reference and regulating voltage amplifier. Since the emitters of the series transistors Q_4, Q_5, and Q_6 are referred to the same line as the regulating amplifier, the series transistors are operated in the common-emitter mode.

Terminal B_1 on the transformer is the common lead between rectifiers CR_5 and CR_6. Resistors R_{40} and R_{41} are the turn-on surge-current limiters

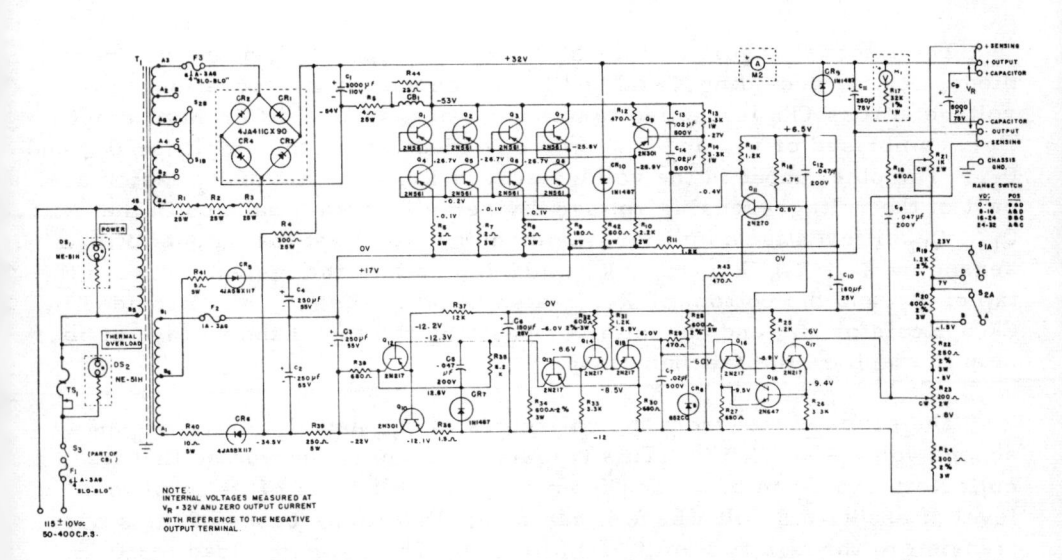

Circuit 8-10 - Convection-cooled wide range power supply

Circuit 8-10, Convection-Cooled, Wide-Range Power Supply (cont'd.)

for these rectifiers. The half-wave rectified voltage placed on C_2 by CR_6 is filtered by R_{39} and C_3 and applied to an internal voltage regulator, comprised of Q_{10}, Q_{12}, Q_{13}, Q_{14}, Q_{15} and reference CR_8, whose function it is to develop a stable 12 volts for operation of the reference CR_8 and the regulating voltage amplifier composed of Q_{16}, Q_{17}, and Q_{18}.

The 6-volt breakdown-diode reference, CR_8, is operated by current from R_{28} and sets the voltage on the bases of Q_{15} and Q_{16} which are the input transistors for the internal and main regulating amplifiers, respectively. Transistors Q_{14} and Q_{15} are connected as a balanced differential amplifier which compares the potential from CR_8 with the potential from the voltage divider R_{32} and R_{34}. These transistors strive for the balance which occurs when the anode of CR_8 and the negative end of R_{34} reach -12 volts and both bases rest at -6.0 volts.

The amplified error signal in the collectors of Q_{14} and Q_{15} is applied to the base and emitter of the second voltage amplifier, Q_{13}. The output from Q_{13}, taken across its collector load resistor, R_{35}, is passed by the emitter follower, Q_{12}, to the base of the series regulating power transistor, Q_{10}, which regulates only the internal power flowing into the -12 volt line. Dissipation is minimized in Q_{12} by operating it within the 0.5 volt forward drop of silicon diode CR_7 which carries current from R_{38}. The cathode of CR_7 is also the most negative stable voltage in the entire circuit and therefore makes a desirable return point for R_{35}, the collector load for Q_{13}. Capacitors C_5 and C_6 stabilize the feedback loop of this internal regulator.

The internal regulator then stabilizes the operation of the reference diode, CR_8, by providing a stable 12 volt source for its operation. The voltage across CR_8 is applied to one input of a second differential amplifier, comprised of Q_{16} and Q_{17}, which is identical to the amplifier Q_{15} and Q_{14}. The other input is the voltage across R_{24} and part of R_{23} which are part of the voltage-sensing voltage divider. To obtain balance in the Q_{16} - Q_{17} differential amplifier, the circuit will adjust the output voltage sensed by R_{21}, R_{19}, R_{20}, R_{22}, R_{23}, and R_{24} so that the voltage between the tap of R_{23} and the bottom of R_{24} is also equal to the 6.0 volts provided by CR_8. Resistor R_{18} and capacitor C_8 shunting the top of the voltage divider help to stabilize the feedback loop.

The error signal from the collectors of Q_{16} and Q_{17} drive Q_{18} as a second voltage amplifier. This transistor is an n-p-n type so that its collector output can be raised from the -9.5 volt level of its input to a level of about -0.5 volt where it can be applied to the bases of the series transistors through two emitter followers. The collector load for Q_{18},

Circuit 8-10, Convection-Cooled, Wide-Range Power Supply

R_{16}, receives a positive voltage from C_4 with additional filtering by R_{11} and C_{10}. Capacitor C_{12} also helps to stabilize the feed-back loop.

The error signal proceeds through the first emitter follower, Q_{11}. The collector of Q_{11} must be supplied from a negative source, and, since the -12 volt line is the only stable one, it is used with resistor R_{43} to minimize the collector dissipation. The signal then passes to a power-transistor emitter follower, Q_8, which drives the paralleled bases of the series power transistors, Q_4, Q_5, and Q_6. Resistor R_9 provides about 95 mAdc of base turn-off bias to these three transistors from the bias voltage across C_4. Resistors R_6, R_7, and R_8 help to divide the output current equally through the three series transistors.

Since the voltage and power that must be handled by the series regulating element are greater than can be tolerated by the three power transistors Q_4, Q_5, and Q_6, a two-level high voltage stack is constructed with Q_1, Q_2, and Q_3 respectively in series with Q_4, Q_5, and Q_6. The voltage across the entire stack is measured by the divider R_{13} and R_{14}, and precisely half the entire voltage is placed on the base of emitter follower Q_9. Transistor Q_9 then drives, and resistor R_{10} provides the turn-off bias for a second emitter follower, Q_7, which not only drives the paralleled bases of Q_1, Q_2, and Q_3, but also establishes the collector voltage for Q_8. By this means, the collector-to-emitter voltages of all six series power transistors and the currents flowing through each are kept closely equal.

These power transistors then function as the series regulating element in the negative line from the bridge rectifier comprised of CR_1, CR_2, CR_3, and CR_4. The voltage drop across, and the heat dissipation in, these power transistors is kept low by switching the a.c. output from the top two windings of transformer T_1 with switches S_{2_B} and S_{1_B} which are ganged with switches S_{2_A} and S_{1_A} located in the voltage-sensing divider. In each of the four combinations of positions these switches may have, the voltage-sensing divider is adjusted to perform over one of four 8-volt bands in the range of zero to 32 volts, and the transformer a.c. output is adjusted to be the minimum necessary to permit the regulator to function at full load current and low line voltage.

Resistor R_4 draws a bleeder current of about 110 mAdc to 220 mAdc from the output to help the series transistors retain control at high temperatures. Diode CR_9 prevents a reversal of output voltage greater than about -1 volt under the action of the bleeder current or an external inductive load. Diode CR_{10} and resistor R_{42} help keep the series transistors biased off during turn-off of the supply, when the normal turn-off bias from C_4 has collapsed, and thereby prevent output voltage surges during turn-off.

Circuit 8-10, Convection-Cooled, Wide-Range Power Supply (cont'd.)

    This power supply is designed for operation with local or remote sensing. When used with local sensing, all three positive output terminals are connected together, as are all three negative terminals. There are several methods of obtaining remote sensing, but the best method is to carry separate wires from each of the six terminals to the load where all positive and all negative leads are again connected together. It may be desirable in the case of long leads to provide an external capacitor directly at the load in place of C_9.

    The specifications for this supply state that the line regulation, for input variations from 105 to 125 volts, and load regulation, for load variations between zero and 2 amperes, will be better than 0.15% or 20 mVdc, whichever is greater. Ten-volt line transients will not cause the output voltage to depart from the regulation specification. The output voltage is also within regulation specifications within 50 μs after a sudden application or release of a 2 ampere load. Its internal d.c. resistance is less than 0.025 ohms and its ripple and noise is less than 1 mVdc r.m.s. Either output terminal may be connected to the chassis ground. Protection from overload or short circuit is provided by a magnetic circuit breaker, CB_1, which opens the line switch, S_3. A thermostat is also provided which will turn off the supply if the ambient temperature exceeds safe levels.

CIRCUIT 8-11

HIGH-TEMPERATURE, 120-VOLT SILICON REGULATOR

Texas Instruments, Inc.

This regulator, constructed entirely from n-p-n silicon transistors, can operate in ambient temperatures between -55°C and +85°C while delivering up to 600 mAdc at a fixed output voltage between 110 and 130 volts. Its load regulation is about 0.008% and its input regulation is about 0.016%. A d.c. output resistance of 0.02 ohms is attained, and the output ripple is less than 5 mVdc, peak to peak.

This is an example of a common-emitter type of series regulator since the regulating voltage amplifier is referred to the emitter of the series transistor, Q_1. The first stage of the voltage amplifier is the differential amplifier comprised of Q_5 and Q_6. The voltage reference, formed by three 5-volt breakdown diodes in series, is applied to the base of Q_6. This -15 volt input is matched by a similar voltage applied to the base of Q_5 by the voltage divider R_4, R_5, and R_6. These resistances are so proportioned that the regulator must deliver a 120-volt output in order to attain

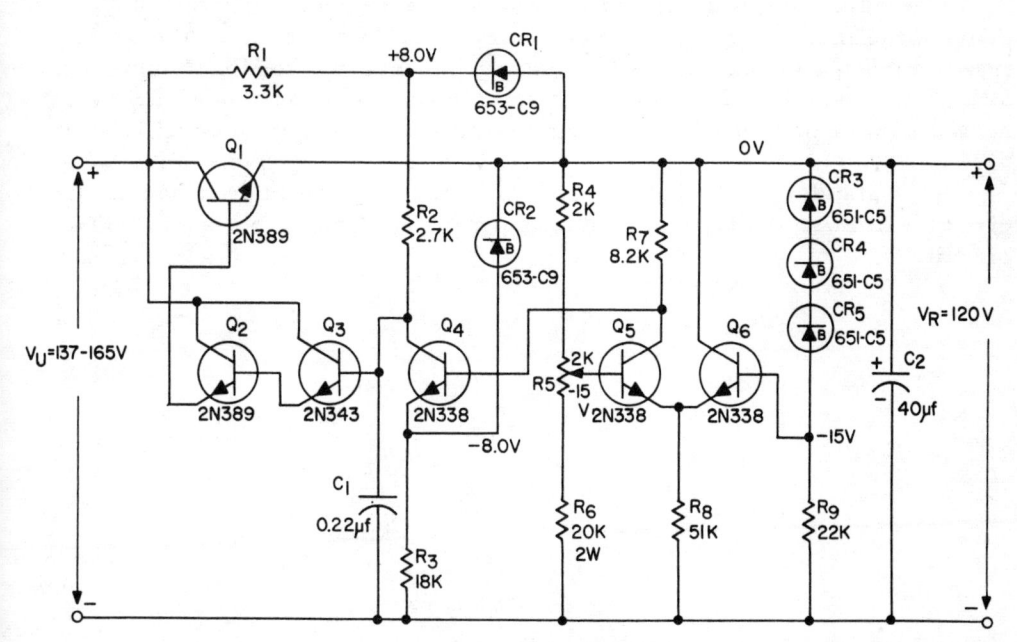

Circuit 8-11 - High-temperature, 120-volt silicon regulator

Circuit 8-11, High-Temperature, 120-Volt Silicon Regulator (cont'd.)

a -15-volt input to Q_5. Adjustment of R_5 permits variation of the output voltage V_R over the range of 110 to 130 volts.

Because the output voltage range is so restricted, it is possible to use it to supply power for the voltage amplifier so that no auxiliary or internal power supply is necessary.

The error signal output from Q_5 is applied to a second voltage amplifier, Q_4. The emitter of Q_4 is held at a reasonably constant -8.0-volts by breakdown diode CR_2 which is operated by current from R_3. The small impedance of CR_2 causes some emitter degeneration and loss of gain in Q_4, but this effect is not serious.

The output signal from Q_4 must be able to go positive with respect to the emitter of Q_1 in order to provide a forward bias to the base of Q_1. A positive voltage return for the collector load, R_2, is created by breakdown diode CR_1 which operates at about 8.0 volts with current from R_1. The filtering action of R_1 and CR_1 reduces the amount of ripple and unregulated voltage variations which are introduced into the regulating amplifier through R_2.

The output of Q_4 is applied to two emitter followers, Q_3 and Q_2, which provide the increased current necessary to drive the base of the series transistor, Q_1. Capacitor C_1 is necessary to prevent feedback instability at high frequencies. It is said that any amount of capacitance may be connected across the output of the regulator without danger of oscillation.

As it stands, this regulator is not protected against overload or short circuit. Almost any type of protection may, and should be, added to this circuit.

CIRCUIT 8-12

SILICON CONTROLLED RECTIFIER REGULATOR

General Electric Company*

This unusual circuit, which combines the functions of rectification and voltage regulation, is able to deliver up to 15 amperes at rectifier stud temperatures as high as 100°C and regulated output voltages between about 44 and 85 volts. The output voltage will vary less than a volt with changes

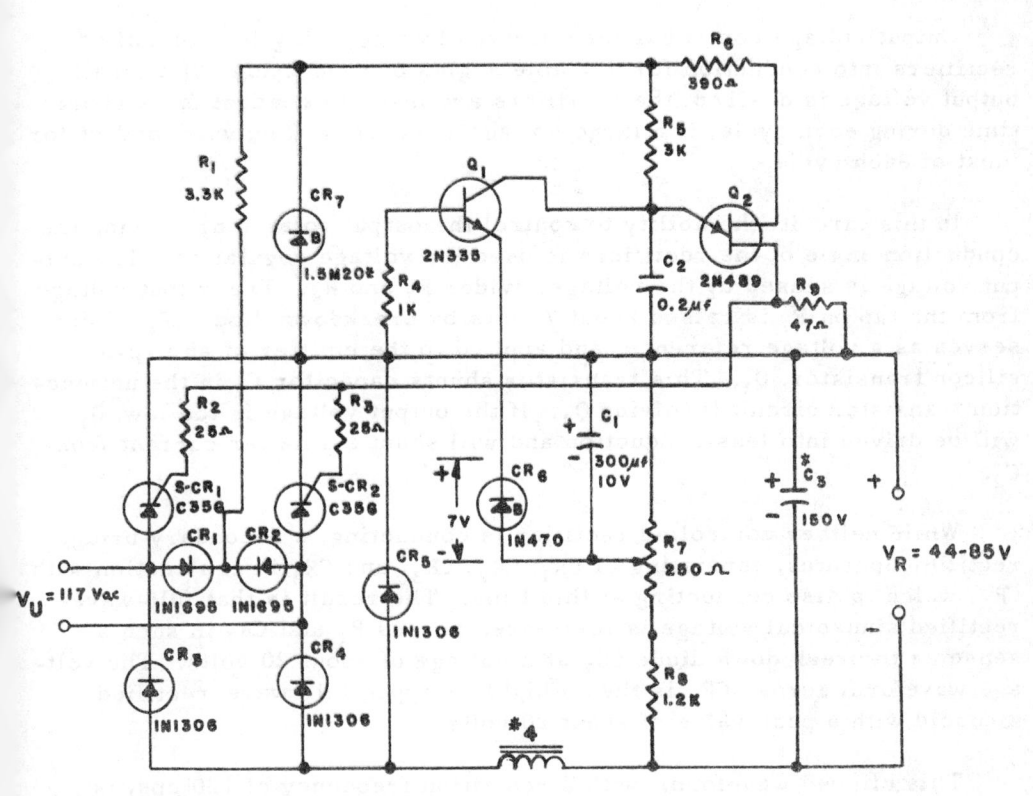

*L_1 & C_3 MAY BE ANY VALUES DETERMINED BY OUTPUT RIPPLE REQUIREMENTS.

Circuit 8-12 - Silicon controlled rectifier regulator

---

*Notes on the Application of the Silicon Controlled Rectifier, Report ECG-371-1, December 1958.

Circuit 8-12, Silicon Controlled Rectifier Regulator (cont'd.)

in the line voltage from 94 volts to 140 volts or with changes in load current from zero to 15 amperes.

The circuit is similar to that described in Section 1.B.4 of this chapter and illustrated in Figure 8-0.14(A). As was explained there, silicon-controlled rectifiers SCR_1 and SCR_2 are two arms of a bridge rectifier with CR_3 and CR_4. If SCR_1 or SCR_2 fail to conduct, it will cause the bridge to be non-conducting. While this state persists, any output current will flow through CR_5 and discharge the stored energy in L_1. When either of the controlled rectifiers is triggered into conduction, power will again flow from the bridge to the load and L_1 will be recharged.

Output voltage control is then derived by triggering the controlled rectifiers into conduction for variable angles of conduction. If a small output voltage is desired, the rectifiers are made to conduct for a short time during each cycle; if a large voltage is desired, they will conduct for most of each cycle.

In this circuit, the ability to control the output voltage by varying the conduction angle of the rectifiers is used for voltage regulation. The output voltage is sensed by the voltage divider R_7 and R_8. The output voltage from the tap on R_7 is raised about 7 volts by breakdown diode CR_6, which serves as a voltage reference, and applied to the emitter of an n-p-n silicon transistor, Q_1. This transistor shunts capacitor C_2 in the unijunction transistor circuit involving Q_2. If the output voltage is too low, Q_1 will be driven into less conduction and will shunt a smaller current from C_2.

While neither controlled rectifier is conducting, a secondary bridge rectifier operates, consisting of CR_1, CR_2, CR_3, and CR_4 in conjunction with CR_5, which is also conducting at this time. The result is that full-wave-rectified sinusoidal voltage is impressed across R_1 and CR_7 in such a sense as to break down diode CR_7 at a voltage of about 20 volts. The voltage waveform across CR_7 is then a highly-clipped full-wave-rectified sinusoid with a peak value of about 20 volts.

This clipped waveform, with a repetition frequency of 120 cps, is applied through R_5 to capacitor C_2. If the shunting effect of Q_1 is slight, C_2 will quickly charge to the voltage which fires the emitter of Q_2, causing the unijunction transistor to break down and pass a current from R_1 through R_6 and into R_9. The resulting positive pulse on R_9 is applied to both controlled rectifiers and will cause the one which is forward-biased to conduct, in this case rather early in the cycle so that a large output is developed.

Circuit 8-12, Silicon Controlled Rectifier Regulator (cont'd.)

If, on the other hand, the shunting effect of Q_1 is heavy because the output voltage is too large, C_2 will charge more slowly and cause firing of Q_2 at a later instant in the cycle. The controlled rectifiers will then be pulsed on later in each cycle, and since they automatically turn off like thyratrons when their voltage reverses, they will conduct for a shorter interval in each cycle and produce less output. In this way, the regulating feedback loop is constructed.

The values of L_1 and C_3 are not important, since they have no great effect on the operation of the controlled rectifiers. They may be chosen from consideration of the output ripple requirements.

CIRCUIT 8-13

DUAL-REGULATOR, WIDE-RANGE POWER SUPPLY

Electronic Research Associates, Inc.

This unusual power supply features two separate regulators in series, the first being an efficient, but slow, magnetic regulator which crudely regulates its output voltage to be only slightly larger than the output from the second regulator, which is a conventional inefficient, but fast, transistor regulator. The net result is an efficient high-speed regulator which combines the advantages of both magnetic and transistor types. The magnetic regulator also provides current-limited overload and short-circuit protection by permitting its output voltage to collapse during overloads.

The circuit presented here is the ERA Model TR36-8M which is capable of supplying up to 8 amperes at voltages between 0.75 and 36 volts. A line voltage change from 105 to 125 volts causes an output voltage change of ±0.05% or 30 mVdc, whichever is greater. Its load regulation amounts to a change of no more than 35 mVdc from zero to 8 amperes load. The output is stable within 50 mVdc over an 8-hour interval after an initial stabilization period. The use of a magnetic regulator restricts the permissible line frequency to either 60 or 400 c.p.s.

In the schematic diagram of Circuit 8-13, the magnetic regulator is represented by transformers T_1 and T_2 in conjunction with capacitor C_1.

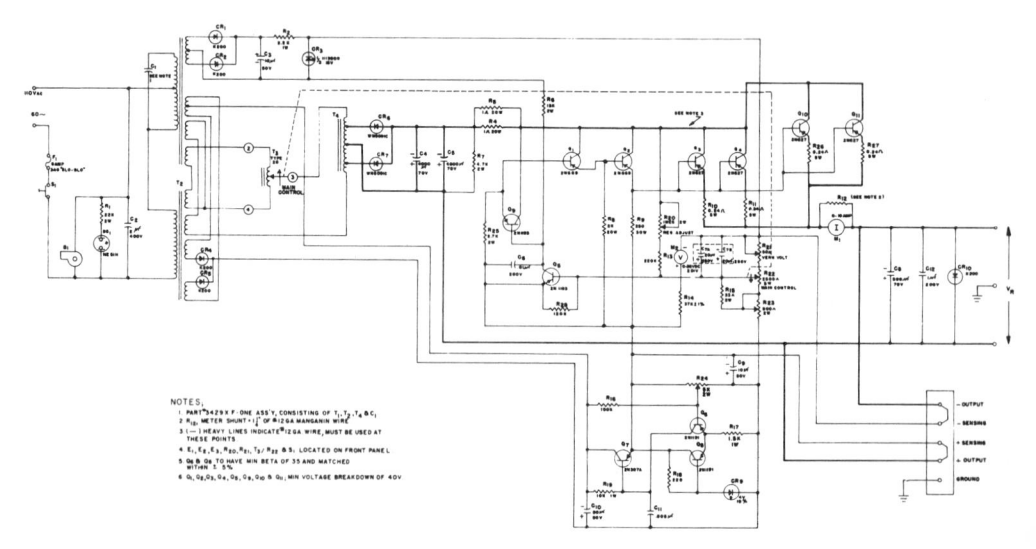

Circuit 8-13 - Dual-regulator, wide-range power supply

Circuit 8-13, Dual-Regulator, Wide-Range Power Supply (cont'd.)

A portion of the stabilized output from T_1 and T_2 is applied to the variable transformer, T_3, which is ganged with the voltage control potentiometer, R_{22}. In this way, the autotransformer T_4 receives a stabilized input voltage from T_1, T_2, and T_3 which develops a full-load d.c. voltage across C_4 and C_5 which is only slightly larger than the required d.c. output voltage from the power supply. Since the series-regulating power transistors can operate with collector-emitter voltages as low as one or two volts, the difference between the voltages on C_4 and at the output can be made quite small without impairing performance but with the important advantage that heat dissipation in the transistor regulators is greatly reduced.

Two additional windings on transformers T_1 and T_2 supply internal power for operating the regulating amplifier. Rectifiers CR_1 and CR_2 develop 15 volts across CR_3, whose positive terminal is connected to the negative output line and whose negative terminal supplies the collector load for Q_5. Rectifiers Q_4 and Q_5 develop a voltage across C_{10} which is regulated by Q_6, Q_7, and Q_8 with voltage reference CR_9 and becomes a stable reference voltage across C_9.

The output across C_9 operates the 4 volt, ±10% voltage reference breakdown diode, CR_9, through resistor R_{18}. The 4 volt reference is then one input to the differential amplifier comprised of Q_6 and Q_8, while the other input is provided by R_{24} which senses the output voltage. The collector output from Q_6 is applied to the base of power transistor Q_7 which then series-regulates power into the internal load across C_9. The voltage across C_9 is determined by the setting of R_{24}, since regulating action will tend to maintain the voltage between the wiper and positive end of R_{24} at the same voltage as developed by CR_9. Resistor R_{16} will cause the regulated voltage across C_9 to decrease slightly with an increase in line voltage as a means of improving the line regulation of the power supply by compensation.

The regulating amplifier, Q_5, is constructed on the positive sensing line, while the regulating power transistors are placed in the negative output line. The configuration then operates its series transistors as emitter-followers. The voltage sensing system and regulating amplifier actually operate independently of the main power circuits and are connected to them only by the sensing connections which may be made locally at the supply output terminals or remotely at the load itself.

The emitter of Q_5 is tied directly to the positive sensing line so that its base will operate a few tenths of a volt below this line. A current will flow from the voltage across C_9 through the network R_{23}, R_{22}, and R_{15} and into the base circuit of Q_5. This current would easily bias Q_5 off if regulating action did not cause an output to develop which draws precisely the

Circuit 8-13, Dual-Regulator, Wide-Range Power Supply (cont'd.)

same current, less the necessary base current in Q_5, through resistor R_{21} and out of the base circuit of Q_5.

Since the base voltage on Q_5 and the voltage across C_9 are both constant, the current through R_{23} is also constant. The output voltage will then be approximately equal to this current times the resistance of R_{21} plus that between the wiper and negative end of R_{22}. Hence, output voltage varies linearly with the resistance of R_{22} and it is feasible to gang this control with a linear-taper variable transformer, T_3. Capacitors C_{7_A} and C_{7_B} bypass R_{21} and R_{22} and maximize the a.c. regulating gain.

The collector output from Q_5 passes to emitter follower Q_9, which is also a small transistor. Transistor Q_9 then drives emitter follower Q_1 which in turn drives emitter follower Q_2, both of which are high-gain, low-current power transistors. Finally, Q_2 drives the paralleled bases of Q_3, Q_4, Q_{10}, and Q_{11} which are the power regulating transistors. Resistances R_{10}, R_{11}, R_{26}, and R_{27} tend to equalize the current division through the four transistors so that each will carry a maximum of two amperes.

No short circuit protection is provided in the transistor regulator aside from resistances R_4 and R_5. These will limit the short circuit current to about 80 amperes when the series transistors are saturated. As capacitors C_4 and C_5 quickly discharge, the load on the magnetic regulator increases past its normal load limit. The output voltage from the magnetic regulator then collapses and reduces the current which will flow into the output. When the short circuit, or overload, is removed, the magnetic regulator will resume its normal operation and will cause the supply output voltage to rise to its normal level.

When a load current is applied, the collectors of the series regulating transistors will move positively because of the source impedance of R_4, R_5 and the rectifier circuit of CR_6 and CR_7. An adjustable amount of this positive change is fed into the base of Q_5 where it tends to bias Q_5 to a smaller collector current, which in turn allows the collector to become more negative and bias the power transistors to a lower voltage drop. The result is a controllable decrease in output resistance by a compensation method. Resistance R_{20} is given such a range that the output regulation may be adjusted to be positive, zero, or even negative.

CIRCUIT 8-14

TRANSISTOR CONSTANT-CURRENT REGULATOR

Minneapolis-Honeywell Regulator Company*

The current regulator shown here is capable of controlling currents from less than 50 mAdc to greater than 6 amperes, depending on the

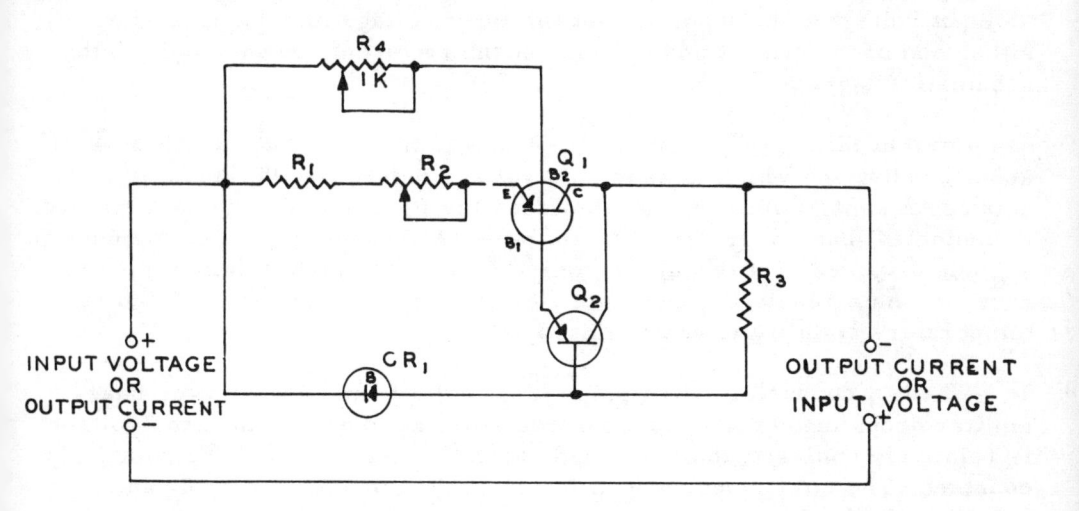

REGULATED CURRENT	RESISTANCE VALUES			DIODE	USE TRANSISTOR TYPE	
	R_1	R_2	R_3	CR_1	Q_1	Q_2
50-100 ma	50Ω 1W	150Ω 2W	4700Ω ½W	SV7	2N540, H45, H200E	2N43, H45, 2N540
100-200 ma	25Ω 2W	75Ω 4W	4700Ω ½W	SV7	2N540, H45, H200E	2N43, H45, 2N540
200-500 ma	10Ω 4W	30Ω 4W	4700Ω ½W	SV7	2N540, H45, H200E 2N574, 2N575	H45, 2N540
500-1000 ma	5Ω 10W	15Ω 10W	4700Ω ½W	SV7	2N540, H45, H200E 2N574, 2N575	H45, 2N540
1A-3A	2Ω 25W	7Ω 25W	3300Ω ½W	SV7	2N540, H45, H200E 2N574, 2N575	H45, 2N540
3A-6A	0.75Ω 50W	3Ω 50W	1000Ω ½W	SV7 SV806	H200E, 2N574 2N575	H45, 2N540

Circuit 8-14 - Transistor constant-current regulator

*Application Notes AN3, August 15, 1958.

383

Circuit 8-14, Transistor Constant-Current Regulator (cont'd.)

transistor type used. Typical regulation of a 1.0 ampere current with a 28 volt source is better than 2% for load changes from zero ohms to maximum resistance, which is 20 ohms. Either conventional power transistors may be used, or the H200E tetrode may be utilized for a substantial increase in the maximum temperature at which regulation can be maintained.

Basically, this circuit is merely a very high impedance placed between an input voltage and a load, so that the input voltage may be applied at either end of the circuit and the load at the other end, as indicated in the schematic diagram.

Current through R_3 operates the breakdown diode, CR_1, which develops about 7 volts, and which is used as a voltage reference. Transistor Q_2 acts as an emitter follower which drives the base of Q_1. If Q_1 is a tetrode, as indicated, base #1 is driven while base #2 absorbs the collector junction I_{CO} and disposes of it through R_4 which is adjusted for a minimum output current. As a result, I_{CO} need not flow through the base #1 lead and high-temperature stability is vastly improved.

When Q_1 conducts, it will place the 7 volts from CR_1, less the base-emitter drops in Q_1 and Q_2, across resistors R_1 and R_2. Since the voltage is relatively constant, the current through the resistors will be relatively constant. The emitter current in Q_1 is equal to the current in R_1 and R_2, but the collector current is smaller by the amount of the base #1 current. However, Q_2 receives the base #1 current and transmits α_{fb}, or nearly 1.0 times this current, back to the collector circuit of Q_1 so that the constant current from R_1 and R_2 is delivered almost intact to the load.

CIRCUIT 8-15

HIGH VOLTAGE REGULATOR WITH SERIES STACK

Motorola, Inc.

This regulator provides a fixed 150-volt output to loads as high as 150 mAdc with about 0.05% line or load regulation. Its input voltage is supplied by a d.c. to d.c. power converter which has 5% regulation and which is, in turn, supplied from a primary power source having voltage variations as high as ±23%.

The design of high-voltage transistor regulators is complicated by the limited collector-emitter voltage ratings of power transistors and by the poor temperature coefficients of high-voltage breakdown diode voltage references. The voltage limitation is overcome in this regulator by using several power transistors in a series stack, while a stable voltage reference is obtained from a low-voltage breakdown diode.

The minimum input voltage is 180 volts, which permits an average of no less than 7.5 volts across each power transistor in the series stack,

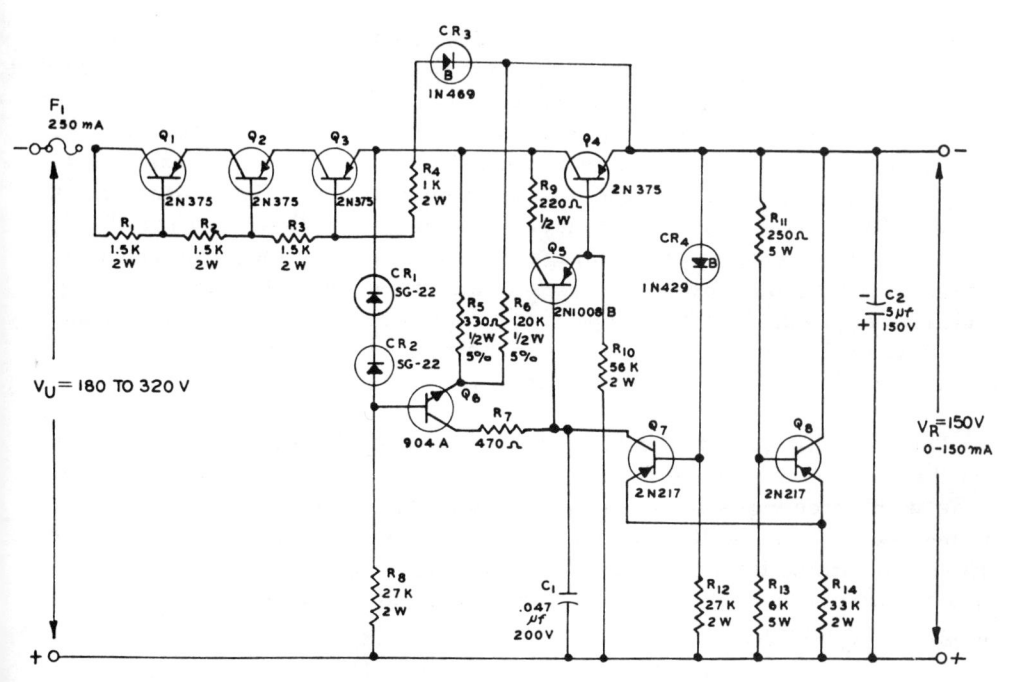

Circuit 8-15 - High voltage regulator with series stack

385

Circuit 8-15, High Voltage Regulator With Series Stack (cont'd.)

Q_1, Q_2, Q_3, and Q_4. Of course, it must be assumed that this minimum input voltage occurs at full-load and when the primary voltage source is 23% low. Then, when the primary voltage is 23% high at no load, the input voltage to the regulator will rise to about 320 volts, placing about 42 volts across each of the four series transistors. Power transistors are readily available with collector-emitter voltage ratings higher than this value. Each may pass a full-load current of 150 mAdc plus another 38 mAdc which operates the regulating amplifier, so that it must dissipate 7.9 watts maximum.

Transistor Q_4 is the only one in the stack which is directly controlled by the regulating amplifier. Control for transistors Q_1, Q_2, and Q_3 is obtained from the voltage divider R_1, R_2, R_3, R_4, and CR_3 whose function it is to divide the 170 volts which may exist across the stack equally between the four transistors. The base current in each transistor is composed of two components; one is the normal current which is a function of collector current, and the other is the I_{CO} of the collector junction which is a function of temperature. These currents are in opposition and each can attain levels between 2 and 3 mAdc. In order that these base currents will not seriously alter the outputs of the voltage divider, the resistances in the voltage divider are selected so that about 30 mAdc flows through it at maximum input voltage and about 4.5 mAdc at the minimum input voltage.

Part of the regulating amplifier is operated from the collector-emitter voltage of Q_4 so that it is necessary that this voltage be maintained at low input voltages. This is accomplished by the 5-volt breakdown diode, CR_3, which holds the base of Q_3 at least five volts away from the emitter of Q_4 if there is any current able to flow through CR_3. Hence, the emitter of Q_3 and collector of Q_4 is held at a minimum of about five volts from the emitter of Q_4. Resistor R_4 is reduced in value from that used for R_1, R_2, and R_3 to compensate for the voltage drop across CR_3 and produce equal voltage drops across the four series transistors when the maximum input voltage is applied. Use of CR_3 permits regulation with inputs as low as 180 volts, while omitting CR_3 and increasing R_4 to 1.5 kilohms causes the regulator to fail at about 200 volts input.

The 150-volt output is divided by R_{11} and R_{13} to a level comparable with the reference voltage from CR_4, which is a 6.2-volt breakdown diode whose voltage changes less than 0.01% per degree centigrade. Comparison between the 6.2-volt reference and the divided output is made by the differential amplifier comprised of Q_7 and Q_8. The differential connection tends to balance out the base-emitter voltage changes with temperature changes in Q_7 and Q_8.

Circuit 8-15, High Voltage Regulator with Series Stack (cont'd.)

However, at high temperatures the I_{CO} of these transistors can become quite large and, since these currents will flow into the base circuits, a base voltage change will result which is proportional to the base circuit resistance. Since the resistance of CR_4 is small, the voltage change at the base of Q_7 is also small, but both the resistance and voltage change will be relatively high at the base of Q_8. This source of temperature drift is minimized by making R_{11} and R_{13} as small as feasible; in this case they carry about 25 mAdc, or possibly 200 times the maximum expected I_{CO} in Q_8. It is also important that R_{11} and R_{13} have the same temperature coefficient of resistance.

The error signal output is taken from the differential amplifier at the collector of Q_7. The collector load of Q_7 consists of the base of Q_5 and the constant-current source, Q_6, which supplies a current of 2 mAdc to Q_7 at a high impedance and thereby maximizes the voltage gain of Q_7. Basically, the constant forward voltage drop across two silicon diodes, CR_1 and CR_2, is placed across R_5 by the emitter of Q_6 with a small and essentially constant base-emitter voltage drop. Hence, the voltage across, and the current through, R_5 is relatively constant. The emitter of Q_6 is then driven from a high impedance while the base sees a low impedance. This is equivalent to a grounded-base connection which displays a very high collector dynamic impedance.

A silicon transistor is used for Q_6 because no germanium n-p-n unit was available with the required high-voltage rating. If Q_6 is ever forced into voltage breakdown for more than a few seconds, as might happen during turn-on, Q_7, Q_5, and possibly Q_4 will be damaged or destroyed. Therefore, a minimum voltage rating of 60 volts, and preferably 80 to 100 volts, is necessary on Q_6. Resistors R_7 and R_9 also help to limit these surges. Resistor R_6 was determined experimentally to improve the regulation against input voltage changes. As the input increases, the voltage across Q_4 increases, driving the emitter of Q_6 slightly positive through R_6. This in turn decreases the collector current from Q_6 and allows the collector of Q_7 to rise positively, carrying the bases and emitters of Q_3 and Q_4 in the same direction. The value of R_6 chosen balances this tendency to decrease the output voltage against the normal tendency to increase the output when the input voltage rises.

Transistor Q_5 is an emitter follower which reduces the base current requirements of Q_4 from about 2 mAdc to about 40 μAdc in the collector of Q_7, thereby permitting Q_7 to develop a large voltage gain. It must be a 300-mW type having a collector diode breakdown voltage of 60 volts. Resistor R_{10} provides a turn-off bias current for Q_4 which permits the regulator to retain control at high temperatures and light output loads.

Circuit 8-15, High Voltage Regulator with Series Stack (cont'd.)

The only components necessary for feedback stabilization are C_1 and C_2.

This regulator will operate in ambient temperatures as high as 60°C with transistor mounting base temperatures as high as 70°C. The measured output voltage temperature coefficient was 0.029% per degree centigrade. It could be reduced by using equal temperature coefficients in R_{11} and R_{13}. The measured output impedance ranges from 0.2 ohms at 400 c.p.s. to a peak of about 5 ohms at about 15 kc, above which it drops again to about an ohm at 300 kc. Overload protection is provided by a 250-mAdc fuse, F_1, which is in series with the collector of Q_1.

CIRCUIT 8-16

500-VOLT REGULATED SUPPLY

Motorola, Inc.

    This unusual regulator illustrates another method of producing a high-voltage output without the use of a high-voltage series stack of transistors. In this case, the high-voltage output is sensed and regulation is performed at a low-voltage, high-current point while the regulated output is transformed to the desired high-voltage, low-current level with a d.c. to d.c. converter. The particular design shown here produces a fixed 500-volt output at load currents up to 150 mAdc with an input regulation of 0.2%, a load regulation of 0.8%, and a voltage stability within ±2.5 volts over the ambient temperature range of -40°C to +60°C. It operates from the standard military voltage of 28 volts d.c. and can tolerate variations between 20 and 30 volts. No auxiliary power supplies are required to operate the regulator.

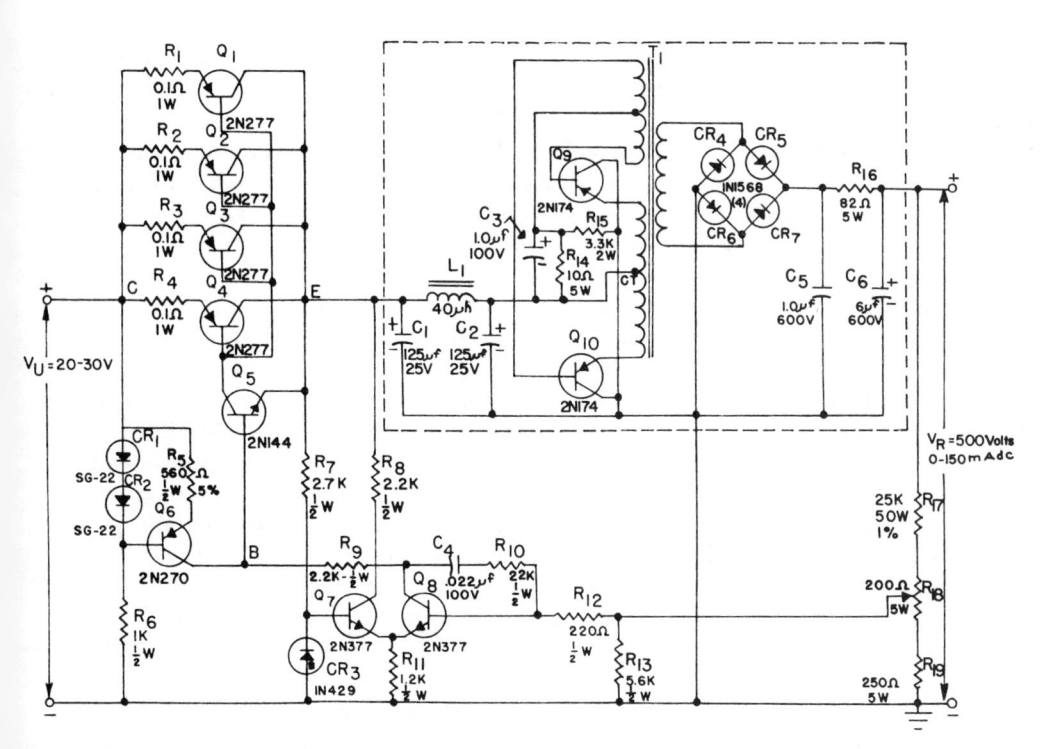

Circuit 8-16 - 500-Volt regulated supply

Circuit 8-16, 500-Volt Regulated Supply (cont'd.)

The d.c. to d.c. power converter, shown within the dashed line, is a standard 75-watt design. The transformer is wound to deliver a 500-volt d.c. output when the input voltage to the converter is about 18 volts. The only modification necessary in the converter is the addition of a filter comprised of C_1, L_1, and C_2 which prevents the converter switching transients from disturbing the regulator. As is customary performance from d.c. to d.c. converters, an overload will cause the converter to cease its switching and thereby protect itself and its regulator from damage.

A portion of the output voltage is sensed by the voltage divider, R_{17}, R_{18}, and R_{19}, and delivered to the base of Q_8. Resistor R_{12} provides a bit of protective isolation for the base of Q_8, while R_{13} acts as a base return in the event that the voltage divider is disconnected from the regulator. This portion of the output voltage is compared with the 6.2 volts from CR_3, a 0.01%/°C reference breakdown diode, in the differential amplifier composed of Q_7 and Q_8. The collector of Q_7 has a load resistance, R_8, which merely reduces the collector dissipation, while the amplifier output is taken from the collector of Q_8.

The impedance presented to Q_8 by the constant-current source, Q_6, and transistor Q_5 is sufficiently high that little gain is lost by placing R_9 in series with Q_8 as a collector-dissipation reducing device. The forward bias voltage developed across the two silicon diodes, CR_1 and CR_2, by current from R_6 is placed by Q_6 across resistance R_5. This causes about 2 mAdc to flow into the emitter and out of the collector of Q_6 which displays the high dynamic impedance of a grounded-base transistor and thereby maximizes the voltage gain developed by Q_8.

At first sight, Q_5 might seem to be a confused emitter follower. Actually, it functions with the paralleled p-n-p power transistors, Q_1, Q_2, Q_3, and Q_4, in such a way as to synthesize a high-power n-p-n transistor which has a current gain equal to the product of that of the paralleled transistors and the current gain of Q_5. The collector, emitter, and base leads of the synthesized n-p-n transistor appear at the points in the circuit marked "C", "E", and "B", respectively. This part of the circuit may be viewed as a two-stage grounded-emitter amplifier which has its output, the p-n-p collectors, tied directly to the input emitter in Q_5 for 100% voltage feedback. By means of this degenerative feedback, the p-n-p collectors are made to follow the base voltage of Q_5 in the same way that an n-p-n emitter would follow its base.

When analyzed in this manner, the regulator circuit may be seen to be an emitter-follower type of series regulator, since the regulating amplifier applies its control between collector and base of the synthesized n-p-n power transistor.

Circuit 8-16, 500-Volt Regulated Supply (cont'd.)

Transistor Q_5 is a medium-power n-p-n type, while Q_1 through Q_4 are high-power p-n-p units. An additional p-n-p unit might be added in parallel for a 90-watt output, or one removed for a 60-watt converter. As is customary, resistances R_1 through R_4 equalize the division of current through the paralleled transistors. At full load, the series transistors pass about 6 amperes to the converter which operates at 85% efficiency.

No provision is made for turn-off bias current for Q_1 through Q_4. It is stated in the original report that the converter draws about one ampere with no output load and that this is about ten times the combined high-temperature I_{CO} of transistors Q_1 through Q_4 so that no turn-off bias is necessary. However, this circuit configuration will multiply the I_{CO} of these transistors by their current gain and this could quite easily exceed an ampere at high temperatures. Therefore, at least a small turn-off bias current seems desirable for Q_1 through Q_4 to prevent current-gain multiplication of their I_{CO} leakage currents.

Feedback stabilization is obtained with C_4 and R_{10}.

While this technique of high-voltage regulation may seem awkward, there are a number of advantages to be noted. The life of the series transistors should be very good because of their low collector-emitter voltage. Turn-on and off surge problems are also minimized because of the low voltage levels in the regulator. A high degree of regulation can be maintained on one output of the converter, and if other output windings are added to the transformer they will also be regulated to a lesser degree. The advantage of overload protection provided by the converter is also important.

It is equally feasible to supply this regulator from a rectifier operating from the 117-volt power line through a transformer. A single capacitor filter will provide sufficiently low ripple. In addition to the same advantages just mentioned, such a system would also permit emergency operation from a 28-volt battery or electrical system.

CIRCUIT 8-17

MINIATURE SILICON REGULATOR MODULE

Power Sources, Inc.

This small all-silicon module contains all the semiconductor components for a regulated power supply capable of delivering as much as 1.5 amperes at voltages between 100 and 140 volts. It is built on a heat sink which may be used to dispose of the internal heat directly by convection or to conduct this heat to another heat sink. It is intended for modular use in military equipment and systems. The user need provide only a power transformer and filtering components suitable to his particular application to complete a regulated power supply using one of these modules. Other modules in the same series are available with output voltages in the range between 25 and 190 volts.

The particular module presented here, a 100 to 140 volt version of Model PS8001, is able to provide load regulation of less than 0.1% and regulation against ±10% input voltage changes of 0.02%. At 25°C heat sink temperature, it can dissipate up to 50 watts in its series regulator, although at 125°C this must be derated to about 20 watts. Its temperature coefficient is typically less than 50 parts per million per degree Centigrade. The output ripple is no more than 0.2% of the ripple at its d.c. input.

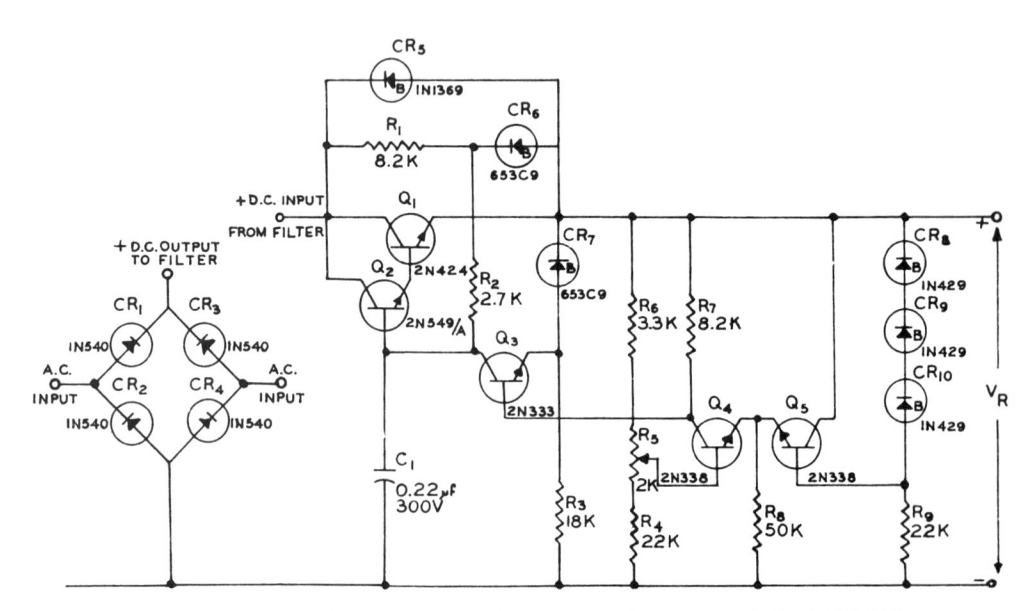

Circuit 8-17 - Miniature silicon regulator module-PS8001

Circuit 8-17, Miniature Silicon Regulator Module (cont'd.)

The user is responsible for adjusting the input voltage range, load current, and heat sink temperature so that the transistor junction temperatures and series transistor collector-emitter voltage do not become excessive under any conditions.

The unit is ruggedly constructed. It can withstand a shock of 100 G for 6 milliseconds, a vibration of 15 G over the frequency range from zero to 2,000 c.p.s., and an acceleration of 100 G.

The regulator utilizes silicon n-p-n transistors and silicon breakdown diodes to achieve high-temperature operation. The voltage reference is a series of three 1N429 reference diodes which develops a total of about -18.6 volts at the base of Q_5 with reference to the positive output line. This voltage is matched at the base of Q_4 by voltage regulator action through the voltage divider comprised of R_4, R_5 and R_6. Adjustment of R_5 will cause the output voltage to vary in such a manner as to maintain -18.6 volts at the base of Q_4.

The output from this differential amplifier, Q_4 and Q_5, is taken from the collector of Q_4. Resistor R_7 supplies the collector current for Q_4 and the base current for Q_3, which is the second stage voltage amplifier. The emitter of Q_3 is held at about -8.0 volts by breakdown diode CR_7, which is operated by current from R_3. The small degeneration caused by the impedance of CR_7 in the emitter circuit is neither serious nor easily reduced.

The collector output from Q_3 is applied to the base of Q_2 which is an emitter follower in the "Darlington connection" with Q_1. Collector current for Q_3 and base current for Q_2 is provided by R_2 and a positive 8.0 volt source generated by breakdown diode CR_6 with current through R_1. Capacitor C_1 prevents high-frequency oscillation and instability. Transistor Q_1 is the main power regulating unit which is capable of dissipating as much as 50 watts at 25°C. Breakdown diode CR_5 will conduct when the voltage across Q_1 reaches about 56 volts and will thereby protect Q_1 from voltage surges which might otherwise destroy Q_1.

PART 9

POWER CONVERTERS

Section 1. Design Philosophy

The transistor power converter consists of an over-driven, push-pull, transformer-coupled transistor oscillator with one or more load windings added to the transformer.

A. Basic Transistor Oscillator

The basic oscillator circuit is shown in Figure 9-0.1. The transistors operate in a push-pull circuit, with the transformer windings arranged to provide positive feedback from the collector of each transistor to its emitter. The operation of the circuit can best be described with the aid of the transformer B-H curve in Figure 9-0.2. Assume transistor Q_1 is non-conducting, transistor Q_2 is conducting, and the transformer core saturated at the point "P" on the B-H curve.

When Q_1 starts to conduct, a voltage is developed across the primary winding, inducing a voltage in the feedback windings that drives the emitter of Q_1 positive and causes increased conduction. Thus, the current increases rapidly until the collector of Q_1 is driven into the saturation region of its characteristics. When this occurs, the primary voltage can no longer increase and a condition of quasi-stable equilibrium is maintained. During this equilibrium period, the voltage drop in the

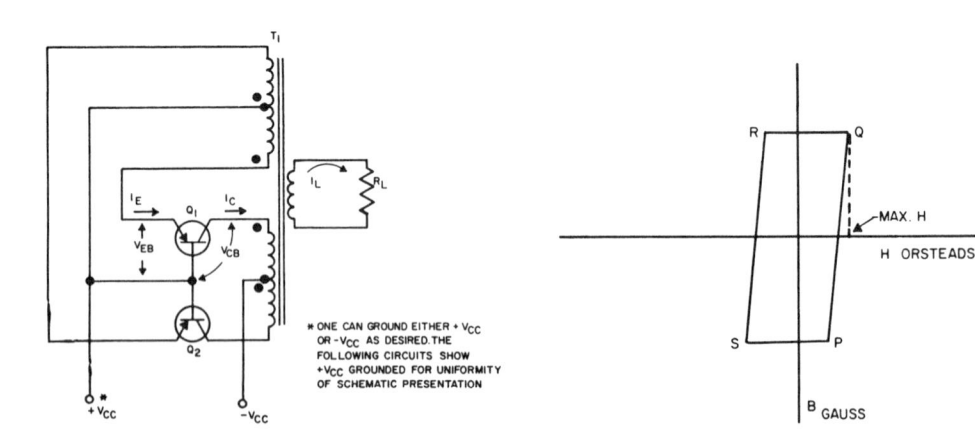

Fig. 9-0.1 - Basic oscillator

Fig. 9-0.2 - Transformer B-H curve

transistor from collector to emitter is very small, and essentially the full supply voltage -V_{cc} appears across that half of the transformer primary.

As long as Q_1 can supply a collector current equal to the total of reflected load current, reflected emitter current, and transformer exciting current, this condition will be maintained. With a resistive load, the reflected load current and reflected emitter current remain almost constant. The transformer exciting current is small as long as the core remains nonsaturated.

With a constant voltage across the transformer primary, the magnetic flux must increase according to the relation

$$v = N \frac{d\phi}{dt}. \qquad 9\text{-}0.1$$

Thus, during the flat portion of the voltage wave, the core flux increases from point "P" to point "Q" on the B-H curve. There will be a slight increase in the exciting current requirements as per the B-H loop characteristics in Figure 9-0.2, but this represents only a small fraction of the total collector current and can be readily supplied by transistor Q_1.

Eventually, at point "Q", the transformer core reaches saturation and the required exciting current increases rapidly to a value greater than can be supplied. As a result, the primary voltage decreases, reducing the emitter voltage and decreasing the collector current. Thus, transistor Q_1 turns off regeneratively, ending the half cycle. As the flux collapses from point "Q" to point "R", voltage is induced in the winding which biases transistor Q_2 to conduction, thereby initiating the next half cycle. The operation is identical to the first half cycle with the exception that Q_2 conducts until the core is driven into negative saturation, represented by point "S" on the B-H curve. After the core saturates, the flux collapses from point "S" to point "P" and the full cycle is complete.

Typical collector voltage and current waveforms are shown in Figure 9-0.3. These pertain to a circuit operating with a properly designed transformer working into

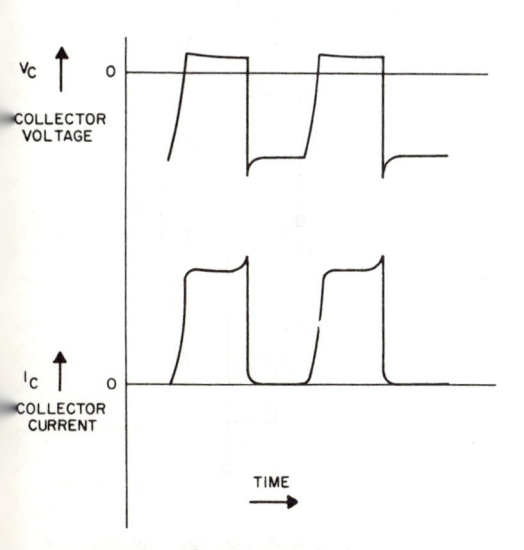

Fig. 9-0.3 - Converter transistor waveforms

optimum load. The waveforms are desirably square, with the exception of
the spike at the end of the current waveform, which represents the additional exciting current required by the transformer as the core approaches
saturation. Since the collector of the transistor is saturated during its
complete conduction period, the transistor dissipation is very low.

B. Other Connections for Basic Circuits

The basic circuit described above and shown in Figure 9-0.1 is a
common base circuit. Its major advantage is that with battery supplies of
small voltages, the transistor voltage drop can be more than compensated
for by the voltage developed by the base-to-emitter drive winding. The
major disadvantage of this type of circuit is that the base-to-emitter winding on the transformer must carry the total collector current for one half
the cycle, and in converters using modern high current transistors this
may be many amperes. This makes the wire size required for the winding
quite large, thus making the transformer winding awkward.

Two types of common collector circuits are shown in Figure 9-0.4 (A)
and (B). The basic grounded collector circuit (A) requires a drive winding

(A) STANDARD CONNECTIONS   (B) AUTOTRANSFORMER CONNECTIONS

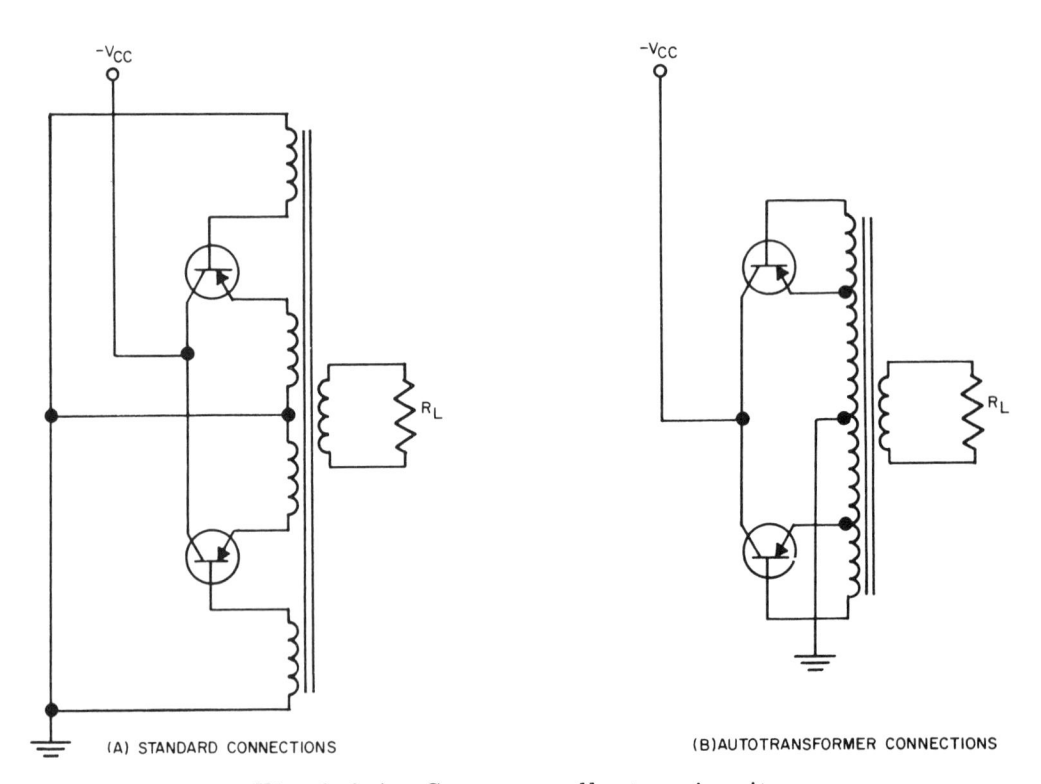

Fig. 9-0.4 - Common collector circuits

with slightly higher number of turns than is on the primary or emitter-to-emitter winding. This can be eliminated by using the modified circuit (B), where the drive winding is taken from the emitter-to-emitter winding by use of an autotransformer connection. This circuit has a great advantage, with respect to transistor cooling, when used in a system that has the negative side of the battery grounded to the vehicle or system ground. The collectors of most transistors are mounted directly on the transistor base plate, and in this type of system the transistor can be attached directly to the vehicle frame without electrical insulators, which are thermal insulators as well.

The common emitter circuit, shown in Figure 9-0.5, is probably the most common circuit in use today, and it obtains its popularity from the fact that the drive requirements are less by a factor of the transistor gain than in the common base circuit. The drive current, in general, is in the milliampere rather than ampere region, as is found in the common base circuit.

C. Starting Circuits

None of the three basic oscillator circuits as described above will oscillate readily or, in some cases, they will not oscillate at all, unless some form of starting circuit is used. The starting problem is most severe when the oscillator is to be started under full load and at low temperature. Many techniques to start oscillation may be used, and one of the most common circuits is shown in Figure 9-0.6. This circuit

Fig. 9-0.5 - Common emitter circuit

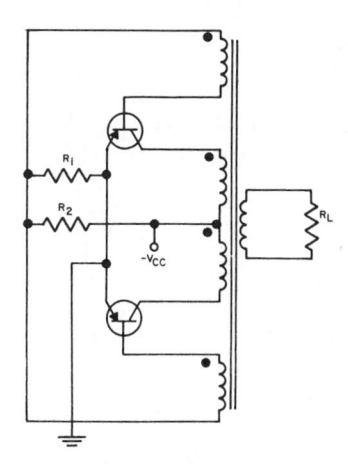

Fig. 9-0.6 - Resistor self-starting circuit

is used often because the only components added to the oscillator circuit are resistors. These resistors make up a low impedance bleeder circuit which bias the transistors to conduction before oscillations start. The major disadvantage to this type of circuit is that when it is used on higher powered converters the power dissipation in the resistors becomes quite high. This becomes a major problem when the converter is to be packaged in a minimum volume, and is to operate at high ambient temperatures.

A second starting circuit is shown in Figure 9-0.7. This circuit uses a diode and a resistor for starting. It is more costly than the circuit using only resistance, but it requires less dissipated power and is less temperature-dependent. The circuit operates in a similar fashion to that using only resistance, but when power is first applied, the bases of the oscillator are driven negative by the full supply voltage which starts oscillation rapidly. As soon as base current flows, the diode clamps the base return to ground.

A third self-starting system is shown in Figure 9-0.8. This circuit uses an extra negative supply consisting of winding 7-8-9 and the two diodes. This circuit operates similar to the other self-starting circuits except that R_1 can be adjusted to limit the base-to-emitter current to a specified value.

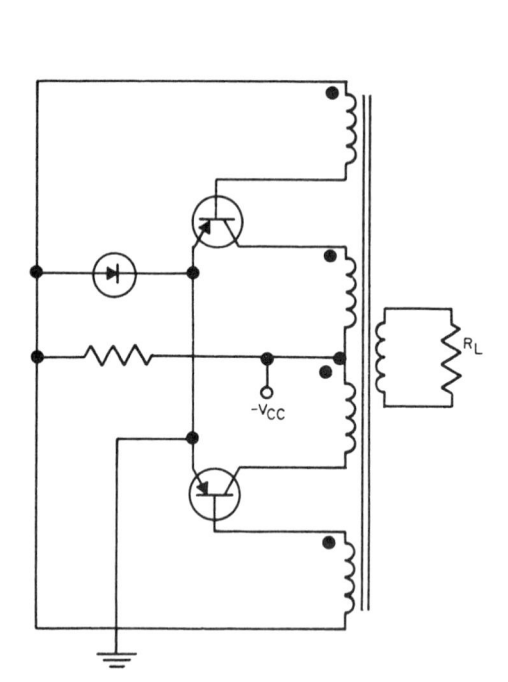

Fig. 9-0.7 - Diode self-starting circuit

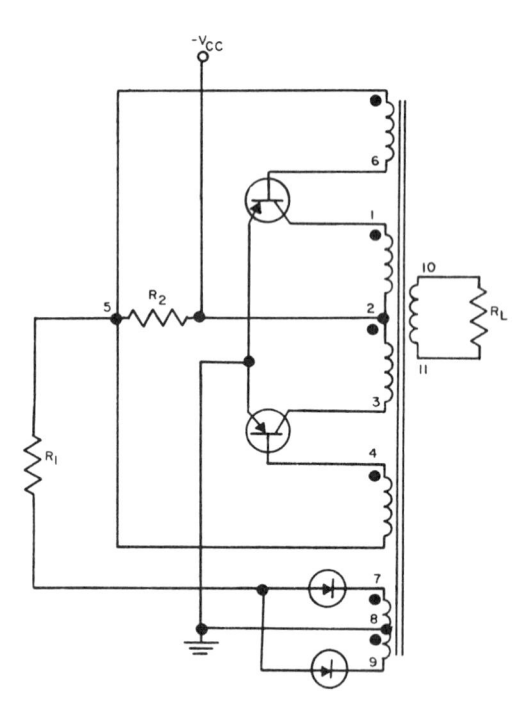

Fig. 9-0.8 - Negative supply self-starting circuit

D. Basic D.C. to D.C. and D.C. to A.C. Power Converters

Figure 9-0.9 shows d.c. to d.c. and d.c. to a.c. converters of a practical nature. Both converters are identical up to the load which is added to the output winding of the transformer. Since the output waveform from the transformer is basically a square wave, some problems are encountered if one tries to drive systems designed for sine wave inputs. This difficulty arises from different peak, average and r.m.s. factors for the two waveforms. Using r.m.s. values for a standard:

	Square Wave	Sine Wave
RMS	1.0	1.0
Peak	1.0	1.4
Average	1.0	0.9

Using the above data, one can see that if a vacuum tube power supply, designed to be operated from a sinusoidal input, were supplied power from a square wave converter, and the r.m.s. value of heater or filament voltage was maintained at the same value, the d.c. output would depend on the type of rectification system used. For an averaging rectification system (choke input filter), the d.c. output voltage would be high by the factor 1.0/0.9, and, if the filter were a peak filter (capacitor input), the d.c. output would be low by the factor 1.0/1.4. Since these figures are based on theoretical waveforms for systems without loss, the actual deviations are not as great, and many power supply systems, particularly choke

Fig. 9-0.9 (A) - A.C. output power converter

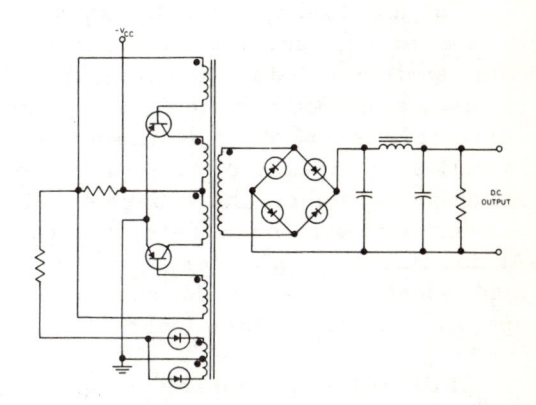

Fig. 9-0.9 (B) - D.C. output power converter

input filter systems, can be driven from square wave converters without any modifications at all. It is also interesting to note that, as shown in Figure 9-0.10, a rectified theoretical square wave gives a pure d.c. voltage as compared to a pulsating d.c. voltage for a sine wave. This indicates that, for the same filter system, the ripple from rectified square waves will be much less than that obtained when the same applied voltage of the sine wave form is used. Thus, for the same reduction, less filtering is needed for the square wave system than for the equivalent sine wave system.

Many types of motors can be driven by square waves as well as sine waves. In most cases the motor acts like a filter and uses only the fundamental component of the square wave. From the Fourier analysis of a square wave, the peak value of the fundamental is $4/\pi$, and, since the r.m.s. value of a sine wave is 0.707 of the peak, the square wave voltage must equal the rated sine wave voltage times the factor $\pi/4 \times 0.707$. Thus, for example, a motor that requires 115V a.c. sine wave would require approximately 128V a.c. square wave for the same speed and torque output.

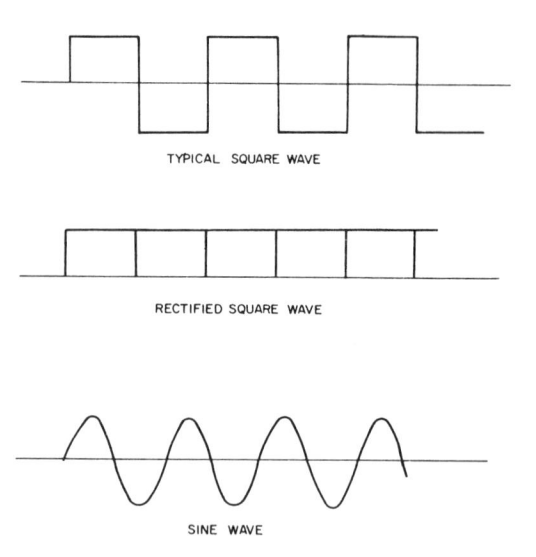

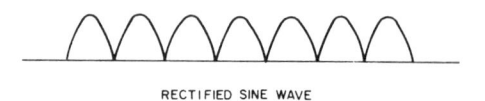

Fig. 9-0.10 - Comparison of rectified sine and square waves

The d.c. to d.c. converter shown in Figure 9-0.9 (B) is the same as the d.c. to a.c. converter shown in Figure 9-0.9 (A) but with rectifier and filter system added to the output transformer winding. The type of filtering used after rectification greatly affects the converter operation. The converter, as any other oscillator, will not operate properly with large inductive or capacitive loads. If a choke input filter is used with a large inductive reactance, the voltage developed across the inductor can cause current to flow through the rectifiers after the normal cycle is complete. At this time the converter is switching and it sees a very low impedance load, which causes oscillations to cease. Thus, the converter will tend to operate in a mode which looks like a continuous-running blocking oscillator.

If the filter is a capacitor input system, and the value of the capacitive reactance is too low, the converter will start very slowly or not start at all, and the current required to be supplied by the transistors may be in excess of the transistor current ratings. There is a greater variety of capacitive reactances than inductive reactances into which the converters will operate properly, so that capacitor input filter systems are almost always used as converter filters.

E. De-Spiking Networks

Quite often spikes will appear across collector-to-emitter of the switching transistor as it turns off, as shown in Figure 9-0.11. This spike may be close to, or exceed, the value of voltage allowed across the transistor and steps should be taken to keep this to a minimum. These spikes are generated in the circuit, or by the load, and are not spikes appearing on the converter power source which will be discussed later.

One of the greatest sources of spikes is the converter transformer. In general, every effort should be made to have a transformer with high-coupling and low-leakage reactance.

A small value of capacitance can be placed across the primary of the transformer, as shown in Figure 9-0.12. This gives a storage device which absorbs the energy released when the transformer flux collapses during switching. The value of capacitance is best selected experimentally

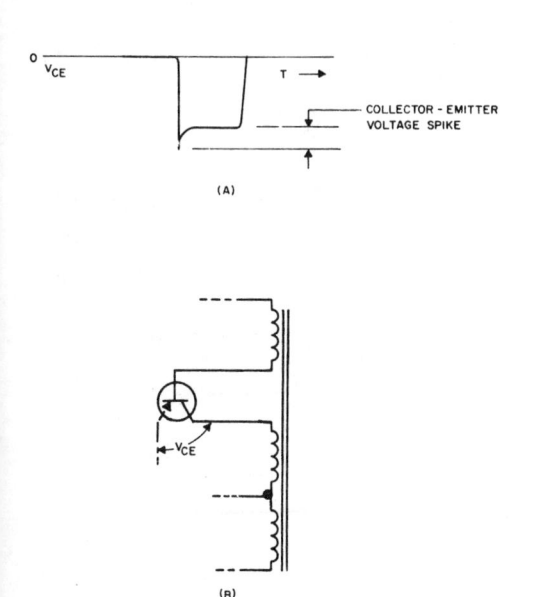

Fig. 9-0.11 - Spiking across transistors in power converter applications

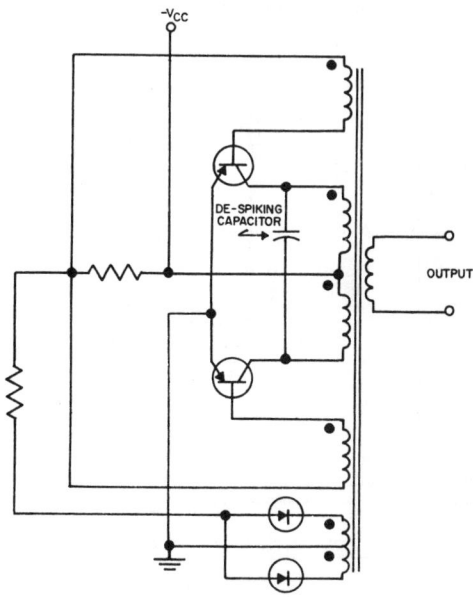

Fig. 9-0.12 - De-spiking circuit with collector-to-collector capacitor

for minimum spike. This circuit is very useful under severe environmental conditions, even though the capacitor may have to be of the energy-storage type. It is not necessary, however, to use a large value of capacitance such as found in the electrolytic capacitor.

Another circuit is shown in Figure 9-0.13. In this circuit, large values of capacitance effectively clamp the base circuits and prevent the collector of the transistor that is turning "on" from going positive.

A third system of de-spiking is shown in Figure 9-0.14. In this circuit, power Zener diodes are connected across the transistors. If a voltage spike appears that exceeds that allowed across the transistor, the Zener diode breaks down and the spike energy is dissipated in the diode.

One, or a combination of more than one, of these circuits may be necessary to limit the transient voltage across the transistor to a safe limit.

F. <u>Trasistor Selection</u>

The voltage developed across the "off" transistor in a full-wave converter circuit is twice the supply voltage, plus any transient spike. Thus,

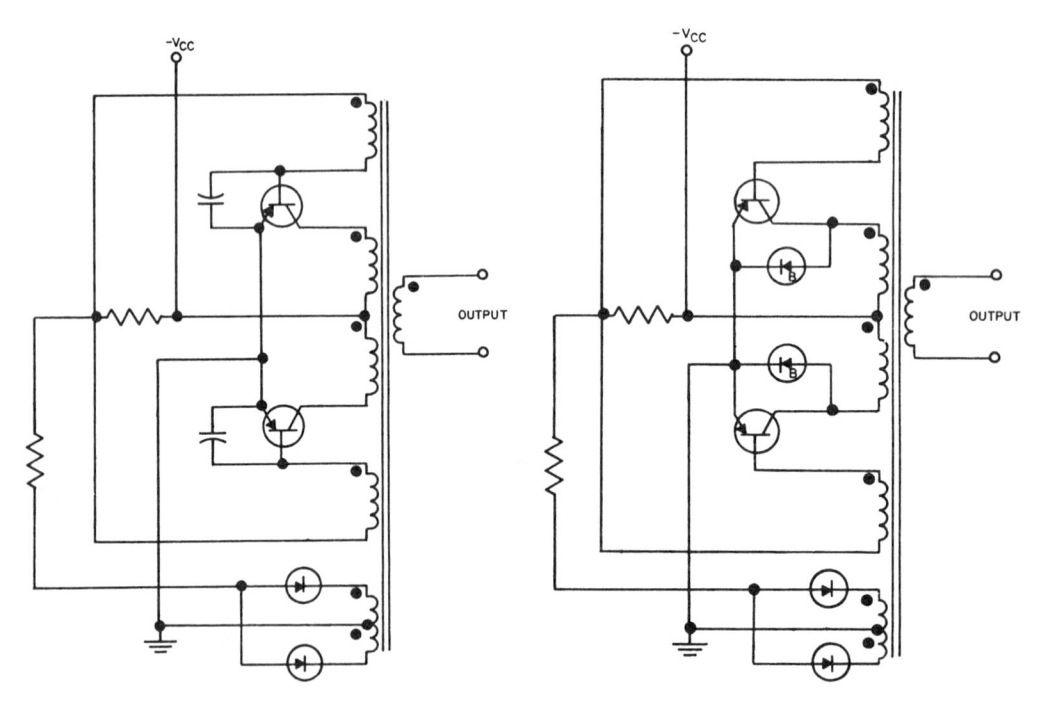

Fig. 9-0.13 - De-spiking circuit with base clamping capacitors

Fig. 9-0.14 - De-spiking circuit with Zener diodes

a transistor should be selected having an open base rating that is twice the supply voltage, plus a safety factor. Typical values are as follows:

Supply Voltage	Collector-Diode Voltage
24 - 28V	60 - 80V
12 - 14V	30 - 40V
6 - 8V	20 - 40V

The transistor collector current is approximately the current measured in series with the supply voltage. For design purposes, the current can be estimated from the following formula

$$I = \frac{\text{Watts Out}}{\text{Input Voltage} \times \text{Efficiency}}.$$ 9-0.2

Depending upon the circuit, the usual efficiencies lie between 60% - 80%.

Another important transistor parameter is its saturation resistance (or saturation voltage). Since the I^2R loss in this resistance represents power dissipated in the transistor, this value should be as low as possible. In general, transistors with high collector current ratings have the lowest saturation resistance.

With most germanium transistors, gains are sufficiently high and saturation resistances sufficiently low so that transistor selection within a particular type is not generally necessary. With some of the new high power silicon transistors, however, matching at the rated converter current for gain and saturation resistance may be a necessity.

G. Transformer Selection and Design

The transformer, in general, is the most important item in the power converter. For proper operation, the transformer should have close-coupling, low-leakage reactance and have a core material that saturates sharply. Many types of core material have been used in converters, including standard transformer laminations, cut and uncut "C" cores of grain oriented steel, ferrite cores, tape wound nickel-iron materials, just to name a few. The best type of material for most applications is tape wound toroids of 50-50 nickel-iron sold commercially as Orthonol (1), Deltamax (2), Hypernik V (3). This material has a high saturation flux density, square hysteresis loop (or B-H loop) and a small core loss. In toroidal form, with proper winding techniques, very close coupling can be achieved. Another advantage of this core material is that the core characteristics are affected very slightly by temperature over the range where the transistor can be used. When extra tight coupling is desired, the windings may be bi-filar and tri-filar wound.

Unlike most transformer designs, the frequency is selected for optimum operation of the power converter, transistor and transformer. The lower limit of frequency is in general determined by the allowable size of the magnetic components or load frequency requirements. The upper frequency, in most cases, is limited by the frequency cutoff characteristics of the transistors, and the core loss of the magnetic material used in the transformer. In accordance with these considerations, most of the higher powered converters operate from 60 c.p.s. to about 2000 c.p.s. ultrasonic oscillator frequencies at high power levels and with a high efficiency will probably be possible shortly, with the introduction of high frequency power transistors and high saturation flux density, square loop, non-temperature sensitive ferrite core materials.

The normal transformer equation

$$v = 4.44 \times 10^{-8} \times A \times S.F. \times f \times B \times N \qquad 9\text{-}0.3$$

must be modified by changing the constant to 4×10^{-8} for square wave operation.

A = core area in square inches,

S.F. = steel stacking factor,

f = frequency of converter operation,

B = saturation flux density in lines per square inch,

N = number of turns, and

v = voltage across the winding.

It must be remembered that twice the supply voltage, minus the transistor drop, appears across the transformer. Since most of the data given on the square loop toroidal cores are expressed with area in square centimeters and flux density in gauss, the equation can also be expressed as follows:

$$\frac{N}{v} = \frac{25 \times 10^6}{A_{cm^2} \times S.F. \times f_{c.p.s.} \times B_{gauss}} . \qquad 9\text{-}0.4$$

Output transformers, or driver transformers in more complicated converter circuits, are designed in similar fashion, except that a flux density well below saturation is used in the transformer design.

H. Modified and Improved Circuits

The simple converter may be modified or improved depending upon its specific application. In converters delivering power over about 50 watts, it is often desirable to use separate oscillator and output transformers as shown in Figure 9-0.15. Only the small oscillator transformer (T_1) saturates and therefore the necessary extra current at saturation is small compared to the load current.

Another useful modification of the basic converter is shown in Figure 9-0.16. In this circuit, the basic converter drives another push-pull stage. The drive to the output stage is such as to turn the transistors "on" hard, or "off" as in the oscillator. The circuit is often used when the output stage is to drive a motor or some other highly reactive or varying load.

This type of circuit is also used when it is desirable to frequency-regulate a low power stage and, on occasion, a series of stages may be

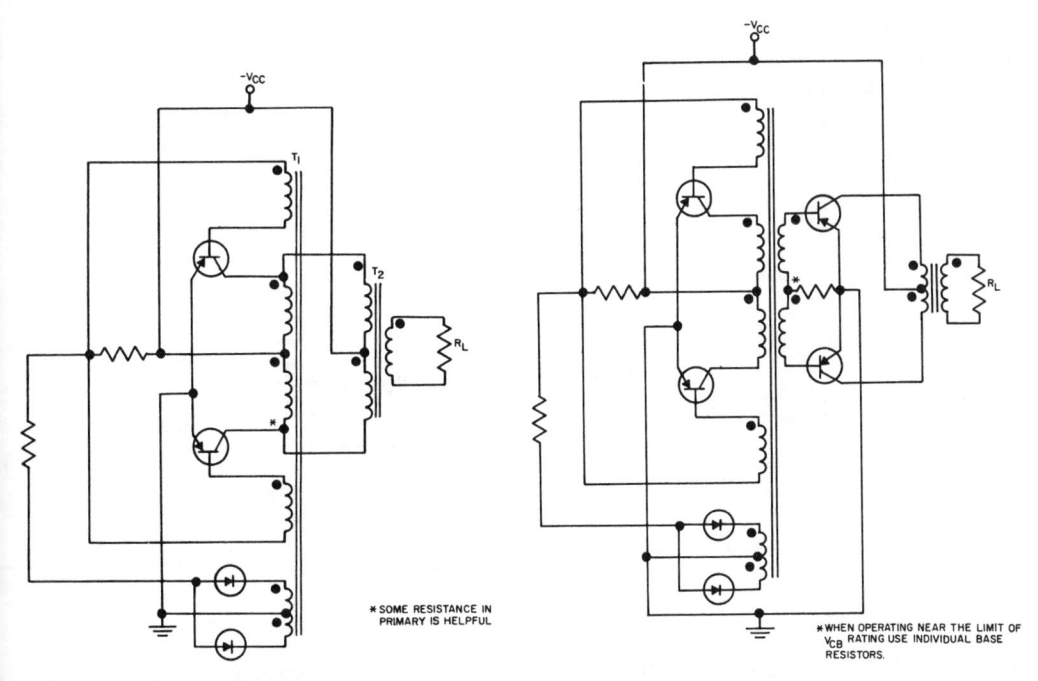

Fig. 9-0.15 - Converter with separate saturating transformer

Fig. 9-0.16 - Converter and driven stages

used to go from a milliwatt control stage to an output stage of several hundred watts.

Another type of output stage can be added to the basic converter circuit, as shown in Figure 9-0.17. In this circuit the output stage is connected in a bridge arrangement. The output transformer required is single-ended and the back voltage across the transistors is the supply voltage instead of twice the supply voltage. This circuit is useful in driving reactive loads or varying loads where excessive spiking would occur in a full wave circuit. The circuit requires twice the drive power and twice the number of output transistors as does the full wave circuit.

Figure 9-0.18 shows a circuit for operating a series of converters using the same transformer core for high voltage operation. No voltage equalization is required, for the magnetic circuit demands that the voltages divide equally. Modifications and additions to this circuit indicate that converters with input systems of high voltage d.c. and high power levels can be constructed using the presently available commercial transistors.

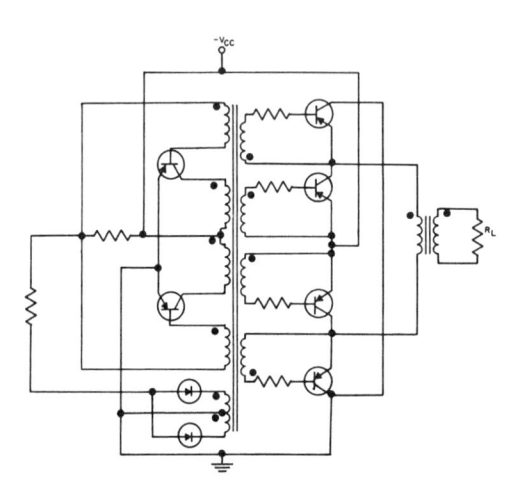

Fig. 9-0.17 - Converter with driven bridge connected output stage

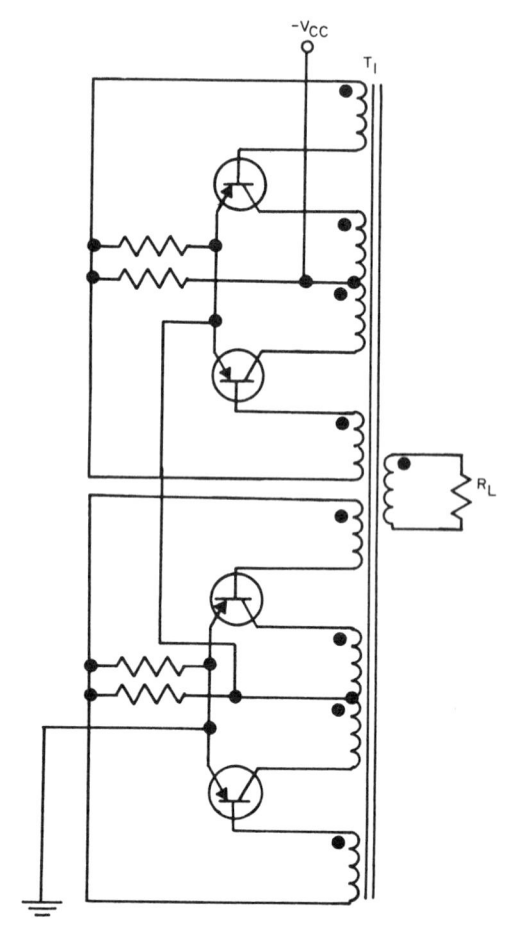

Fig. 9-0.18 - Series connected converters for high d.c input voltages

I. Regulators and Protective Circuitry

In both the d.c. to a.c. and d.c. to d.c. converters, once the output voltages are obtained, many types of standard regulators may be used. Two types of regulators however, are especially suited to the power converters. The first regulator, as shown in Figure 9-0.19, is very useful with low power converters. In the base-controlled regulator, the transformer primary voltage is rectified and fed back through an amplifier in such a fashion that transistor Q_3 acts as a variable resistance to keep the voltage across the transformer primary constant.

Another type of regulator, as pictured in Figure 9-0.20, is a full-wave magnetic amplifier type. A magnetic amplifier can be added to the bridge circuit or separate oscillator circuit for d.c. regulation, as well as to the full wave circuit. The operation of the three different types of magnetic amplifier regulators is the same, a portion of the d.c. output voltage is fed to a differential amplifier with a reference. The error voltage is amplified and used to control the control winding of the magnetic amplifier. Typical waveforms are shown in Figure 9-0.21. Unlike the base control regulator, which dissipates the extra power in the oscillator transistors, the magnetic amplifier looks like a high impedance, using a portion of the cycle and thereby limiting the average voltage delivered to the load. In this type of system, it is possible to have d.c. regulated output power converters with efficiencies greater than 80%. This makes an excellent regulator for converters that must be packaged for minimum size and operate at high temperatures.

The basic converter will be damaged if the input should be connected in the wrong polarity. A diode in series with the input, as

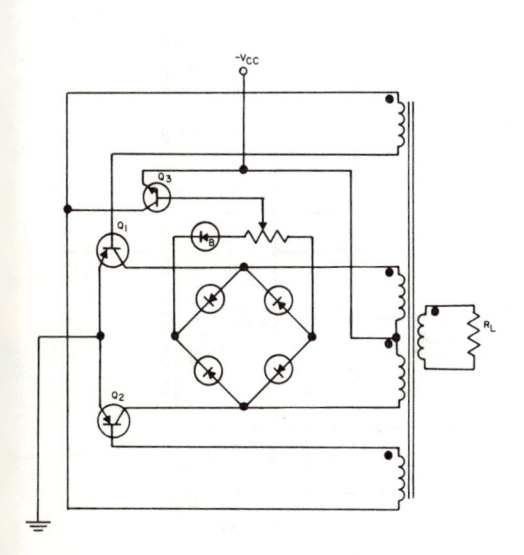

Fig. 9-0.19 - Basic base-controlled regulator

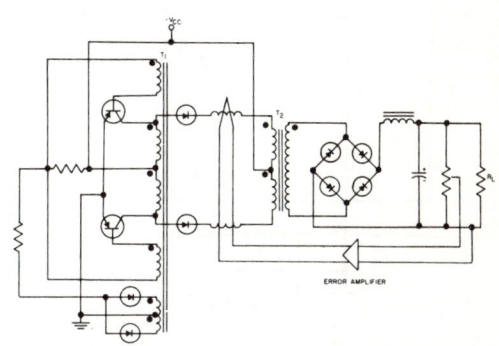

Fig. 9-0.20 - Basic magnetic amplifier controlled amplifier

shown in Figure 9-0.22, will limit the current to a few mills or less under a wrong polarity, but when the converter is operating properly the diode will be a major factor in limiting the converter efficiency. This is because of the large voltage drop that occurs across silicon diodes.

Another reverse polarity protective circuit is shown in Figure 9-0.23. This circuit uses a transistor in the base bias circuit and under reverse polarity conditions the transistor is biased off and the oscillator base current is limited to a safe value. When the converter is operating properly, a very small voltage appears across the protective transistor, but it must handle the total oscillator base current. Under a wrong polarity condition, it must have a breakdown rating exceeding the supply voltage.

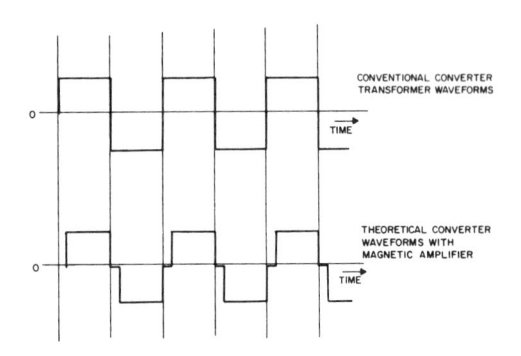

Fig. 9-0.21 - Transformer waveforms with magnetic amplifiers

As with most oscillators, the basic power converter will stop

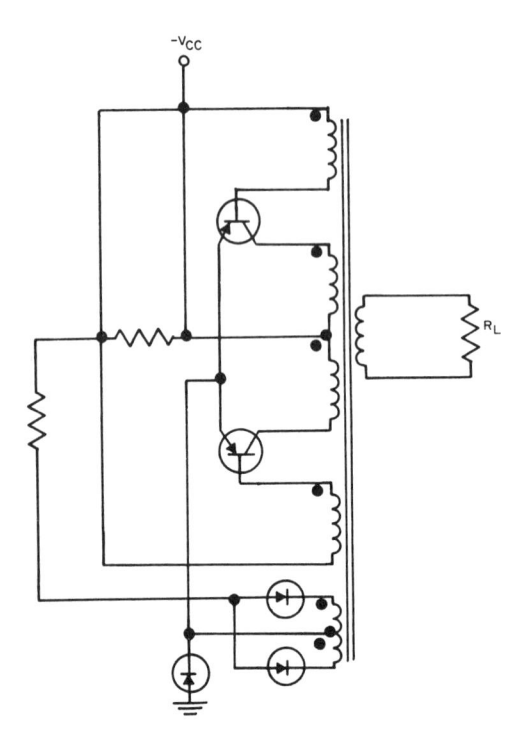

Fig. 9-0.22 - Reverse polarity protection with input diode

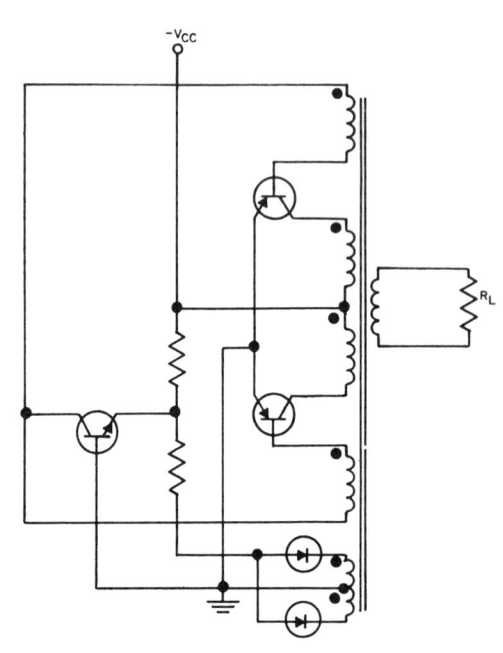

Fig. 9-0.23 - Reverse polarity protection with base control

oscillating if the output is short circuited. If care is taken to adjust the base bias conditions, the power dissipated in the oscillator transistors can be limited to a safe value under short circuit conditions, making it possible to short the output of the converter for long periods without any damage to the converter. Once the short circuit is removed, the oscillator functions normally.

By proper component selection, use of proper biasing for optimum operating and short circuit conditions, and use of reverse polarity protection, power converters can be designed and fabricated with extremely long life expectancies.

J. Power Converter Thermal Considerations

In operating a power converter of any type, care must be taken to insure that the transistors are operating properly electrically, and that adequate thermal cooling is provided by conduction or convection. The transistor junction temperature must not be exceeded for reliable converter operation. To observe the transistor operating in the circuit, a good d.c. coupled oscilloscope should be used to measure the voltage across the transistor during the "on" time. This voltage (which should be similar for both transistors) times the d.c. input current gives the power dissipated in the two oscillator transistors. Another source of transistor heating can be caused by excessive current in the emitter base circuit due to excessive drive from the transformer drive winding. Since the winding is shunted by the emitter base diode, the current should be measured by inserting a small resistor (0.01 to 0.1 Ω) in series with the base lead. If excessive current should be found, it can be limited by either inserting larger resistors in series with the bases or inserting a resistor in the common base return. The resistors should be adjusted to limit the current to some normal value.

One of the greatest causes of failure of the power converter is the presence of spikes or excessive ripple on the input supply to the converter. These spikes can be of very short duration and might not be seen on a good oscilloscope, but they can exceed the breakdown voltage of the oscillator transistors and cause transistor failure. This is particularly true when a battery which energizes the converter also supplies motor loads, or is charged by a generator with a slow response time regulator. It is also quite likely to occur on converter power sources that are rectified from the a.c. power system. Slow response systems, such as magnetic amplifier regulated supplies, can cause high line spikes.

Power converters, especially d.c. to a.c. units, will work well into pure resistive loads, but care should be taken when trying to operate into other types of loads, including nonlinear resistive loads, such as light bulbs. A normal light bulb can have a cold resistance which is 1/10th of the hot resistance. Such operation can cause excessive transistor current in some types of converters, and result in transistor failure.

REFERENCES

1. Magnetics, Inc., Butler, Pennsylvania

2. Arnold Engineering Company, Marengo, Illinois

3. Westinghouse Electric Corporation, Pittsburgh, Pennsylvania

BIBLIOGRAPHY

Bright, Pittman, and Royer, "Transistors as On-Off Switches in Saturable-Core Circuits," ELECTRICAL MANUFACTURING, December 1954.

Bright, R. L., "Junction Transistors Used as Switches," AIEE TRANSACTIONS, pp 155-156, 1955.

Royer, G. H., "A Switching Transistor D-C to A-C Converter Having an Output Frequency Proportional to Input Voltage," AIEE TRANSACTIONS - COMMUNICATIONS AND ELECTRONICS, Vol. 74, p. 322, 1955.

Smyth, R. R., "Transistors as Power Conversion Devices," paper delivered at IRE-AIEE Conference on Transistor Circuits, University of Pennsylvania, 18 February 1955.

Smyth, R. R., and Shorr, M. G., "Transistorized Power Sources for D-C to A-C and D-C to D-C Conversion," ELECTRONIC DESIGN, November 15, 1956.

Uchrin, G. C., and Taylor, W. O., "A New Self-Excited Square Wave Oscillator," PROC. IRE, 43, 99, 1955.

"Development of Power Transistor Circuitry," Final Report, U. S. Army Signal Corps, Contract DA-36-039-sc-63072, May 1956.

"Engineering Data and Application Notes," Delco Radio Division, General Motors Corporation.

CIRCUIT 9-1

D.C. TO D.C. POWER CONVERTER

Technical Operations Contributed by R. R. Smyth

Circuit 9-1 is a simple but highly reliable d.c. to d.c. power converter for both military and commercial applications. This unit, with proper mechanical mounting, can be operated over a temperature range of -50°C to 85°C.

The adjustment of the circuit for optimum performance can be obtained by adjusting R_1 and R_2. R_1 is made as large as possible and still have the circuit start reliably at the lowest temperature of operation and under full load (1.5K to 10K).

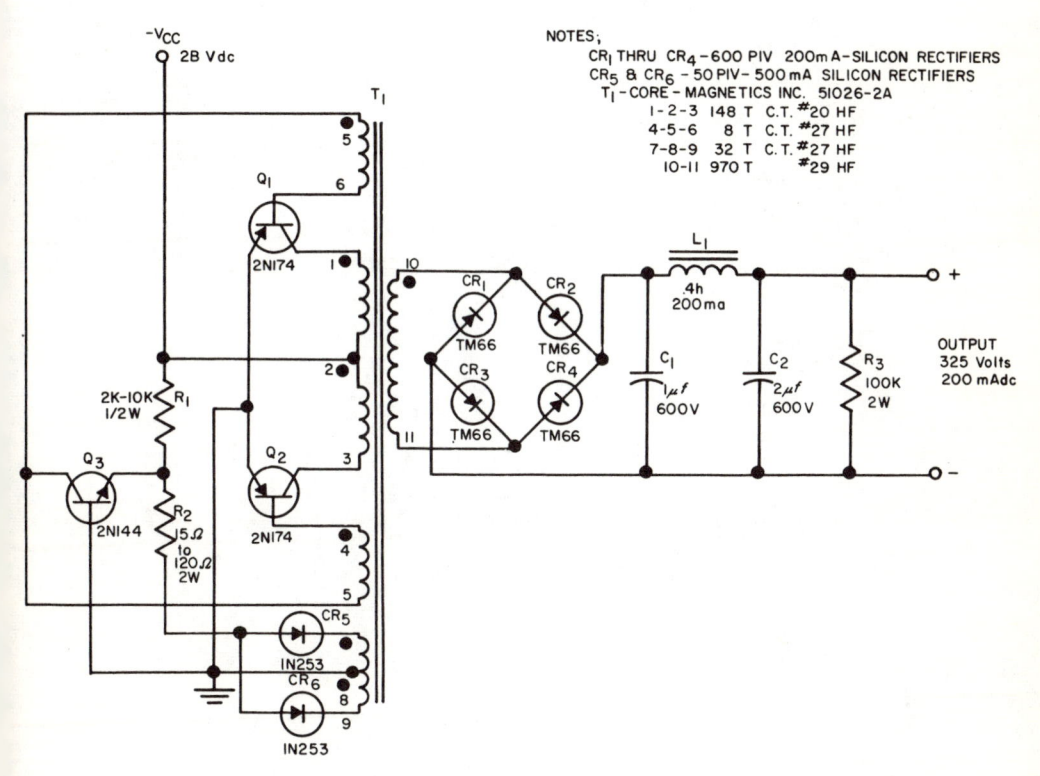

Circuit 9-1 - D.C. to d.c. power converter

Circuit 9-1, D.C. to D.C. Power Converter (cont'd.)

R_2 (15 to 100 Ω) is adjusted until the transistors are just saturated at slightly over maximum current. Under these conditions, when the output is shorted, oscillations will cease, making the converter short circuit proof. The addition of Q_3 makes the converter reverse polarity proof also

CIRCUIT 9-2

MAGNETIC AMPLIFIER REGULATED D.C. TO D.C. CONVERTER

Technical Operations							Contributed by R. MacMillan

    Circuit 9-2 is a magnetic amplifier regulated d.c. to d.c. converter designed to supply power to a missile telemetering system.

    The regulator operates in the following manner. The output of the 180V is sampled through the divider consisting of R_4, R_5, and CR_{16}. A portion of this voltage appears at the base of Q_5. The base of Q_4, the other transistor in the differential amplifier, is held at a fixed voltage by the Zener diodes CR_{11} and CR_{12}. If the output voltage tends to increase because the supply voltage rises, Q_5 conducts more heavily, increasing the current through the magnetic amplifier control winding T_1, reducing the voltage across the primary of the output transformer T_3, to restore the output voltage to its original value.

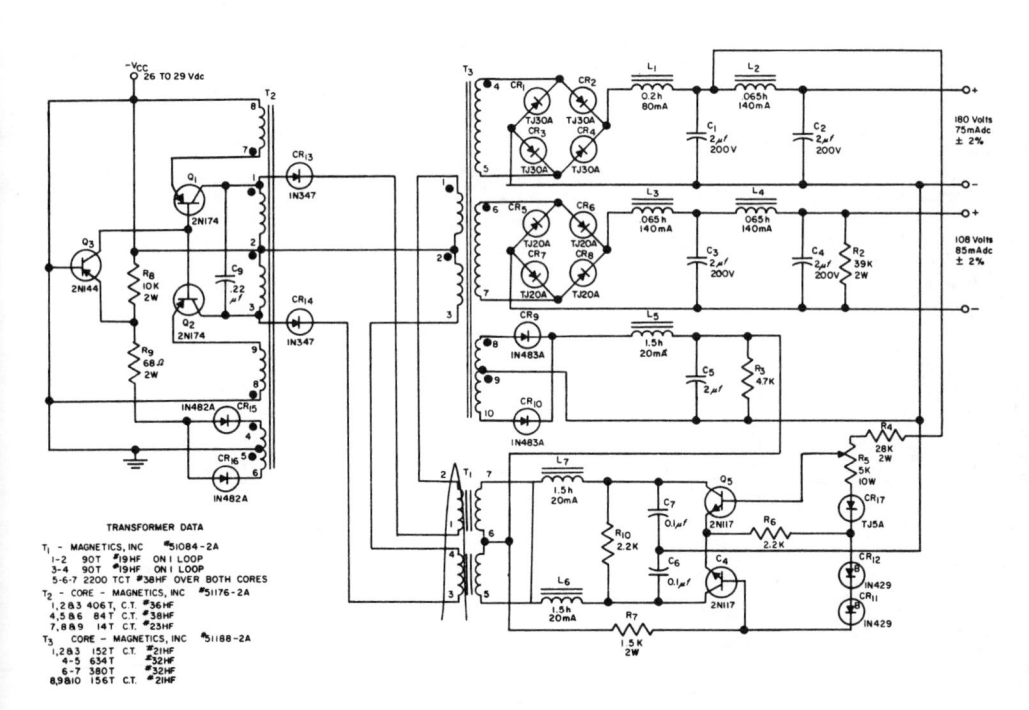

Circuit 9-2 - Magnetic amplifier regulated d.c. to d.c. converter

Circuit 9-2, Magnetic Amplifier Regulated D.C. to D.C. Converter

Since this system tends to keep the a.c. voltage across the primary of the output transformer constant, the second output voltage (108V d.c.) is regulated at the same time.

This circuit is also reverse polarity proof by the use of Q_3 which limits the converter back current when the converter is connected to the wrong polarity.

CIRCUIT 9-3

D.C. TO D.C. CONVERTER USING THE BASE CONTROL METHOD OF OUTPUT REGULATION

Technical Operations   Contributed by R. R. Smyth

Circuit 9-3 is a d.c. to d.c. converter using the base control method of output regulation. The operation of the circuit is as follows.

If the output voltage increases due to an increase in the d.c. input voltage, the voltage across the primary of the output transformer must increase. This in turn increases the drop across R_3 which biases Q_4 in such a fashion as to decrease the base current of Q_3. Since the base return for the oscillator transistors is also the collector current of Q_3, the oscillator transistors are current limited and the original increase in input

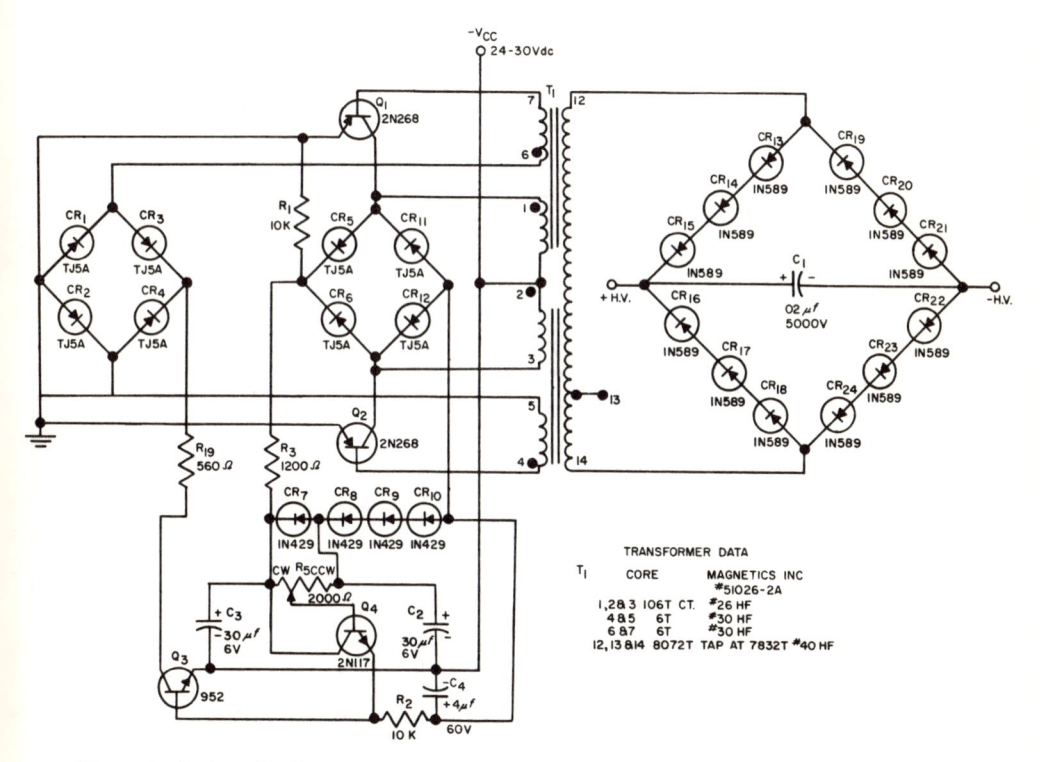

Circuit 9-3 - D.C. to d.c. converter using the base control method of output regulation

Circuit 9-4, D.C. to A.C. Converter with 400 C.P.S. Square
          Wave Output (cont'd.)

diode at minimum input voltage and R_4 is adjusted to give the proper oscillating frequency output.

The frequency will hold to a few percent of its set value over the normal 25 to 32V d.c. input voltage variations.

CIRCUIT 9-5

D.C. TO D.C. CONVERTER, 130 WATTS OUTPUT WITH ZENER DIODE TRANSIENT SURGE PROTECTION

Texas Instruments   Contributed by R. G. McKenna

    Circuit 9-5 is a d.c. to d.c. converter which incorporates Zener diodes as transient voltage protection devices. One major source of difficulty with this type of converter, when protection devices are not

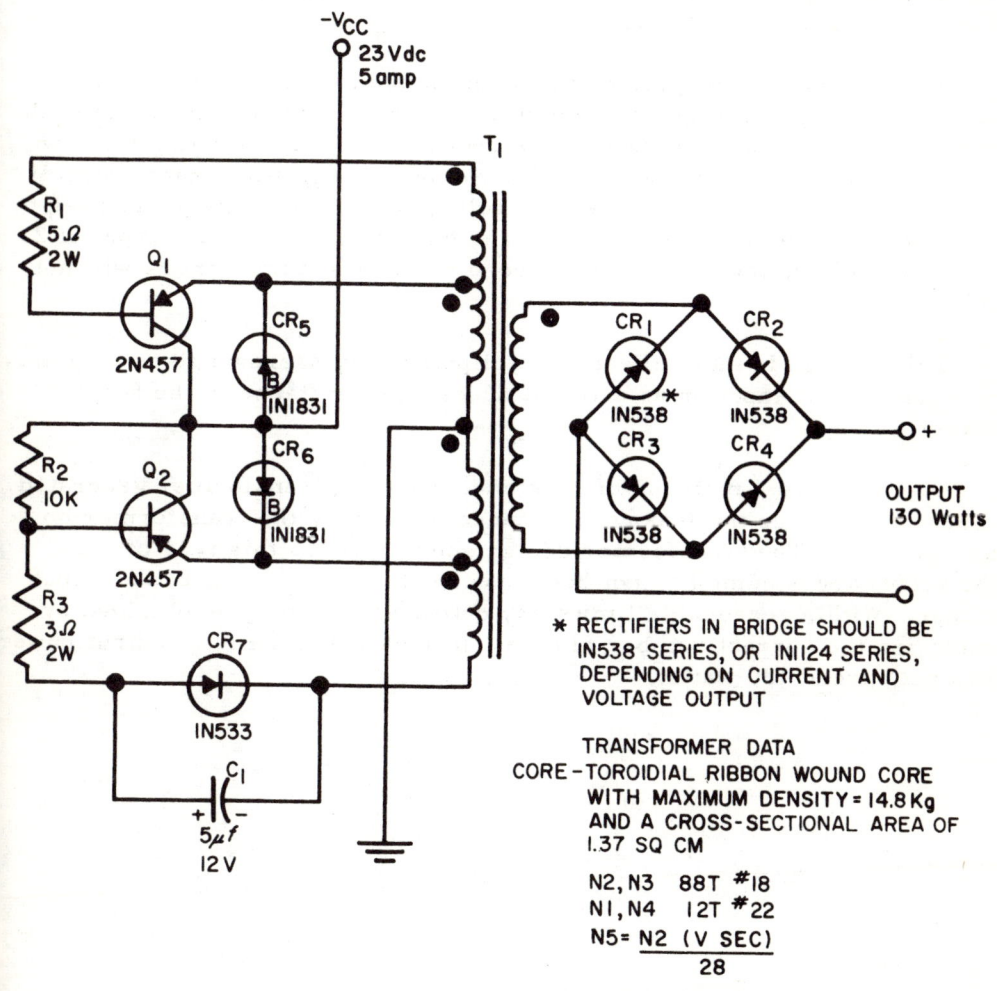

* RECTIFIERS IN BRIDGE SHOULD BE 1N538 SERIES, OR 1N1124 SERIES, DEPENDING ON CURRENT AND VOLTAGE OUTPUT

TRANSFORMER DATA
CORE - TOROIDIAL RIBBON WOUND CORE WITH MAXIMUM DENSITY = 14.8 Kg AND A CROSS-SECTIONAL AREA OF 1.37 SQ CM

N2, N3 88T #18
N1, N4 12T #22
$N5 = \dfrac{N2 \; (V \; SEC)}{28}$

Circuit 9-5 - D.C. to d.c. converter, 130 watts output with Zener diode transient surge protection

Circuit 9-5, D.C. to D.C. Converter, 130 Watts Output with Zener
    Diode Transient Surge Protection (cont'd.)

employed, is failure of the switching transistors due to excessive collector-to-emitter back voltage. In normal operation, the back voltage on the "off" transistor will be at least twice the input voltage. If transient voltages of very short duration appear on the input these must be added to the instantaneous back voltage on the transistor.

Another common source of spiking results from the finite leakage inductance of the output transformer. Either or both of these sources may generate spikes of sufficient magnitude to cause the back voltage rating of the transistor to be exceeded. When this occurs it is almost certain to cause transistor failure.

The present circuit provides protection against this source of transistor failure by shunting Zener diodes from collector-to-emitter of each transistor. Zener diodes shown have breakdown voltage ranging from 56 to 60 volts which is less than the back voltage rating of the transistor. If spikes of greater than the breakdown voltage occur, they will be clipped by the Zener. In the absence of spikes, and operating at the rated input of 28 volts, the back voltage should not exceed 56 volts and the Zeners will not conduct.

This system is most effective in clipping very narrow spikes originating either on the input or as a result of leakage reactance in the transformer.

One of the best methods of reactance reducing transformer generated spikes is to reduce the leakage reactance by winding the transformer on a toroidal core. The transformer design shown utilizes this type of core. The secondary winding has not been specified in terms of an exact output voltage. Within reasonable limits any output voltage may be obtained by employing the correct number of turns, provided that the total output power does not exceed 130 watts.

CIRCUIT 9-6

D.C. TO D.C. CONVERTER, 50 WATTS, 400 C.P.S.

Delco Radio

Circuit 9-6 was selected to represent the simplest possible converter circuit. This particular unit was designed for 6 volts input and is operated in the common collector mode. The use of the common collector circuit has definite advantages for most mobile systems, since the negative input lead is normally grounded. This permits the collectors to be tied right to the chassis thereby eliminating additional thermal drops associated with the insulating mica washers.

```
TRANSFORMER DATA
CORE - H-42 "C" CORE
FEEDBACK NEXT TO CORE
FEEDBACK;   54T C.T. #26
SECONDARY   786T #27
PRIMARY     36T C.T. #15
```

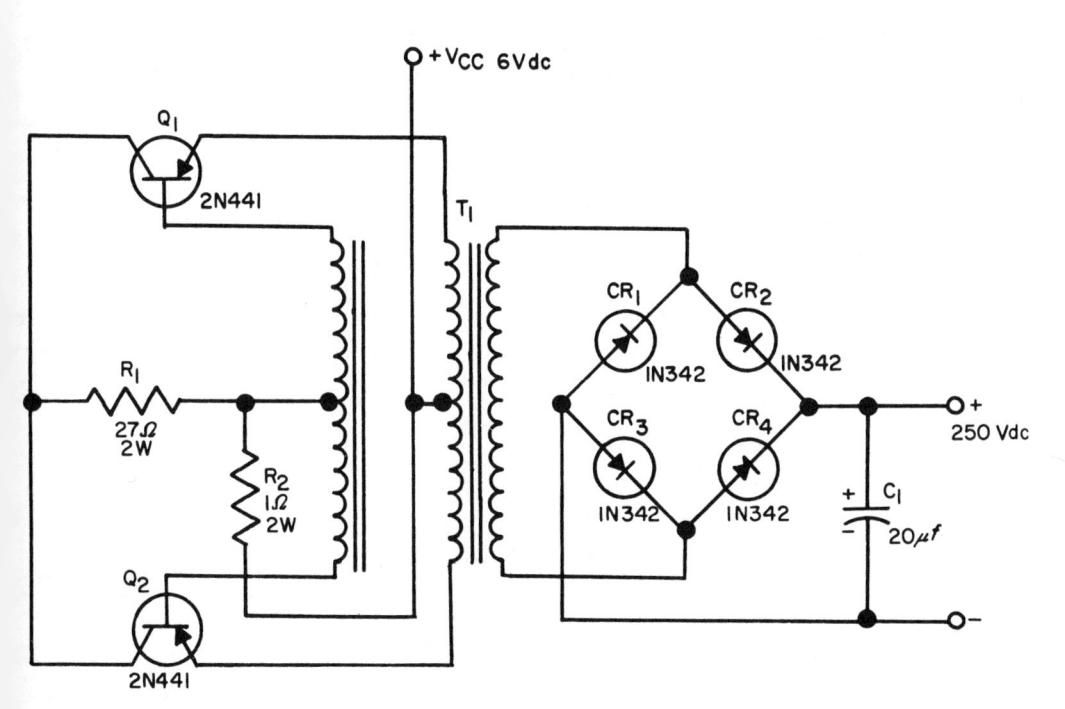

Circuit 9-6 - D.C. to d.c. converter, 35 watts, 400 c.p.s.

Circuit 9-6, D.C. to D.C. Converter, 50 Watts, 400 C.P.S. (cont'd.)

The simplest possible starting has been achieved by simply biasing the center tap of the base winding supply below the positive voltage. The values of the two biasing resistors shown are adequate for a 35 watt load. If the load power is to be increased, the values of these resistors must be decreased accordingly.

The major disadvantage of this simplified circuit lies in the method of obtaining starting bias. As shown, the circuit will not be stable in the event that the load becomes short-circuited. Methods for overcoming this problem are given in the following selected circuits.

The output d.c. voltage is obtained through a full wave bridge rectifier with a capacitor input filter. The diodes should have a peak inverse rating of at least 300, and preferably, 400 volts. The capacitor input filter will provide superior converter operation, although the exact value of the capacitor is not critical. As shown, with a 20 mfd capacitor, the ripple should be considerably less than 1% of the d.c. output. If additional filtering is required, standard L-C sections can be added.

The transformer has been designed on a C core. In constructing this transformer care should be exercised in minimizing the leakage inductance to avoid generating high spikes which could appear across the transistors and cause failure. Any of the standard techniques used to minimize leakage inductance may be employed. In particular, care should be exercised in obtaining a minimum gap when the core is banded.

CIRCUIT 9-7

SERIES CONNECTED D.C. TO D.C. CONVERTER

Delco Radio

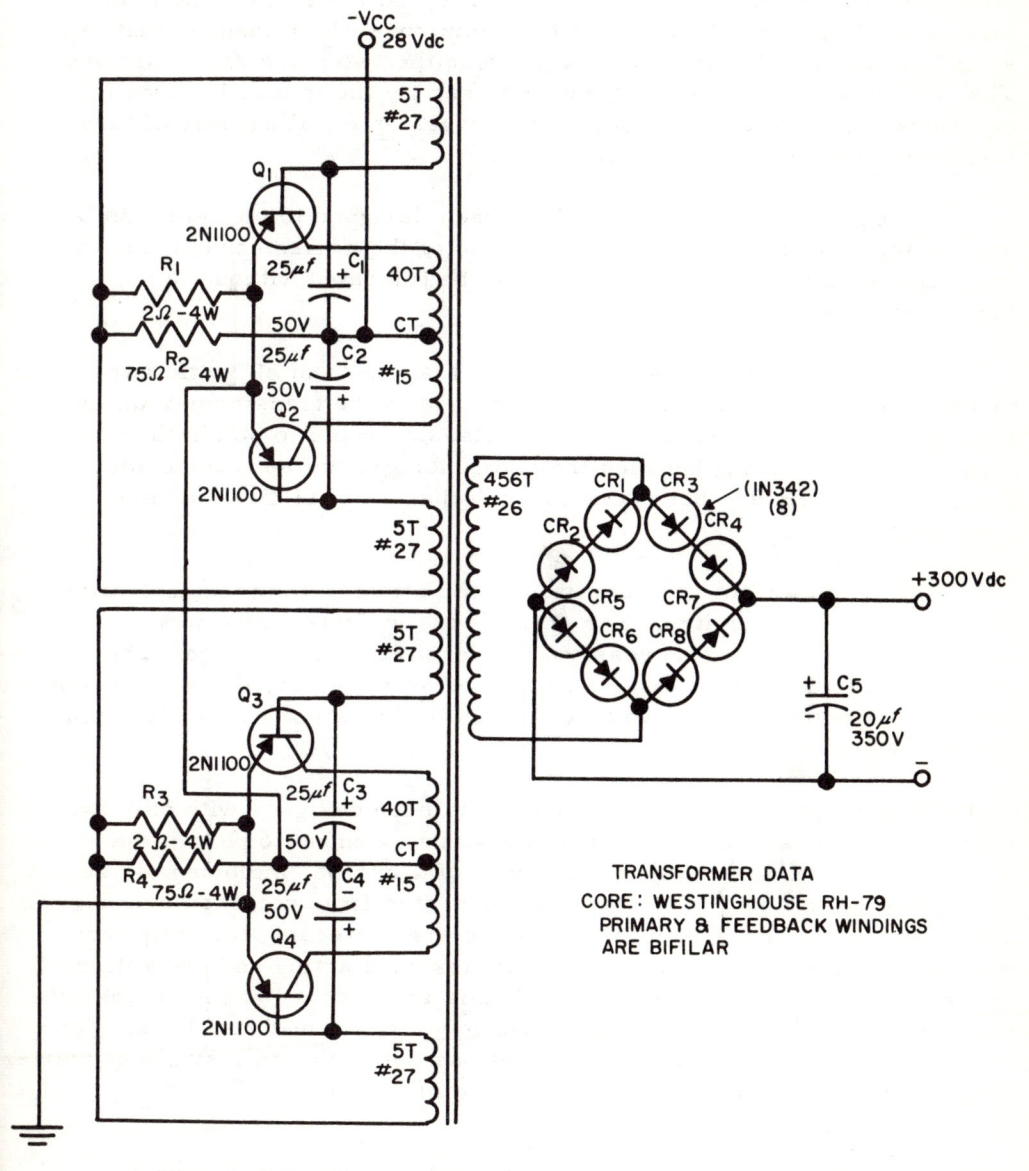

Circuit 9-7 - Series connected d.c. to d.c. converter

Circuit 9-7, Series Connected D.C. to D.C. Converter (cont'd.)

Circuit 9-7 is a series connected d.c. to d.c. converter. This type of circuitry has many important applications where it is desirable to operate from higher d.c. supply voltages than is permitted by the collector breakdown rating of the available transistors, or where reliable protection against input transients is desired. As shown, the circuit consists of two standard common emitter converters connected in series. The transformer windings are all wound on the same core which insures that the supply voltage must divide equally between the two pairs of transistors. The reason is quite simple. In order to satisfy the magnetic circuit equations, equal voltages must appear across an equal number of turns for any winding on a single core.

This particular circuit shown has been designed to operate from a normal input voltage of 28 volts. Because of the series connection, however, transients as high as 80 volts may be tolerated without damage to the transistors.

While the circuit can withstand input transients of 80 volts, care must be exercised to minimize leakage inductance in the transformer which could generate damaging spikes. It is also desirable to obtain the maximum possible coupling between the two windings. In the transformer design shown, this has been done through the use of bi-filar windings on the transformer.

Another important consideration is the effect of the transient voltages on the secondary circuitry. For sustained transients on the input, there will be a corresponding increase in the secondary a.c. voltage. The diodes selected must, therefore, have a peak inverse rating capable of withstanding the maximum expected voltage during the transient. This also applies to the filter condenser.

While the particular circuit shown has been designed with only two series converters, the same process may be extended to permit operation at even higher supply voltages. Circuits of this type employing six series pairs have been built and successfully operated from d.c. inputs as high as 150 volts. As the number of series pairs are increased, the frequency must be reduced. This must be done to assure starting and saturation of the transistors at the same instant. If this is not done, voltage transients will appear across the transistors that have the slowest rise time. Furthermore, the various circuit modifications applicable to a single converter stage may also be employed with this type of circuitry.

PART 10

SMALL-SIGNAL NONLINEAR CIRCUITS

Section 1. Design Philosophy

A. Introduction

The circuits commonly known as modulators, mixers, converters, detectors, frequency multipliers, and frequency dividers will be covered in this part. They have in common the fact that they depend upon some nonlinearity for their action. However, the circuits to be considered are such that, because of one or more filters, the input and output voltage or currents are essentially sinusoidal; they are to be differentiated, therefore, from waveform circuits such as multivibrators, which will be covered elsewhere.

B. Modulators

A modulator will be defined here as a device that varies some property of a high-frequency wave in accordance with the amplitude of a low-frequency wave. Amplitude modulation will be of principal interest, as the application of transistors in frequency-modulation equipment has been relatively slight. When frequency modulation is desired using semiconductor devices, a variable-capacitance diode is normally used to vary some parameters in a tuned circuit. One frequency modulator operating in the broadcast band is included among the selected circuits, however.

It is convenient to consider two types of amplitude modulators, high level and low level. High level modulation is used with a Class C amplifier in which practically the entire collector-supply voltage is switched across the tuned circuit at the amplifier output. The collector-supply voltage is then varied in accordance with the modulating wave. Excellent linearity can be obtained because of the good bottoming characteristics of the transistor, and the efficiency of the transistor Class C amplifier can be very high at medium or low frequencies. Unfortunately, high-frequency transistors with power capabilities greater than a few milliwatts are not available, and therefore there has been little application of this type of modulation. In low-level modulation, a bias is varied rather than the collector-supply voltage. If the carrier is small in comparison to the emitter bias, the modulation process can be studied in terms of the variation of transistor gain with emitter current. Linear modulation is possible if the correct operating point is chosen - in typical small-signal transistors the emitter current for minimum envelope distortion may be in the range from 0.1 to several milliamperes. To estimate the gain of the modulator at the chosen operating point, it should be recalled that the transconductance of the transistor is very nearly 40 I_E, where I_E is the average emitter current.

C. Mixers and Converters

Mixers and converters are normally used to translate a high signal frequency to a lower or intermediate frequency where more efficient amplification or improved selectivity is possible. If the local heterodyning frequency is generated within the mixer, the device is commonly called a converter. Although the use of a single transistor as both mixer and oscillator is economical, and is therefore common in broadcast-type receivers, there are disadvantages to combining both functions in a single semiconductor. Thus, a converter is limited in the maximum amount of signal that can be applied because full AGC cannot be used. If it were applied, the oscillator would cut out somewhere in the AGC range. Also, at the higher frequencies, the variation of local-oscillator frequency with AVC would cause difficulty, particularly in narrow-band systems.

In general, the design of a mixer will be much the same as that of an RF amplifier for use at the same frequency. Thus, the input impedance of the mixer will be essentially the same as that of the transistor (h_{11_e}) at the quiescent bias point. Since the collector circuit is tuned to a frequency different from the input, the effects of the normal feedback parameters need not be considered. The conversion gain will be roughly 6 to 8 db less than the gain as an amplifier. As an approximation, if the emitter current is just 100% modulated by the local-oscillator signal, the conversion conductance can be assumed to be 10 I_E. The IF output impedance will be very nearly that of the transistor as an amplifier with the input shorted. The noise figure will normally be 6 to 10 db higher than would be obtained with the transistor as an amplifier. However, at frequencies so high that the performance as an amplifier is poor, there will be little difference in noise figure between the same transistor used as an amplifier or a mixer.

The usual value of local-oscillator injection voltage for a germanium transistor mixer is in the range from 30 to 100 millivolts. Fortunately there is a wide region in which the rate of change of gain with injection is small. A poor noise figure may be obtained with an excessively high injection voltage and, also, a poor noise figure may be obtained if the average emitter current is chosen excessively high.

It is normally undesirable to use a transistor as a mixer at a frequency higher than that at which it might normally be used as an amplifier. Although it might be convenient to picture the transistor mixers as a diode mixer and intermediate frequency amplifier combined in a single device, the variation of amplifier transconductance with instantaneous emitter current gives a truer picture of its operation. In addition, it should be stated that hole-storage effects in the emitter-base junction will present efficient rectification at frequencies much above f_{ab} anyway.

In a transistor mixer, it is convenient to apply the local-oscillator voltage between emitter and ground, while the high-frequency signal is applied between base and ground. To prevent degeneration at the intermediate frequency, the IF emitter-base impedance should be kept as low as possible.

D. Detectors

A detector is used to extract the desired signal from a modulated wave. To do so it is necessary to choose a suitable nonlinearity and combine it with a satisfactory filter so that the desired output frequencies are obtained. Unfortunately a simple square-law nonlinearity is unsatisfactory because harmonics of the modulation frequency are produced. Normally a diode detector is run at a sufficiently high level to obtain essentially linear operation, so that this type of distortion is minimized. The same considerations apply to a transistor detector which can be considered as a diode detector followed by a transistor amplifier. To obtain good rectification efficiency, the diode must be operated at the most nonlinear portion of its characteristic. Unfortunately this corresponds to a low value of emitter current, which may be a poor operating point for the amplifying portion of the transistor. Thus, detector design is a compromise. With germanium transistor detectors, the optimum emitter current is of the order of 50 microamperes, and the gain, defined as the ratio of audio output power to available sideband power, may be 18 or 20 db with a 20,000 ohm audio load. The input resistance at the base averaged over a complete cycle will be higher than the no-signal value. This is fortunate because it tends to compensate for the square-law distortion produced in the rectifying junction. Thus, the transistor detector will change from square-law to linear operation at a somewhat lower power level than the diode detector.

One of the principal reasons for using the transistor as a detector is that d.c. control power is available for an AGC system. This is helpful because the controlled stages of a transistor receiver require AVC power, principally because of the low-impedance networks which are necessary for operation point stabilization.

If the transistor cutoff frequency is not high in relation to the signal frequency, the emitter-base junction will suffer from hole-storage effects and the rectification efficiency will be poor. If the signal frequency must be high, better results will be obtained with a separate diode detector. This will also then permit the choice of a more favorable operating point for the transistor amplifier. Silicon detectors make poor low-level detectors because, as amplifiers, silicon transistors have low gain at low emitter currents. This effect is much more pronounced than with germanium units. It would seem highly desirable to use a separate silicon-diode detector in equipment requiring silicon semiconductors. The germanium detector is not without its limitations, however. The 50 microampere operating point indicated above may not be too far removed from the collector cutoff current for some types, particularly at the higher ambient temperatures, and therefore the zero-signal operating condition may have to be carefully stabilized.

E. Frequency Multipliers

The frequency multiplier normally consists of a Class C amplifier with its output tuned to a multiple of the input frequency. The considerations involved in the design of Class C frequency multipliers are much the same as those in connection with Class C amplifiers, with the exception that, since the harmonic content of the collector-current pulse is very sensitive to angle of collector-current flow, the correct angle must be chosen for the desired frequency ratio.

The optimum angle, expressed in degrees, is approximately 180 divided by the order of the harmonic; thus, when doubling, 90 degrees should be used. The collector-circuit efficiency will decrease at the higher ratios and will be given approximately by 100 divided by the order of the harmonic.

Unfortunately the current gain of the transistor will decrease with frequency and current, rendering it difficult or impossible to obtain the desired conduction angles at the higher frequencies. Hence the collector-current pulses may be broadened because of the random times taken for holes to diffuse through the base region, and the tops of the pulses may be rounded because of the decrease of α_{fe} with emitter current. These effects are difficult to evaluate and therefore, as a practical matter, an experimental approach is usually necessary in order to obtain the best operating conditions. Assuming that the base of the transistor is driven, with a time constant such that self-bias is developed, the conduction angle can be varied by changing the amplitude of the drive and by changing the time constant. The load impedance can be varied at the collector circuit to effect the optimum transfer of power. If the use of a chain of frequency multipliers is contemplated, the design of each stage must be such that each stage must be capable of driving the following one. The use of push-pull stages is advantageous with odd frequency ratios, while the push-push connection is helpful with even ratios. Such use of two transistors per stage will double the power output and will also simplify the filtering requirements since the number of collector-current pulses per unit time is doubled.

F. Frequency Dividers

The frequency dividers of the type under consideration in this section will produce a sinusoidal voltage or current at a submultiple of a sinusoidal input frequency. Although waveform devices such as multivibrators or counters could be used in the intermediate stages of such a device, with filtering to produce a sinusoidal output, this discussion will be limited to circuitry that is essentially sinusoidal in nature. It must be recognized, however, that some nonlinearity must be present to effect division. As a practical matter, a synchronized oscillator is usually involved in the generation of subharmonics. In one method, a voltage of frequency nf is injected into an oscillator operating at a frequency f. It is assumed that a voltage at $(n-1)f$ is generated within the oscillator so that a beat of frequency f can be produced between nf and $(n-1)f$. Such a simple system is usually termed a locked oscillator and is frequently useful when the frequency ratio n is small. A more complicated system can be devised in which the functions of frequency multiplication, beating or phase comparison, and oscillator phase control or locking, are carried out in separate portions of the circuit. Although such a device is capable of good performance at large frequency ratios, it is complicated, and will not be considered here.

A second type of comparatively simple and very useful frequency divider is the so-called regenerative or Miller type, which does not contain a free-running oscillator as such, but does contain an oscillating loop which is completed by the input frequency f, which is applied to a mixer. The output of the mixer is tuned to a frequency f/n, which then drives a frequency multiplier whose output is at $(n-1/n)f$. The mixer is driven by $(n-1/n)f$ and by the original input f, and therefore will have an output at f/n. The system will oscillate if the loop gain is greater than unity, and cannot produce a spurious output frequency other than harmonics of the output frequency, if the filtering is such that the loop gain is less than unity at all other frequencies. The use of a balanced modulator for the mixer will also help to prevent the generation of spurious output frequencies, particularly if the frequency ratio is only 2, where direct feedback at $f/2$ could occur to produce self oscillation at $f/2$. Some dividers of this type are not self-starting, particularly when using divide-type mixers and large-ratio frequency multiplication ratios. With these, special starting devices must be employed.

BIBLIOGRAPHY

Dewitt, David, and Rossoff, Arthur L., "Transistor Electronics," McGraw Hill, New York, 1957.

Shea, Richard F., "Transistor Circuit Engineering," John Wiley & Sons, Inc., New York, 1957.

Wolfendale, E., "The Junction Transistor and Its Applications," The Macmillan Co., New York, 1958.

Zawels, J., "The Transistor as a Mixer," Proc. I.R.E., March 1954.

CIRCUIT 10-1

LOCKED-OSCILLATOR FREQUENCY DIVIDER

Sulzer Laboratories, Inc.   Contributed by Peter G. Sulzer

The simple locked-oscillator frequency divider is capable of providing reliable frequency division at small frequency ratios. It consists of a grounded-base-oscillator, with the synchronizing voltage applied between base and ground. Successful operation of this oscillator as a frequency divider requires that harmonics be produced at frequencies near the input frequency. It has been found experimentally that f_{ab} should be several times the input frequency if reliable locking is to be accomplished.

The circuit shown employs d.c. stabilization to assure reliable starting. The impedance of the synchronizing-voltage source should not greatly exceed 1000 ohms. The optimum input is 0.7 volts r.m.s.; however, variation from 0.5 to 1.2 volts will have little effect. The output with a 20 volt supply is approximately 2 volts r.m.s. to a 1000 ohm load. Since the output exceeds the drive requirement, these dividers can readily be cascaded.

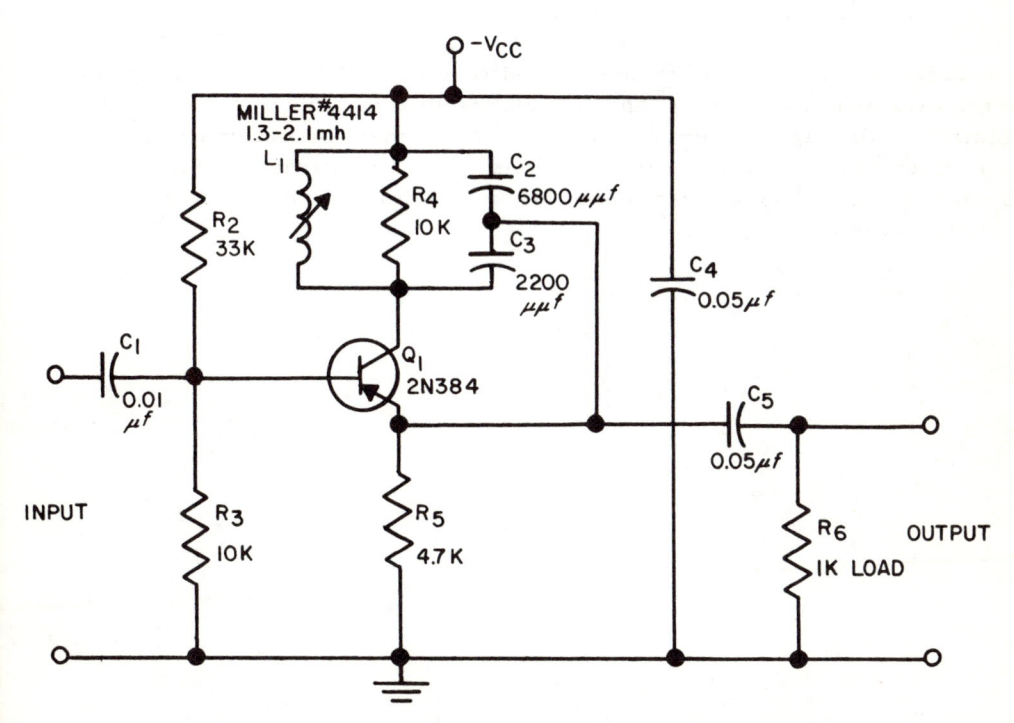

Circuit 10-1 - Locked oscillator frequency divider

Circuit 10-1, Locked-Oscillator Frequency Divider (cont'd.)

If the dividers are to be cascaded, it is suggested that each stage be driven from a tap on the emitter resistor of the preceding stage.

Performance of the circuit at various frequency ratios is given in the table below. In making these tests, the output frequency was kept constant at 100 kc, while the input frequency was increased in steps of 100 kc from 100 kc to 700 kc. The operating bandwidth was measured at a 20 volt supply. The supply-voltage range over which locking was obtained was measured with the input frequency properly centered in the operating band of the oscillator. For all measurements, the input was constant at 0.7 volt r.m.s.

Frequency Ratio	Bandwidth %	Supply-Voltage Range
1	±3.5	1 - 40
1/2	±7.5	2 - 40
1/3	±5	3 - 40
1/4	±6.5	3 - 32
1/5	±1.1	9 - 40
1/6	±1.8	5 - 33
1/7	±1.5	14 - 25

It can be seen that at frequency ratios as great as 1/4 the operation of the circuit is not critical and therefore very reliable results should be obtained. Although systematic tests were not made to determine the effects of transistor parameter variations, it was determined that the performance of the divider was much the same with any one of several transistors of the type shown.

CIRCUIT 10-2

LOW-LEVEL MODULATOR

Transistor Applications, Inc.

Circuit 10-2 is a simple, low-level modulator suitable for use in the early stages of a radio-frequency transmitter. The circuit is a common-emitter amplifier, with a fixed base bias furnished by a voltage divider, R_1 and R_2 and the emitter current determined by an emitter bias resistor R_4. Modulation is accomplished by applying the audio signal to the emitter circuit, which varies the instantaneous emitter-base potential, thereby modulating the transconductance of the transistor.

The optimum carrier level at the base of the transistor is 0.1 volt r.m.s. To produce 100% modulation, 0.12 volt r.m.s. of audio-frequency voltage must be applied to the emitter. The modulator will supply 3 volts r.m.s. of carrier to a 10,000-ohm load. The envelope distortion is approximately 5% at 80% modulation and 20% at 100% modulation.

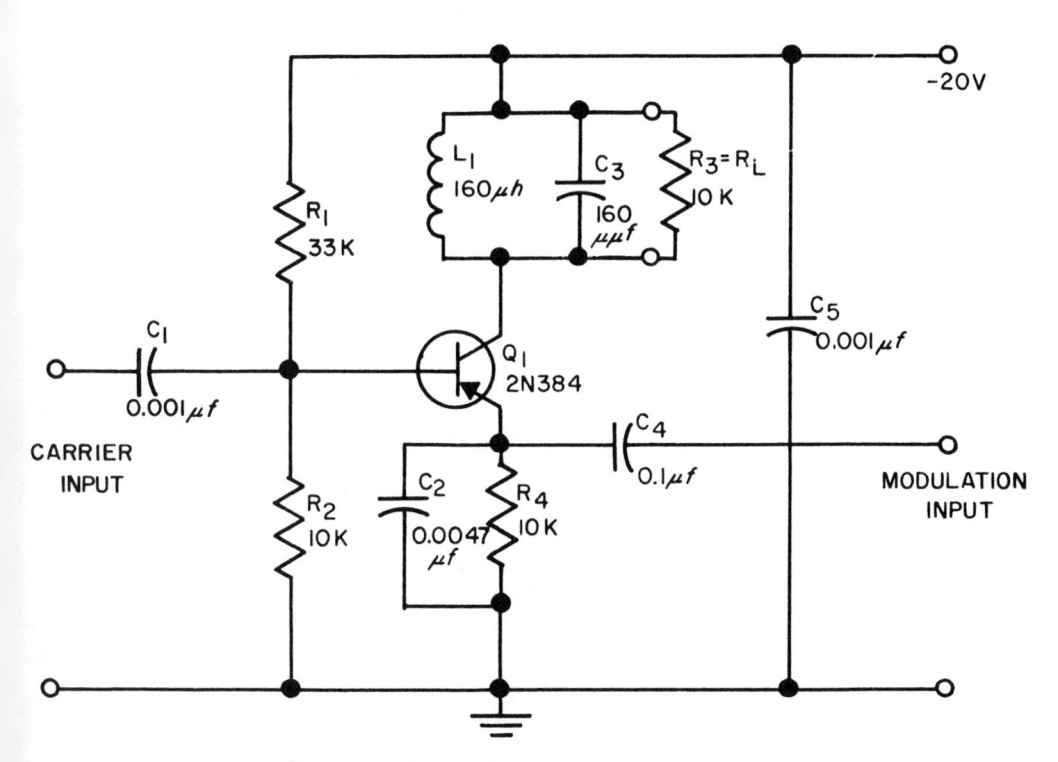

Circuit 10-2 - Low-level modulation

Circuit 10-2, Low-Level Modulator (cont'd.)

With the type 2N384 transistor the radio frequency response of the modulator is essentially flat up to 30 mc. The audio-frequency response is determined principally by the loading of the audio-frequency source by the 0.0047 μf emitter-bypass capacitor. This emitter-bypass capacitor is sufficiently large for carrier frequencies down to one megacycle. If lower carrier frequencies are to be used, more capacitance should be used. If, however, this imposes too great a burden upon the audio-frequency driver, the modulator can be made push-pull as far as the radio-frequency inputs and outputs are concerned, while the emitters are driven in parallel with the modulating voltage. If balanced operation is attained, the bypass capacitor can be omitted, with a resulting improvement in high-frequency response.

CIRCUIT 10-3

FREQUENCY MODULATOR AND OSCILLATOR

General Motors Corp.   Contributed by Paul W. Wood

The frequency modulator employs the Miller effect to vary the equivalent capacitance across a tuned circuit. In the circuit diagram a complete oscillator-modulator combination is shown. Transistor Q_2 operates as a Colpitts oscillator at a frequency of approximately 1 mc. Feedback is from the emitter through a capacitively-tapped tuned circuit to the base. The operating point is determined by the conventional base voltage divider R_6, R_7 and emitter bias resistor, R_8.

The modulator, Q_1, has an input capacitance at the base which is approximately equal to the emitter-junction capacitance plus the voltage gain multiplied by the collector capacitance. If the gain is varied by modulating the operating point of the transistor, the input capacitance will change, and frequency modulation will be produced. A frequency shift of ±20 kc is obtained at 772 kc with fair linearity. Increased deviation could probably be obtained by adding extra capacitance between the base and collector of Q_1.

(This circuit is reproduced with permission of Electronics Magazine. "Transistorized F-M Oscillator" By Paul W. Wood, Jan. 30, 1959.)

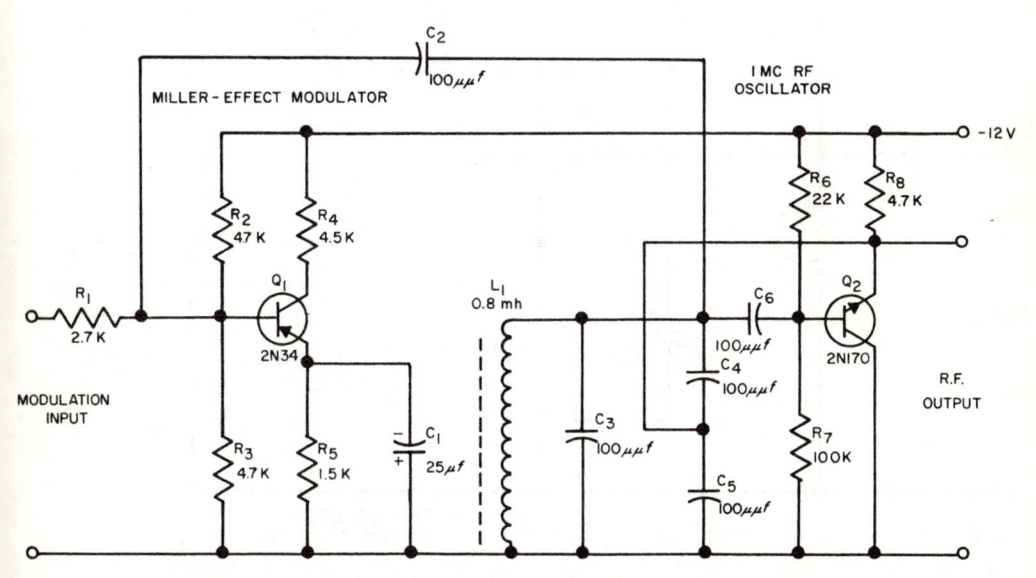

Circuit 10-3 - Frequency modulation and oscillator

CIRCUIT 10-4

FREQUENCY MULTIPLIER

Sulzer Laboratories, Inc. Contributed by Peter G. Sulzer

The frequency multiplier is a simple, low-level Class C amplifier employing the type of d.c. stabilization that is normally used with Class A stages. The zero-signal operating point, 15 volts at 3 mAdc, is chosen well below the maximum collector dissipation of the transistor. The collector voltage is made as high as possible without exceeding the breakdown voltage of the transistor on negative collector swings. The type 2N384 drift transistor was selected for this application because it combines good high-frequency performance with reasonably high collector voltage and dissipation ratings.

In order to determine the capabilities of the multiplier, it was tested with a constant input frequency of 1 mc from a signal generator, which was adjusted to produce the optimum drive amplitude for the particular frequency ratio under test. The load impedance was adjusted by the substitution of various small, high-frequency type carbon resistors until the

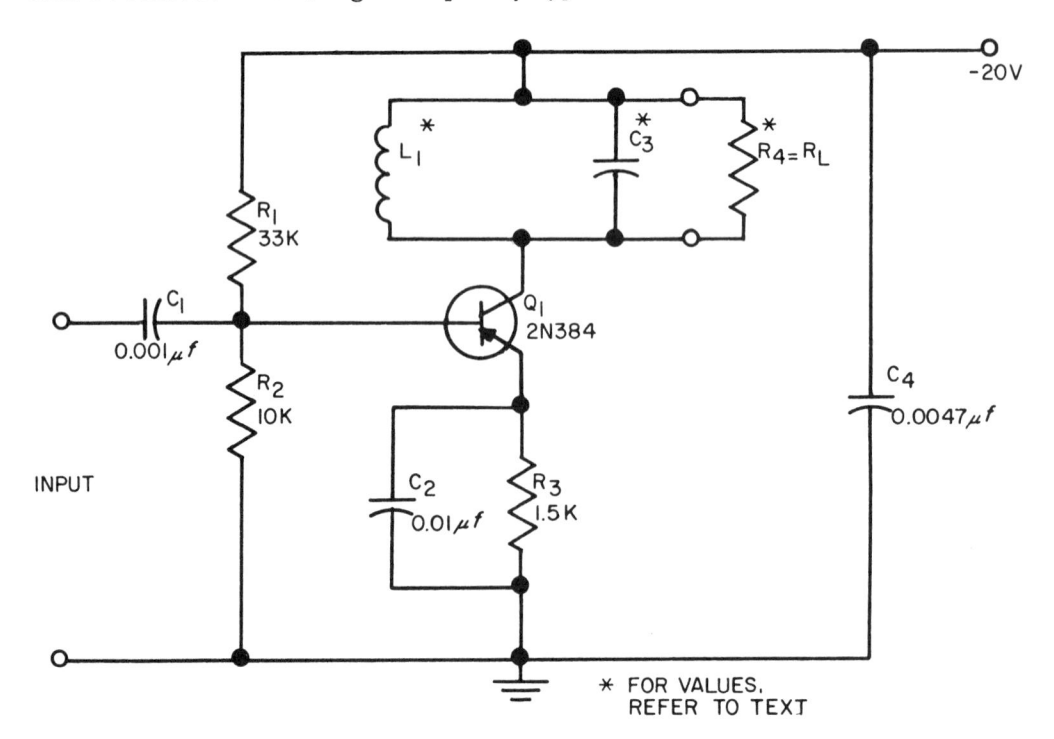

Circuit 10-4 - Frequency multiplier

Circuit 10-4, Frequency Multiplier (cont'd.)

maximum power output was obtained. The tuned circuit LC was kept at a medium value of impedance ($X_L = X_C$ = 300 ohms) so that a loaded Q of 30 or more could be obtained at the higher harmonics. In the following table, f_o is the output frequency, R_L is the load impedance, P_o the power output, N the efficiency, and V_i the r.m.s. input voltage.

f_o mc	R_L Ω	P_o mW	N%	V_i volts
1	3.3K	37	82	0.4
2	6.8K	21	47	1.0
3	10K	14	31	1.1
4	10K	10	22	1.2
5	10K	6.4	14	1.3
6	10K	3.6	8	1.4
7	10K	2.5	5.5	2.0

Straight-through operation is included for purposes of comparison. It will be noted that the efficiencies are reasonably close to the expected values (N = 100/n) at the second, third and fourth harmonics, but become much lower at the higher harmonics. This is probably because of the fact that the minimum collector-current pulse duration that could be obtained was 1/8 microsecond, which corresponds to 1/2 cycle at an output frequency of 4 mc. At higher output frequencies this pulse will bridge an excessive portion of the output-voltage cycle, and will actually extract energy from the tuned circuit during a portion of the cycle.

CIRCUIT 10-5

AUTODYNE CONVERTER

Radio Corporation of America

The autodyne converter is designed for efficient broadcast-band operation with a single transistor. The incoming radio-frequency signal is applied between base and ground, and the intermediate-frequency output is taken at the collector. Feedback for the local-oscillator portion of the converter occurs from collector to emitter through a tuned transformer. Such operation gives a conversion gain 2 or 3 db lower than would be

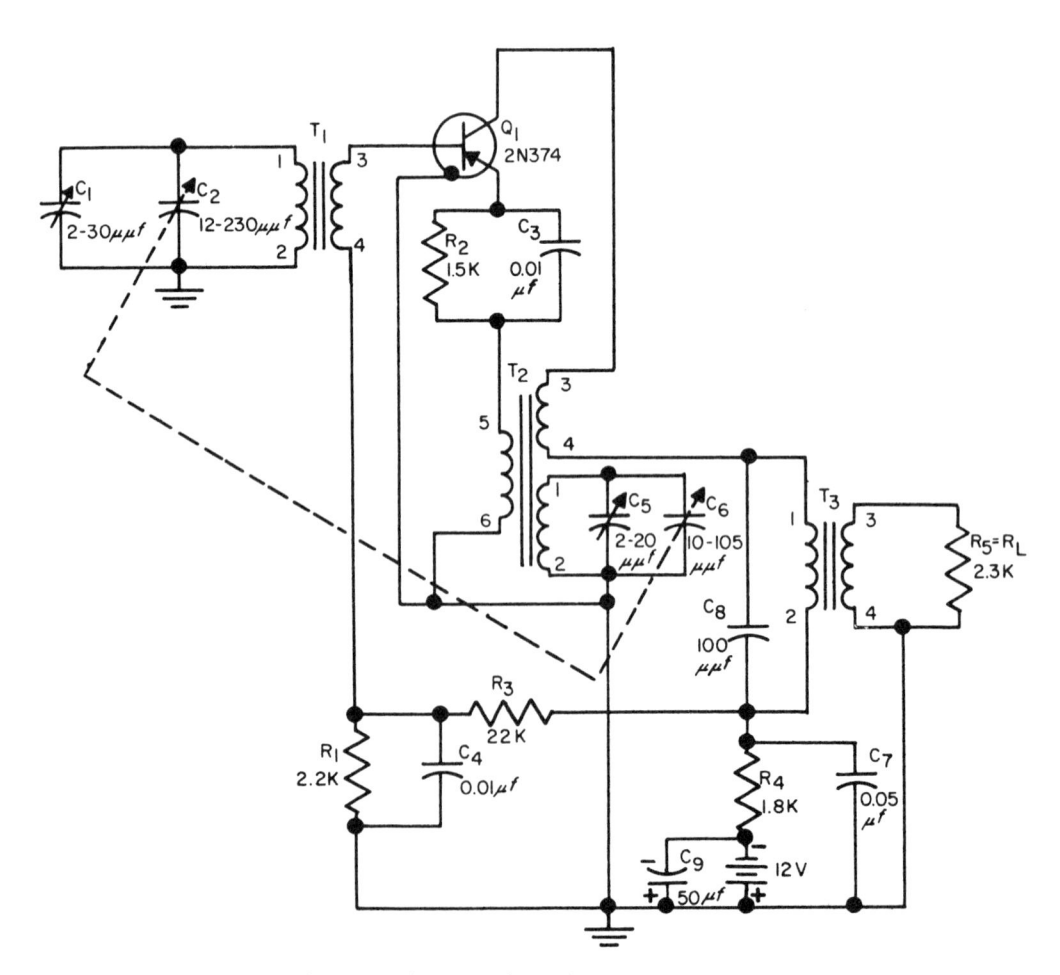

Circuit 10-5 - Autodyne converter

Circuit 10-5, Autodyne Converter (cont'd.)

obtained with the same transistor as a separately excited converter but, in most applications, this is more than offset by the saving in power and components.

In the common-emitter circuit shown, a conversion power gain of 40 db is obtained at one megacycle. The input impedance at this frequency is 2600 ohms, while the output impedance is one megohm. The input impedance varies from 3200 ohms at 500 kc to 1250 ohms at two megacycles. It should be noted that the stated conversion gain is produced with the input circuit matched, but with a great mismatch at the output; thus the transistor worked into a load of 180,000 ohms, only a part of which represents the transformed input impedance of the following transistor intermediate-frequency amplifier.

One of the greatest difficulties encountered with transistor converters is a tendency toward the production of relaxation-type oscillations, particularly over portions of the tuning range and when transistors of the same type are interchanged in a given circuit. This circuit has been designed with relatively low bias impedances and short time constants to decrease this difficulty. In addition, the transistor has carefully controlled characteristics to assure interchangeability. It is seen, however, that AGC is not applied to this converter, and that the operating point is stabilized by emitter-bias resistor R_2.

	*T_1	**T_2	T_3
Primary Tuned Resistance (ohms)	635,000	---	239,000
Primary Input Resistance (ohms)			
With Secondary Terminated	317,000	---	179,000
Secondary terminating resistance (ohms)	2,300	---	2,300
Turns Ratio:			
Terminals 1 and 2 to terminals 3 and 4	---	---	17.7
Core:			
Unloaded Q with transformer mounted on chassis	280 at 1 mc	70 at 1455 kc	72 at 455 kc
Loaded Q with transformer mounted on chassis	140 at 1 mc	---	35 at 455 kc
Primary Inductance (μh)	350	210	---
Material	Ferrite	Ferrite	Powered Iron

*The primary has 98 turns wound on a ferrite rod about 8" long and 11/32" in diameter. The primary winding should be centered on the rod with spacing between turns equal to the thickness of the wire. The secondary has 9 turns wound with no spacing between turns (close wound) on the ground end of the rod. Use #7/41 Litz Wire.

**The primary (terminals 1 and 2) has 120 turns wound with #5/21 Litz Wire. The secondary between terminals 3 and 4 has 4 turns of #34 SSE Wire. The secondary between terminals 5 and 6 has 3 turns of #34 SSE Wire. All windings are bifilar wound. Use a threaded resinite coil form with a length of 3/16" and an outside diameter of 0.235".

Fig. 10-5.1 - Transformer design information

CIRCUIT 10-6

VHF MIXER WITH RF AMPLIFIER-OSCILLATOR

Texas Instruments, Inc.   Contributed by H. F. Cooke

Circuit 10-6 was chosen because it is an excellent example of the application of a germanium tetrode transistor in the VHF range. The gain of the mixer ranges from 4 to 10 db at 215 mc with various type 3N25 transistors. The average noise figure is 10 db at channel 2 and 12 db at channel 13. It is interesting that the noise figure of this type transistor is approximately the same as a mixer, or as an amplifier, over this frequency range. At lower frequencies improved noise figure and gain would be obtained, and the operation of the transistor would be superior as an amplifier. It is also interesting that the addition of the 3N25 as an RF amplifier at VHF TV frequencies will actually degrade the noise performance of the system, but as a practical matter, is worthwhile to obtain increased selectivity and decreased oscillator radiation. The increased gain is more

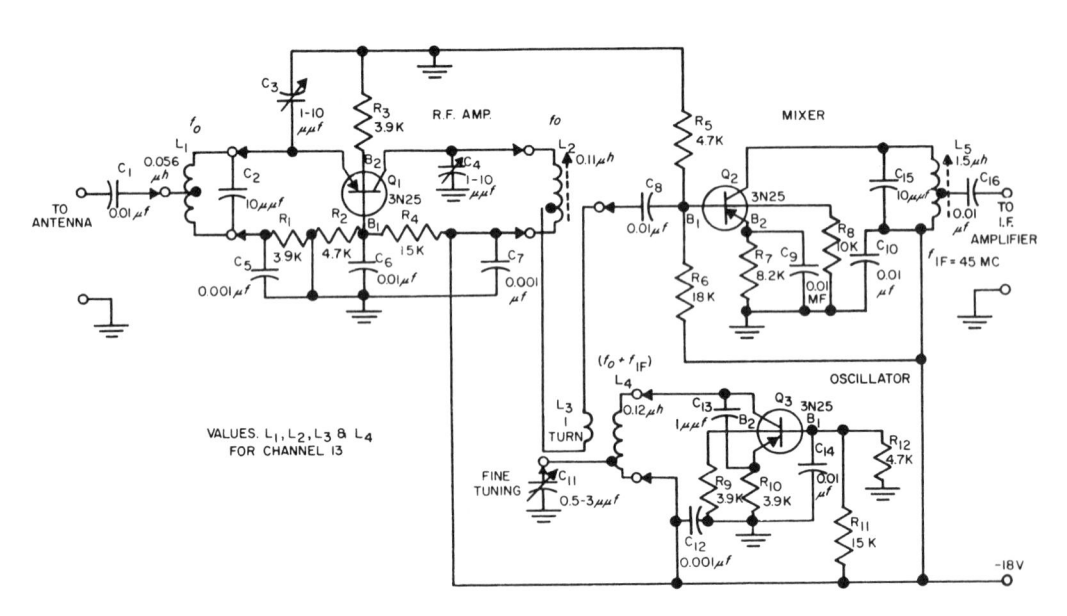

Circuit 10-6 - VHF mixer with RF amplifier and oscillator

Circuit 10-6, VHF Mixer with RF Amplifier-Oscillator (cont'd.)

or less incidental, because it could be obtained more economically at intermediate frequencies.

The mixer is shown link-coupled to the oscillator because this method of coupling was easy to apply and also because injection of the local-oscillator power at the base required slightly less power than at the emitter. The conversion gain is flat over the oscillator-voltage range from 0.07 to 0.1 volt and, as the input impedance is about 125 ohms, the required local-oscillator power is of the order of 40 microwatts. The operating point for the mixer, 0.35 mAdc emitter current and 0.12 mAdc base 2 current, is determined by the conventional type of resistive network. That the operating point is not critical, however, is shown by the fact that the conversion gain decreases only 1 db as the power supply is cut from 15 to 9 volts.

The circuit of the local oscillator and RF amplifier are included because they logically belong with the mixer. The oscillator is of the simple tuned-collector type, with feedback from collector to emitter through the transistor internal impedance plus a small external capacitor. The oscillator will furnish approximately 70 microwatts at 257 mc with a 15 volt supply.

L_1 - Low Q; Antenna may be connected to top of coil with slight loss in gain.

L_2 - Q should exceed 200. Tap point determines tuner BW. Q of C_4 also should exceed 200 at 200 mc.

L_3 - Coupling to L_4 determines oscillator injection. Tight coupling on high channels; very loose coupling on low channels.

L_4 - Q should exceed 200. Extreme care should be exercised with this circuit. Keep losses low. Lossy insulation and fine tuning capacity can prevent oscillation on high channels. Q's should exceed 200 at 250 mc.

Transistor Bias Conditions	I_E	I_{B_2}
Q_1, Q_3	0.8 mAdc	0.20 ma
Q_2	0.35 mAdc	0.12 ma

Fig. 10-6.1 - Transformer and transistor information

The circuit shown in Figure 10-6.2 is included to show the present trend in the design of TV Tuners utilizing the mesa transistor.

This trend is due to the mesa transistor having a better noise figure and gain as well as being a more economical device. Its superiority as a mixer can be attributed to its very low transition capacitance. The latter

Circuit 10-6, VHF Mixer with RF Amplifier-Oscillator (cont'd.)

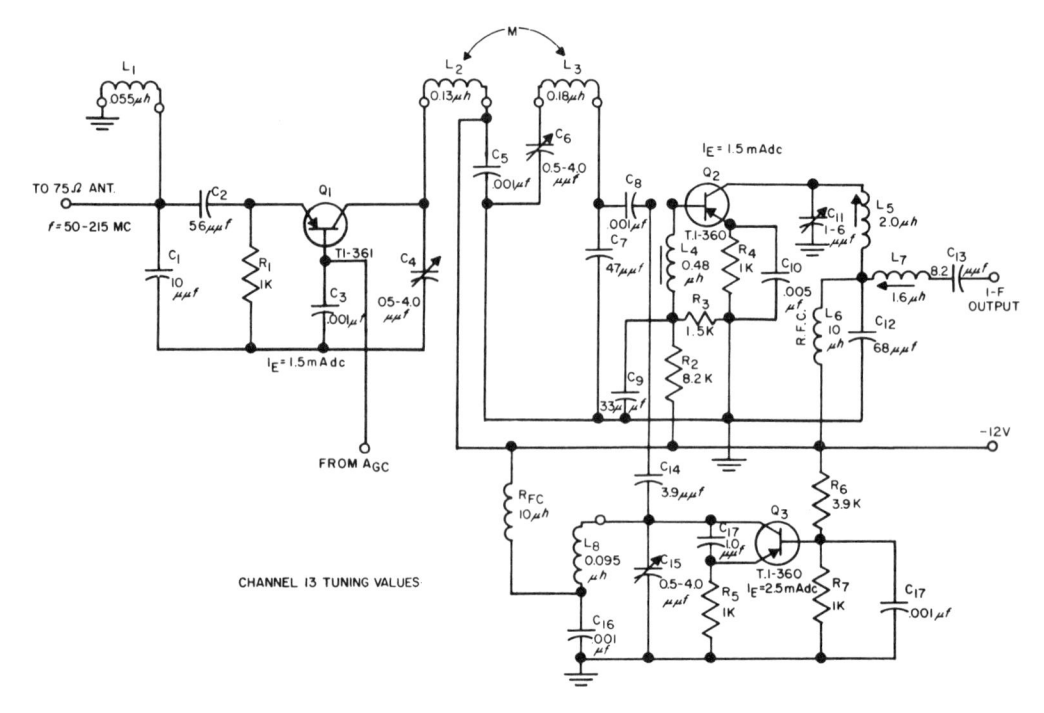

Fig. 10-6.2 - TV tuner with mesa transistors

in combination with $r_b{}'$ form a low pass network, which for a mesa unit has a 3 db point of several thousand megacycles.

On Channel 13, the gain of the tuner was between 20 and 27 db with a noise figure of 7 db, while on Channel 2 the gain of the tuner was between 30 and 40 db with a noise figure of 5 db.

CIRCUIT 10-7

REGENERATIVE FREQUENCY DIVIDER

Sulzer Laboratories, Inc.   Contributed by Peter G. Sulzer

The divider is driven at a frequency of one megacycle and produces an output at 100 kc. The circuit is stable and reliable, and, if properly adjusted, will not oscillate in the absence of an input. The phase stability is good and the operating bandwidth is such that excessively close tolerances in the tuned circuits need not be maintained.

Transistor Q_1 is a mixer that receives an input of 0.7 volt r.m.s. at one megacycle from an external source. The output of the mixer is at 100 kc multiplied X3 twice in Q_2 and Q_3 to produce 900 kc. This signal drives the emitter of Q_1 to produce a frequency difference of 100 kc, which is selected by means of a tapped tuned circuit.

The 4700 ohm resistor is connected in series with the collector of the transistor to suppress a form of negative resistance oscillation that is encountered with high frequency junction transistors when bottoming occurs. It also serves the very useful function of limiting the peak collector current to a satisfactory value. The X3 frequency multiplier Q_2 is driven from a capacitive tap on the 100 kc tuned circuit through a series isolating resistor R_6 which also helps to prevent VHF parasites. The second frequency multiplier, Q_3, produces the 900 kc voltage required for the mixer. The output amplifier, Q_4, is driven by the mixer output, and has sufficient

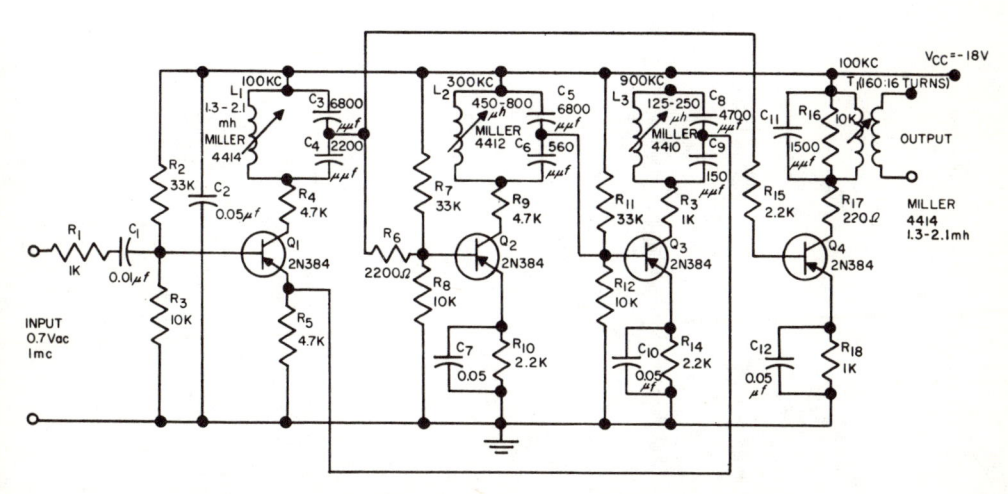

Circuit 10-7 - Regenerative frequency divider

Circuit 10-7, Regenerative Frequency Divider (cont'd.)

gain to deliver 20 milliwatts to a 50 ohm load. The 10,000 ohm resistor R_{16} across the tuned circuit of Q_4 stabilizes the amplifier and also prevents an excessively high voltage from being developed at this point in the absence of a load. The frequency-multiplier stages and the output amplifier also contain series collector resistors for the reasons given above.

    The alignment of the divider is best accomplished by driving the base of each transistor separately at the frequency of the collector tuned circuit. The tuned circuit is then adjusted for maximum response while the input is decreased if necessary, to avoid limiting. An input of 0.7 volt r.m.s. at one megacycle should then be applied to Q_1; the system should oscillate and, as a final step, each tuned circuit should be adjusted to the center of the range over which correct operation is obtained. When properly adjusted, the divider should work as the supply is varied from 5 to 40 volts. Increasing the voltage beyond 40 may cause transistor damage, and should not be attempted. The output should be zero in the absence of an input, except for a small amount of noise. The operating bandwidth should be at least ± 2% at the middle of the supply-voltage range.

CIRCUIT 10-8

SECOND DETECTOR WITH AGC

Raytheon Manufacturing Company   Contributed by W. E. Sheehan
and J. H. Ivers

The second detector is of the Class B type, which provides 10 db of gain as well as a suitable source of AGC voltage. The emitter is grounded and, in the absence of a signal, the base is given a slight negative bias so that the emitter-base diode is just at the point of conduction. Conduction occurs on the negative signal half cycles, and the resulting collector current produces after filtering by C_1, an audio voltage across the 1K

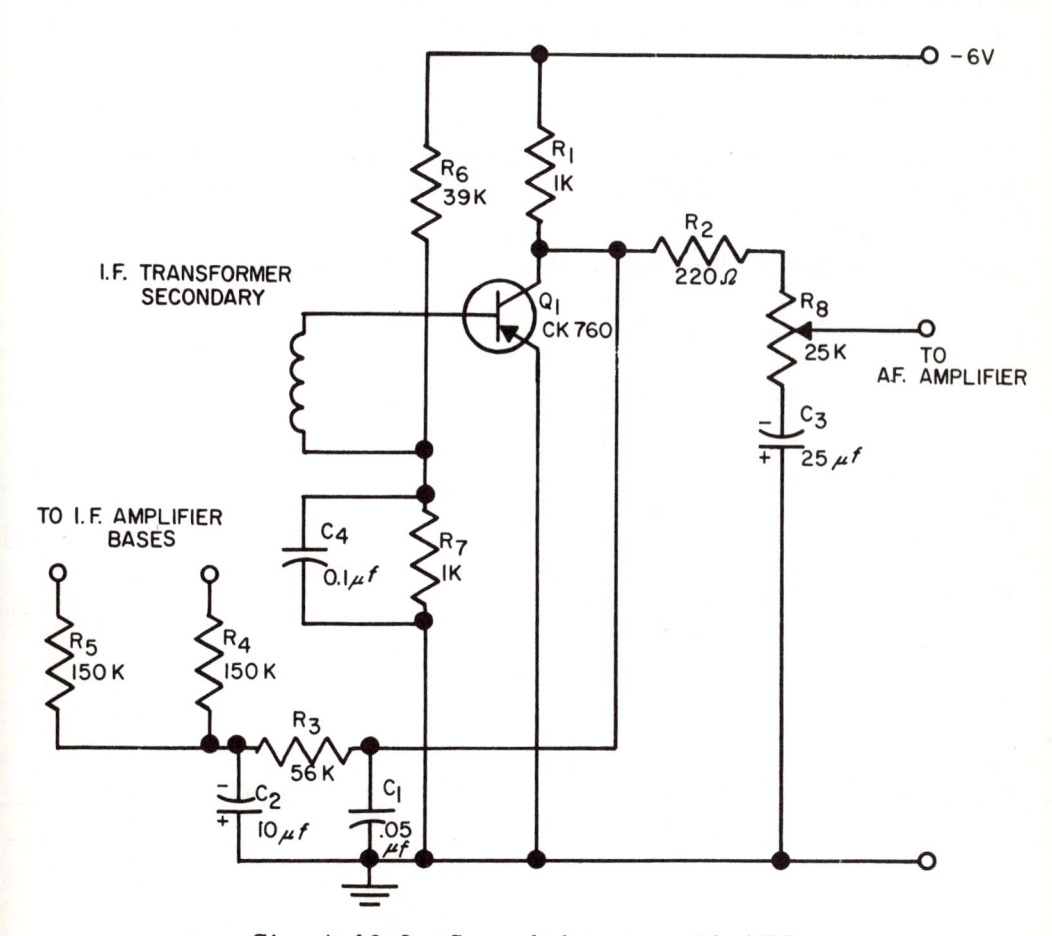

Circuit 10-8 - Second detector with AGC

Circuit 10-8, Second Detector With AGC (cont'd.)

collector-load resistor, R_1. The series resistor, R_2, serves as a filter to keep i.f. voltage out of the audio amplifier.

 The AVC voltage is also obtained from the collector of the detector, and is fed back to the bases of the two i.f. amplifier stages through filter R_3 C_2 and isolating resistors R_4 and R_5. In the absence of a signal, the detector collector operates approximately at the 8-voltage. The resulting i.f.-amplifier base-bias currents flowing through R_4 and R_5 are decreased in the presence of a signal, which decreases the magnitude of the average detector collector voltage. Thus, the i.f. amplifier gain is decreased by the presence of a strong signal. With this simple AGC system, it was found necessary to return the emitter of the first IF stage to a point 3 volts negative with respect to ground to obtain enhanced AGC operation of the first stage, preventing overloading of the second stage or the detector. It will be noted that there is no provision for delayed operation of this AGC.

(This circuit is reproduced with the permission of "Electronics Magazine," "Design of Transistorized High-Gain Portable" by W. E. Sheehan and J. H. Ivers, March 1955.)

Reprinted from Electronics, March 1955; copyright McGraw-Hill, Inc. 1955.

CIRCUIT 10-9

SECOND DETECTOR WITH DELAYED AGC

Radio Corporation of America   Contributed by D. D. Holmes,
                                      T. O. Stanley, and L. A. Freedman

    Circuit 10-9 differs from the previous one principally in the manner in which AGC is applied to the i.f. amplifier and in the fact that there is a delay in the application of the AGC.

    In the absence of a signal, the base and emitter of the transistor operate at approximately the same potential. A signal will produce components of collector current at the i.f. and its harmonics, at the modulation frequency and its harmonics, and also a direct component. In this detector, a well-shielded series-resonant circuit $L_1 C_1$, tuned to 910 kc, is used to shunt the i.f. second harmonic from collector to emitter to prevent pickup by the loop-stick antenna. The audio component of collector current develops a voltage across the volume control, R_4, while the i.f. is rejected by the filter $R_3 C_4$. At an input of 50 millivolts to the detector, 80% modulated, the distortion is approximately 9%, while the distortion is less than 2% at an input of 250 millivolts.

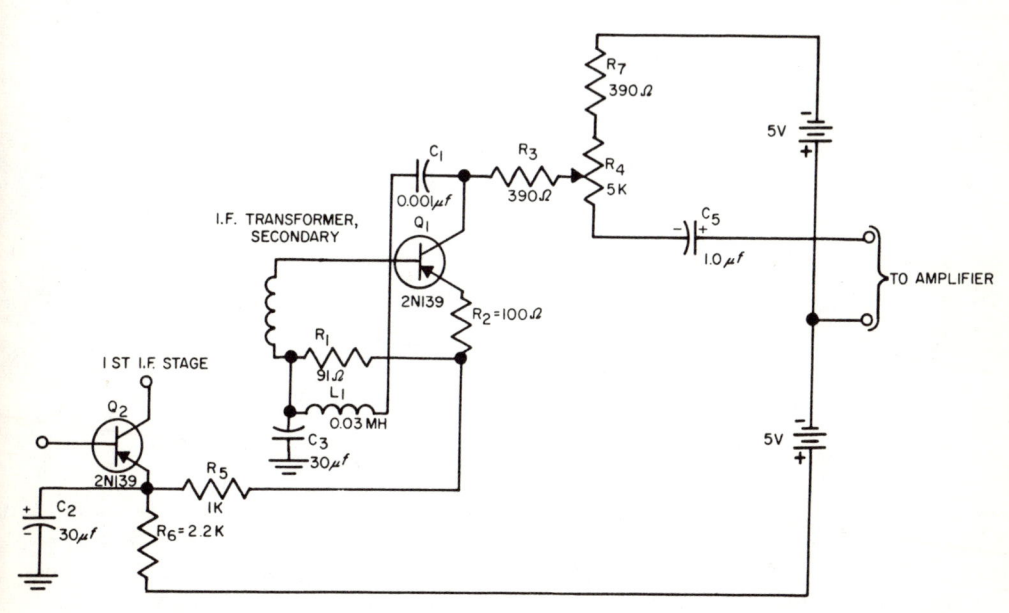

Circuit 10-9 - Second detector with delayed AGC

Circuit 10-9, Second Detector with Delayed AGC (cont'd.)

The AVC is obtained from the d.c. component of the detector emitter current. The d.c. emitter return is through R_6, which is connected in the emitter circuit of the first i.f. stage, and then through R_5 and R_2 to the emitter. A delay in the application of the AVC is furnished by R_2, which produces d.c. negative feedback in the detector. The capacitor C_2 and resistor R_5 serve to filter the audio-frequency components of emitter current, preventing audio-frequency modulation of the i.f. stage. When emitter current flows in the detector, a portion of the first i.f. stage emitter current is diverted to the detector. The zero-signal i.f. stage emitter current is approximately 0.5 mAdc; nearly all of this is shifted to the detector in the presence of a strong signal, decreasing the i.f. gain. An effective AGC range of 45 db is obtained.

(This circuit is reproduced with the permission of the I.R.E., from the article, "A Developmental Pocket-Size Broadcast Receiver Employing Transistors," by D. D. Holmes, T. O. Stanley, and L. A. Freedman, June 1955, Proceedings of the I.R.E.)